# NUTRIGENOMICS AND NUTRACEUTICALS

*AAP Advances in Nutraceuticals*

# NUTRIGENOMICS AND NUTRACEUTICALS

*Edited by*

**Raj K. Keservani, PhD, MPharm**

**Eknath D. Ahire, PhD**

**Shubham J. Khairnar, MPharm**

**Sanjay J. Kshirsagar, PhD**

**Rajesh K. Kesharwani, PhD**

APPLE ACADEMIC PRESS

First edition published 2024

**Apple Academic Press Inc.**
1265 Goldenrod Circle, NE,
Palm Bay, FL 32905 USA

4164 Lakeshore Road, Burlington,
ON, L7L 1A4 Canada

**CRC Press**
2385 NW Executive Center Drive,
Suite 320, Boca Raton FL 33431

4 Park Square, Milton Park,
Abingdon, Oxon, OX14 4RN UK



**Library and Archives Canada Cataloguing in Publication**

Title: Nutrigenomics and nutraceuticals / edited by Raj K. Keservani, MPharm, Eknath D. Ahire, PhD, Shubham J. Khairnar, MPharm, Sanjay J. Kshirsagar, PhD, Rajesh K. Kesharwani, PhD.

Names: Keservani, Raj K., 1981- editor. | Ahire, Eknath D., editor. | Khairnar, Shubham J., editor. | Kshirsagar, Sanjay J., editor. | Kesharwani, Rajesh Kumar, 1978- editor.

Series: AAP advances in nutraceuticals.

Description: First edition. | Series statement: AAP advances in nutraceuticals | Includes bibliographical references and index.

Identifiers: Canadiana (print) 20230592996 | Canadiana (ebook) 20230593011 | ISBN 9781774914380 (hardcover) | ISBN 9781774914397 (softcover) | ISBN 9781003428268 (ebook)

Subjects: LCSH: Functional genomics.

Classification: LCC QP144.F85 N88 2024 | DDC 613.2—dc23

**Library of Congress Cataloging-in-Publication Data**

Names: Keservani, Raj K., 1981- editor.

Title: Nutrigenomics and nutraceuticals / edited by Raj K. Keservani, MPharm, Eknath D. Ahire, PhD, Shubham J. Khairnar, MPharm, Sanjay J. Kshirsagar, PhD, Rajesh K. Kesharwani, PhD.

Description: First edition. | Palm Bay, FL : Apple Academic Press, 2024. | Series: AAP advances in nutraceuticals | Includes bibliographical references and index. | Summary: "This new volume provides an authoritative and comprehensive overview of the field of nutraceuticals and genomics-also called nutrigenomics, a rapidly growing field as the science of nutrition and genomics advances. Nutrigenomics is the study of how a person's diet impacts gene expression across their entire genome. Nutrigenetics is an overlapping topic that studies specific genes rather than the complete genome and how they relate to dietary needs. This volume discusses both nutrigenomic and nutrigenetic techniques to investigate the nutritional basis of some disorders because diet can affect gene expression. It covers the nutrigenomic tools for the prevention of disease conditions such as cancer, viral infection, microbial infection, COVID-19, HIV/AIDS, tuberculosis, hypertension, neurodegenerative disease, obesity, baldness, and anemia. Some nutrigenomic tools can also be used for brain health, weight loss, human nutrition, skin care, and maintaining health during pregnancy. Some important aspects such as the role of genetics and nutritional factors in the development of any disease condition, as well as personalized nutrition and functional foods are discussed. The impact of oxidative stress on human genetics is also explored in detail. It is critical for scientists to be able to investigate the relationship between nutrients, food, and genetics and advanced health or disease reduction. This book will have a significant impact on the study of nutraceuticals, nutrigenomics, and associated qualities that may be useful in the prevention of various diseases. Healthcare professionals and scientists in the nutraceutical sector, DNA technology, and genetic research industry as well as research academics interested in nutrition and genetics related study will find this volume quite useful"-- Provided by publisher.

Identifiers: LCCN 2023055267 (print) | LCCN 2023055268 (ebook) | ISBN 9781774914380 (hbk) | ISBN 9781774914397 (pbk) | ISBN 9781003428268 (ebk)

Subjects: LCSH: Nutrition--Genetic aspects.

Classification: LCC QP144.G45 N8854 2024 (print) | LCC QP144.G45 (ebook) | DDC 612.3--dc23/eng/20231227

LC record available at https://lccn.loc.gov/2023055267

LC ebook record available at https://lccn.loc.gov/2023055268

ISBN: 978-1-77491-438-0 (hbk)
ISBN: 978-1-77491-439-7 (pbk)
ISBN: 978-1-00342-826-8 (ebk)

# ABOUT THE BOOK SERIES: AAP ADVANCES IN NUTRACEUTICALS

**Series Editor:**
**Raj K. Keservani, PhD, MPharm**
Faculty of B. Pharmacy, CSM Group of Institutions, Allahabad, India

**Rajesh K. Kesharwani, PhD**
Nehru Gram Bharati (Deemed to be University), Prayagraj, India

**Anil K. Sharma, PhD, MPharm**
Department of Pharmacy, School of Medical and Allied Sciences, GD Goenka University, Gurugram, India

In the modern era, mankind has witnessed a paradigm shift with respect to fundamental eating behavior. The lack of physical workouts and busy schedules at offices and in households have promoted the consumption of junk foods, which eventually results in numerous diseases and disorders since the nutritional content from fast food is inadequate. Plumpness and obesity have become a global health threat. The leading causes of death from most developing nations are noncommunicable diseases such as cardiovascular diseases, cancer, arthritis, osteoporosis, and liver toxicity. Patients suffering from such lifestyle ailments are a bit apprehensive towards prolonged use of costly modern therapeutics, which encourages instead the use of alternative approaches for management of such diseases and disorders.

The emerging sector of the nutraceutical industry encompasses products derived from nature, dietary supplements, and functional foods. Nutraceuticals are used for treatment and prevention of a broad range of diseases, such as the common cold, arthritis, sleep-related disorders, cancers, cardiovascular complications, metabolic disorders, and others.

The research on nutraceuticals is increasing day by day, considering the beneficial effects of food or food supplements in the management of diverse diseases. The issue of paramount concern is standardization and establishment of clinical efficacy of nutraceuticals, which is in fact a challenge for researchers around the globe.

This book series aims at realizing the significance of the variety of nutraceuticals for human well-being. The books in the series will also emphasize the role of dieticians and nutritionists for the prescription of judicious eating habits. The key food components such as carbohydrates, proteins, and lipids as well as micronutrients (vitamins, minerals) are demonstrated to maintain good health, obviating the need for medicines. Thus, nutraceuticals that are indeed derived from food ingredients are believed to be potential alternative therapeutics.

Asymptomatic diseases/disorders necessitate proper diagnosis. Stressed circumstances have been reported to cause weight loss. Workplace stress is a key variable in the genesis and propagation of metabolic disorders. The nuances of the underlying mechanisms are still to be deciphered. Poor lifestyle trends have been attributed along with the consumption of junk foods in several such instances. Sugar-rich carbonated beverages, the use of contraband drugs, and the consumption of liquor in excess are assigned as risk factors. This book series will cover advances and applications in addition to providing the basics of nutraceuticals.

The books in this series will be valuable resources for industry professionals to design and develop quality products for end users. In addition, the book series will be fortified by ideas of innovation, concept building, manufacturing aspects, quality control, and regulatory status of nutraceuticals.

Ostensibly, business has seen a rise in the nutraceutical consumption in recent years. Pure, safe, efficacious products are the need of the hour globally. There are a number of books on nutraceuticals in the market, yet these are only a few books presenting certain aspects of nutraceuticals therein. This book series aims to offer a holistic view of this promising strategy, which is picking up pace with time. This book series will be a unique endeavor to bring manufacturing, research and development, and marketing strategies under a single umbrella.

## CURRENT BOOKS IN THE SERIES

- **Micronutrients and Macronutrients as Nutraceuticals**
  Editors: Prakash Chandra Gupta, PhD, Sayan Bhattacharyya, MBBS, MD, Nisha Sharma, PhD, MPharm, Rajesh K. Kesharwani, PhD, and Raj K. Keservani, PhD, MPharm
- **Formulations, Regulations, and Challenges of Nutraceuticals**
  Editors: Tingirikari Jagan Mohan Rao, PhD, Rajesh K. Kesharwani, PhD, Raj. K. Keservani, PhD, and Anil K. Sharma, PhD

- **Food Supplements and Dietary Fibers in Health and Disease**
  Editors: Bhushan R. Rane, PhD, MPharm, Raj K. Keservani, PhD, MPharm, Durgesh Singh, DPhil, Nayan A. Gujarathi, PhD, and Ashish S. Jain, PhD
- **Nutraceuticals in Cancer Prevention, Management, and Treatment**
  Editors: Raj K. Keservani, PhD, MPharm Bui Thanh Tung, PhD, Sippy Singh, DPhil, and Rajesh K. Kesharwani, PhD
- **Advances in Flavonoids for Human Health and Prevention of Diseases**
  Editors: Nisha Sharma, PhD, Deepika Saini, PhD, Rajesh K. Kesharwani, PhD, Prakash Chandra Gupta, and Raj. K. Keservani, PhD
- **The Flavonoids: Extraction and Applications**
  Editors: Deepika Saini, PhD, Rajesh K. Kesharwani, PhD, and Raj K. Keservani, PhD, MPharm
- **Immune-Boosting Nutraceuticals for Better Human Health: Novel Applications**
  Editors: Urmila Jarouliya, PhD, Raj K. Keservani, PhD, MPharm Rajesh K. Kesharwani, PhD, Virendra K. Patel, PhD, and Adi D. Bharti
- **Applications of Functional Foods in Disease Prevention**
  Editors: Raj K. Keservani, PhD, MPharm, and Eknath D. Ahire, MSPharm
- **Plant Metabolites and Vegetables as Nutraceuticals**
  Editors: Raj K. Keservani, PhD, MPharm, Bui Thanh Tung, PhD, Rajesh K. Kesharwani, PhD, MPharm, and Eknath D. Ahire
- **Preventive and Therapeutic Role of Vitamins as Nutraceuticals**
  Editors: Khemchand R. Surana, Eknath D. Ahire, Raj K. Keservani, PhD, MPharm, and Rajesh K. Kesharwani, PhD
- **Nutraceuticals and Bone Health**
  Editors: Deepak Sharma, PhD, Madan Mohan Gupta, PhD, Anil K. Sharma, PhD, MPharm, Raj K. Keservani, PhD, MPharm, and Rajesh K. Kesharwani, PhD
- **Flavonoids as Nutraceuticals**
  Editors: Rajesh K. Kesharwani, PhD, MTech, Deepika Saini, PhD, Raj K. Keservani, PhD, MPharm, and Anil Kumar Sharma, PhD
- **Herbals as Nutraceuticals: Their Role in Healthcare**
  Editors: Raj K. Keservani, PhD, MPharm, Rajeshwar Kamal Kant Arya, PhD, and Rajesh K. Kesharwani, PhD

- **Nutraceutical Fruits: Overview and Disease Prevention**
  Editors: Raj K. Keservani, PhD, MPharm
- **Nutrigenomics and Nutraceuticals**
  Editors: Raj K. Keservani, PhD, MPharm, Eknath D. Ahire, Shubham J. Khainar, PhD, Sanjay J. Kshirsagar, PhD, and Rajesh K. Kesharwani, PhD
- **Nutraceuticals in Cardiac Health Management**
  Editors: Raj K. Keservani, PhD, MPharm, Rajesh K. Kesharwani, PhD, Praveen Kumar Jain, PhD, Bhushan R. Rane, PhD, and Nayan A. Gujarathi, PhD
- **Nutraceuticals for the Treatment and Prevention of Sexual Disorders**
  Editors: Raj K. Keservani, PhD, MPharm, Sharangouda J. Patil, PhD, and Ivan Aranha, PhD
- **Nutraceuticals in Obesity Management and Control**
  Editors: Raj K. Keservani, PhD, MPharm, and Alka Lohani, PhD
- **Antioxidants as Nutraceuticals**
  Editors: Nayan A. Gujarathi, PhD, Raj K. Keservani, PhD, MPharm, Rajesh K. Kesharwani, PhD, Bhushan R. Rane, PhD, and Yogeeta Sameer Goyal, PhD
- **The Nature of Nutraceuticals: History, Properties, Sources, and Nanotechnology**
  Editors: Rajesh K. Kesharwani, PhD, Prashant Kumar, PhD, and Raj K. Keservani, PhD, MPharm
- **Nutraceuticals in Insomnia and Sleep Problems**
  Editors: Raj K. Keservani, PhD, MPharm, Sayan Bhattacharyya, PhD, and Rajesh K. Kesharwani, PhD
- **Nutraceuticals in Respiratory and Pulmonary Diseases**
  Editors: Deepika Saini, PhD, Rajesh K. Kesharwani, PhD, and Raj K. Keservani, PhD, MPharm
- **Nutraceuticals in Arthritis and Psoriasis**
  Editors: Meenakshi Jaiswal, PhD, Raj K. Keservani, PhD, MPharm Rajesh K. Kesharwani, PhD, and Swati G. Talele, PhD

# ABOUT THE EDITORS

**Raj K. Keservani, PhD, MPharm**
*Associate Professor, Faculty of B.Pharmacy,*
*CSM Group of Institutions, Prayagraj, India*

**Raj K. Keservani, PhD, MPharm,** is working as Associate Professor in the Faculty of B. Pharmacy at the CSM Group of Institutions, Prayagraj, India. He has more than 12 years of academic (teaching) experience from various institutes of India in pharmaceutical education. He has supervised more than 25 undergraduate students for their research work. He has published over 30 peer-reviewed papers in the field of pharmaceutical sciences in national and international journals, six patents, over 70 book chapters, three co-authored books, and over 40 edited books. He is also active as a reviewer for several international scientific journals. His research interests include nutraceutical and functional foods, novel drug delivery systems (NDDS), transdermal drug delivery/drug delivery, health science, cancer biology, and neurobiology. Dr. Keservani graduated with a pharmacy degree from the Department of Pharmacy, Kumaun University, Nainital, Uttarakhand, India. He received his Master of Pharmacy (MPharm) (specialization in pharmaceutics) from the School of Pharmaceutical Sciences, Rajiv Gandhi Proudyogiki Vishwavidyalaya, Bhopal, India. He earned his PhD degree from IKG Punjab Technical University, Jalandhar, Punjab, India. Dr. Keservani is a life member of the Society of Pharmaceutical Education and Research (SPER).

**Eknath D. Ahire, MSPharm**
*Assistant Professor, Department of Pharmaceutics, MET's,*
*Institute of Pharmacy, Bhujbal Knowledge City, Adgaon, Nashik,*
*Maharashtra, India*

**Eknath D. Ahire, MSPharm**, is currently working as an Assistant Professor in the Department of Pharmaceutics at MET's, Institute of Pharmacy, Bhujbal Knowledge City, Adgoan, Nashik, affiliated to Savitribai Phule Pune University, Pune, India. He is also working for QCS Labs LLP, Mumbai, as Inspector for Nigeria regulatory market pre-shipment inspection under the Earn and Learn Scheme. He has a total of five years of experience as Formulation Scientist II at Torrent Research Centre, Gandhinagar, Gujrat, as well as four years of academic experience. He is also a reviewer for several reputed international journals, including Pharmaceutical Development and Technology (Taylor and Francis), International Journal of Applied Nanotechnology Research (IGI Global), and Exploration of Immunology Journal. He is an editorial board member of the journals Acta Scientific and Community Acquired Infection. He is life member of the Association of Pharmaceutical Teachers of India. He has more than 30 publications (research, review, and book chapters) in national and international indexed journals as well as six textbooks. He has participated and attended several national and international conferences and workshops. He has worked on nanotechnology-based novel drug delivery systems, nutraceuticals, and immunology. He graduated with a BPharm degree from Savitribai Phule Pune University and master's degree (MS Pharm) in Pharmaceutics from the National Institute of Pharmaceutical Education and Research-Ahmedabad (NIPER-A), India. Currently, he is admitted as a PhD Scholar with Savitribai Phule, Pune University, India. He was awarded a NFST (RGNF) PhD Research Scholar fellowship by the Government of India.

**Shubham J. Khairnar, MPharm**
*Assistant Professor, Department of Pharmacology at MET's Institute of Pharmacy, Bhujbal Knowledge City, Adgaon, Nashik, Pune, India*

**Shubham Jagdish Khairnar, MPharm,** is an Assistant Professor in the Department of Pharmacology at MET's Institute of Pharmacy, Bhujbal Knowledge City, Adgaon, Nashik (Savitribai Phule Pune University), Pune, India. He has also three years of academic experience. He has attended several national and international conferences and workshops. He has worked on neuropharmacological evaluation of polyphenols and phenolic acids, such as vanillic acid, sinnapic acid, and ferulic acid, screened for neuroprotective activity. He has several publications in national and international journals, including research, reviews. and many book chapters. In addition, he has several book publications. He obtained his BPharm degree at North Maharashtra University in 2016. He obtained his master's degree (MPharm) in Pharmacology in 2019 from MGV's College of Pharmacy, Panchavati, Nashik (Savitribai Phule Pune University).

**Sanjay J. Kshirsagar, PhD**
*Professor and Principal, MET's, Institute of Pharmacy, Bhujbal Knowledge City, Adgoan, Nashik, affiliated to Savitribai Phule Pune University, Pune, India*

**Sanjay J. Kshirsagar, PhD,** is working as Professor and Principal at MET's, Institute of Pharmacy, Bhujbal Knowledge City, Adgoan, Nashik, affiliated to Savitribai Phule Pune University, Pune, India. He has more than 19 years of experience in the pharmaceutical field. He has guided 50 MPharm students and is currently in charge of eight PhD students. He has published more than 160 research and review articles in well-reputed national/international journals with a cumulative impact factor of 70. He has been awarded with several awards, including Best Research Guide by national-level Pharma Innova, Best Original Paper Award, Pharma Excipients Poster Award from the International AG, and many more. He is acting as member of the board of studies in pharmaceutics of Savitribai Phule Pune University, Pune, and Inspector of the Pharmacy Council of India, India. He has one patent under his name with four others under progress. He has authored many books and book chapters. In addition, he has presented more than 20 papers in national and international conferences. He is also an expert in colon targeted drug delivery system. He completed his PhD at JNTU, Hyderabad, India.

**Rajesh K. Kesharwani, PhD, MTech**
*Associate Professor, Department of Computer Application, Nehru Gram Bharati (Deemed to be University), Prayagraj, India*

**Rajesh K. Kesharwani, PhD, MTech,** working as an Associate Professor in the Department of Computer Application at Nehru Gram Bharati (Deemed to be University), Prayagraj, India. He has more than 11 years of research and nine years of teaching experience in various institutes in India, imparting bioinformatics and biotechnology education. He has received several awards, including the NASI-Swarna Jayanti Puruskar by The National Academy of Sciences of India. He has supervised one PhD and more than 20 under- and postgraduate students for their research work and has authored over 49 peer-reviewed articles, 20 book chapters, and 14 edited books with international publishers. He has been a member of many scientific communities as well as a reviewer for many international journals. He has presented many papers at various national and international conferences. Dr. Kesharwani received his PhD from the Indian Institute of Information Technology, Allahabad, and worked at NIT Warangal for two-semester. He has been a recipient of the Ministry of Human Resource recipient of Development (India) Fellowship and a Senior Research Fellowship from the Indian Council of Medical Research, India. His research fields of interest are medical informatics, protein structure and function prediction, computer-aided drug designing, structural biology, drug delivery, cancer biology, nano-biotechnology, and biomedical sciences.

# CONTENTS

# CONTRIBUTORS

**Eknath D. Ahire**
Department of Pharmaceutics, MET's Institute of Pharmacy, Bhujbal Knowledge City, Adgaon, Nashik, Maharashtra, India

**Sapana P. Ahirrao**
MET's Institute of Pharmacy, Bhujbal Knowledge City, Adgaon, Nasik, Maharashtra, India

**Shubham S. Bagade**
Sandip Institute of Pharmaceutical Sciences, Nashik, Maharashtra, India

**Harshada U. Bagul**
Sandip Institute of Pharmaceutical Sciences, Nashik, Maharashtra, India

**Atul R. Bendale**
Sandip Institute of Pharmaceutical Sciences, Nashik, Maharashtra, India

**Deepak S. Bhambere**
MET's Institute of Pharmacy, Bhujbal Knowledge City, Adgaon, Nasik, Maharashtra, India

**Akash M. Bhoite**
Sandip Institute of Pharmaceutical Sciences, Nashik, Maharashtra, India

**Laxmikant B. Borse**
Sandip Institute of Pharmaceutical Sciences, Nashik, Maharashtra, India

**Manali P. Borse**
Sandip Institute of Pharmaceutical Sciences, Nashik, Maharashtra, India

**Sandhya Borse**
School of Pharmaceutical Sciences, Sandip University, Nashik, Maharashtra, India

**Kishor Danao**
Dadasaheb Balpande College of Pharmacy, Nagpur, Maharashtra, India

**Neelam L. Dashputre**
Department of Pharmacology, MET's Institute of Pharmacy, Bhujbal Knowledge City, Adgaon, Nashik, Maharashtra, India

**Pranali P. Deshmukh**
Department of Pharmacology, MET's Institute of Pharmacy, Bhujbal Knowledge City, Adgaon, Nashik, Maharashtra, India

**Simona E. D'Souza**
Department of Pharmacology, MET's Institute of Pharmacy, Bhujbal Knowledge City, Adgaon, Nashik, Maharashtra, India

**Anil G. Jadhav**
Sandip Institute of Pharmaceutical Sciences, Nashik, Maharashtra, India

**Jayesh D. Kadam**
Department of Pharmacology, MET's Institute of Pharmacy, Bhujbal Knowledge City, Adgaon, Nashik, Maharashtra, India

**Gayatri N. Kanade**
Department of Pharmacology, MET's Institute of Pharmacy, Bhujbal Knowledge City, Nashik, Maharashtra, India

**Rani Kankate**
Department of Pharmaceutical Chemistry, MET's Institute of Pharmacy, Adgaon, Maharashtra, Nashik, India

**Nilesh Karande**
Institute of Pharmaceutical Education and Research, Wardha, Maharashtra, India

**Shubham J. Khairnar**
Department of Pharmacology, MET's Institute of Pharmacy, Bhujbal Knowledge City, Adgaon, Nashik, Maharashtra, India

**Sanjay Kshirsagar**
Department of Pharmaceutics, MET's Institute of Pharmacy, Bhujbal knowledge City, Adgaon, Nashik, Maharashtra, India

**Umesh D. Laddha**
MET's Institute of Pharmacy, Bhujbal Knowledge City, Adgaon, Nasik, Maharashtra, India

**Shaikh M. A.**
Department of Pharmaceutical Chemistry, MET's Institute of Pharmacy, Adgaon, Nashik, Maharashtra, India

**Akhil Nagar**
Sandip Institute of Pharmaceutical Sciences, Nashik, Maharashtra, India

**Vaishali Naphade**
School of Pharmaceutical Sciences, Sandip University, Nashik, Maharashtra, India

**Vasim T. Pathan**
Sandip Institute of Pharmaceutical Sciences, Nashik, Maharashtra, India

**Tambe P. B.**
Department of Pharmaceutical Chemistry, MET's Institute of Pharmacy, Adgaon, Nashik, Maharashtra,, India

**Surabhi B. Patil**
Department of Pharmacology, MET's Institute of Pharmacy, Bhujbal Knowledge City, Adgaon, Nashik, Maharashtra, India

**Utkarsha P. Patil**
Sandip Institute of Pharmaceutical Sciences, Nashik, Maharashtra, India

**Vijayshri Rokde**
R. C. Patel Institute of Pharmaceutical Education and Research, Shirpur, Maharashtra, India

**Rahul R. Sable**
Department of Pharmacology, MET's Institute of Pharmacy, Bhujbal Knowledge City, Adgaon, Nashik, India

**Suraj M. Sarode**
Shree Sureshdada Jain Institute of Pharmaceutical Education and Research, Jamner, Maharashtra, India

**Masood Siddiqui**
Department of Pharmacology, MET's Institute of Pharmacy, Bhujbal Knowledge City, Nashik, Maharashtra, India

**Anjali B. Tajanpure**
Department of Pharmacology, MET's Institute of Pharmacy, Bhujbal Knowledge City, Adgaon, Nashik, India

**Musab Tanzeel**
Department of Pharmacology, MET's Institute of Pharmacy, Bhujbal Knowledge City, Nashik, Maharashtra, India

**Apurva P. Torane**
Department of Pharmacology, MET's Institute of Pharmacy, Bhujbal Knowledge City, Adgaon, Nashik, Maharashtra, India

**Pavan B. Udavant**
Department of Pharmacology, MET's Institute of Pharmacy, Bhujbal Knowledge City, Adgaon, Nashik, Maharashtra, India

**Pradnya H. Vaidya**
University of Bourgogne Franche Comte Master Innovative Drugs, Dijon, France

# ABBREVIATIONS

| | |
|---|---|
| AA | arachidonic acid |
| AD | Alzheimer's disease |
| ADH | alcohol dehydrogenase |
| AGT | angiotensinogen gene variants |
| AGXT | alanine-glyoxylate aminotransferase |
| AHA | American Heart Association |
| ALA | alpha-linolenic acid |
| ALOX5AP | 5-lipoxygenase-activating protein gene |
| AMD | age-related macular degeneration |
| AN | anorexia nervosa |
| ANP | atrial natriuretic peptide gene |
| APOA1 | apolipoprotein A1 |
| APOC3 | apolipoprotein C3 |
| APOE | apolipoprotein E |
| ARFID | avoidant/restrictive food intake disorder |
| ASN | American Society for Nutrition |
| ATRA | all trans-retinoic acid |
| BCC | basal cell carcinoma |
| BED | binge eating disorder |
| BMI | body mass index |
| BN | bulimia nervosa |
| BPA | bisphenol A |
| BRCA1 | breast cancer gene 1 |
| CaBP | calcium-binding protein |
| CBP | CREB-binding protein |
| CHD | coronary heart disease |
| CNV | copy number variations |
| CRP | C-reactive protein |
| CRUK | Cancer Research UK |
| CVD | coronary heart disease |
| CVD | cardiovascular disease |
| CYP | cytochrome p450 |
| DADS | diallyl sulphides |
| DASH | dietary approaches to stop hypertension |
| DHA | docosahexaenoic acid |
| DNA | deoxyribonucleic acid |
| DNAm | DNA methylation |

| | |
|---|---|
| DNMT | DNA methyltransferase |
| DNMTs | DNA methylating enzymes |
| DRVs | dietary reference values |
| DTC-GT | direct to consumer genetic testing |
| dTMP | deoxythymidine monophosphate |
| dUMP | deoxyuridine monophosphate |
| EAAs | essential amino acids |
| EDs | eating disorders |
| EDNOS | eating disorders not otherwise specified |
| EFS | expenditure and food survey |
| EGCG | epigallocatechin gallate |
| EHH | extended haplotype homozygosity |
| eIF2 | eukaryotic initiation factor 2 |
| ELSI | ethical, legal, and social concerns |
| EPA | eicosapentaenoic acid |
| EPHX | epoxide hydrolases |
| ER | endoplasmic reticulum |
| ESPEN | European Society for Clinical Nutrition and Metabolism |
| FA | fatty acids |
| FA | folic acid |
| FAD | flavin adenine dinucleotide |
| FADS | fatty acid dehydrogenases I and II |
| FES | family expenditure survey |
| FEV1 | forced expiratory volume in one sec |
| FFQRNA | food frequency questionnaire RNA |
| FMD | fasting-mimicking diet |
| FTD | fat mass and obesity associated gene |
| GC | gas chromatography |
| GCN2 | general control nonderepressible 2 |
| GEI | genes and environment initiative |
| GI | gastrointestinal |
| GLP-1 | glucagon-like protein-1 |
| GPS | genome-wide polygenic score |
| GRS | genetic risk scores |
| GST | glutathione-S-transferase |
| GSTM1 | glutathione S-transferase Mu 1 |
| GWASs | genome-wide association studies |
| G6PD | glucose-6-phosphate dehydrogenase |
| HAI | integrated household survey |
| HDAC | inhibit simple protein deacetylase |
| HDL | high-density lipoprotein |
| HGP | Human Genome Project |
| HIFs | hypoxia-inducible factors |

| | |
|---|---|
| HIF1 | hypoxia-inducible factor 1 |
| HMG | hydroxy-methylglutaryl |
| HMGCR | HMG- CoA reductase |
| HNE | hydroxynonenal |
| HNF4 alpha | hepatocyte nuclear factor 4 alpha |
| HPA | healthy bone programming |
| HPV | human papillomaviruses |
| IAP | intracisternal A Particle |
| IARC | International Agency for Research on Cancer |
| IGFBPs | IGF-binding proteins |
| IGF | insulin-like growth factor |
| IGF-1 | insulin-like growth factor-1 |
| IGF2 | insulin-like super molecule 2 |
| IL-1 | interleukin-1 |
| IL-6 | interleukin 6 |
| IPA | ingenuity pathway analysis |
| IUGR | intrauterine growth backward |
| IVF | in vitro fertilization |
| Kyr | thousand years |
| LA | linoleic acid |
| LAGB | laparoscopic adjustable gastric banding |
| LC | liquid chromatography |
| LCPUFAs | long chain polyunsaturated fatty acids |
| LDL | low-density lipoprotein |
| LEPR | leptin receptor gene |
| LHPs | lipid hyper peroxides |
| LNCaP | lymph node carcinoma of the prostate |
| LOC | loss of control over eating |
| LXR | liver X receptor |
| Ma | mega annum (a million years) |
| MALDI | matrix-assisted laser desorption ionization |
| MBD | methyl-CpG-binding domain |
| MC4R | melanocortin-4-receptor |
| MD | Mediterranean diet |
| MDA | malondialdehyde |
| mEH | microsomal epoxide hydrolase |
| miRNAs | micro RNAs |
| MMR | mismatch repair system |
| MnSOD | manganese superoxide dismutase |
| MS | mass spectroscopy |
| mtDNA | mitochondrial DNA |
| MTHFR | methylene tetrahydrofolate reductase |
| MTR | methionine synthase |

| | |
|---|---|
| MTRR | methionine synthase reductase |
| Mya | million years ago |
| NAD | nicotinamide adenine dinucleotide |
| NAT | N-acetyltransferase |
| NCD | noncommunicable diseases |
| NCI | National Cancer Institute |
| ncRNA | non-coding RNA |
| NEAA's | non-essential amino acids |
| NFkB | nuclear factor kappa B |
| NFS | National Food Survey |
| NHANES | National Health and Nutrition Examination Survey I Epidemiologic |
| NIEHS | National Institute of Environmental Health Sciences |
| NMR | nuclear magnetic resonance |
| NR | nicotinamide riboside |
| NPY | neuropeptide Y |
| NSP | non starch polysaccharide |
| NTDs | neural tube defects |
| NuGO | Nutrigenomics Organization |
| OGA | go-between metabolism |
| OGC | O-GlcNAcase |
| O-GlcNAc | N-acetylglucosamine |
| OGT | O-GlcNAc enzyme |
| OTC | ornithine transcarbamylase |
| OXPHOS | oxidative phosphorylation |
| PAGE | polyacrylamide gel electrophoresis |
| PARP | poly(ADP-ribose)polymerase |
| PBMCs | peripheral blood mononuclear cells |
| PDGF | platelet-derived growth factor |
| PEM | protein-energy malnutrition |
| PEPCK | phosphoenolpyruvate carboxykinase |
| PKA | protein kinase activity |
| PKC | protein kinase C |
| PKU | phenylketonuria |
| POMC | proopiomelanocortin |
| PPAR alpha | peroxisome proliferators-activated receptor alpha |
| PPAR | peroxisome proliferator-activated receptors |
| PPARGC1A | peroxisome proliferator-activated receptor gamma, coactivator one alpha |
| PR | protein-restricted |
| PUFA | polyunsaturated fatty acid |
| RARE | retinoic acid reactive elements |
| RD | rumination disorder |

| | |
|---|---|
| RDAs | recommended daily amounts |
| RDAs | recommended dietary allowances |
| RNI | reference nutrient intake |
| ROS | reactive oxygen species |
| SAH | S-adenosylhomocysteine |
| SAM | S-adenosyl essential amino acid |
| SAMe | S-adenosyl-L-methionine |
| SCC | squamous cell carcinoma |
| sHE | soluble epoxide hydrolases |
| SLC6A4 | serotonin transporter gene |
| SNP | single nucleotide polymorphism |
| SNVs | single nucleotide variations |
| SPC | soy phytochemical concentrate |
| SREBP | sterol regulatory element-binding proteins |
| SREBP1C | sterol regulatory element binding protein 1C |
| SULTs | sulfotransferases |
| TC | transfer chain |
| TD | Tajima's D |
| TET | ten-eleven translocation |
| TNF | tumor necrosis factor |
| TSG | tumor suppressor gene |
| TPMT | thiopurine S-methyltransferase |
| TPP | thiamin pyrophosphate |
| T1D | type 1 diabetes |
| T2DM | type 2 diabetes mellitus |
| UDP | uridine diphosphate |
| USOs | Underground Storage Organs |
| VAT | visceral adipose tissue |
| VDR | vitamin D receptor |
| VLDL | very low-density lipoprotein |
| WHO | World Health Organization |
| 5-LO | 5-lipoxygenase |
| 8-OHdG | 8-OH deoxyguanosine |

# PREFACE

The evidence suggesting a direct correlation between greater genome/epigenome damage and an increased risk of bad health outcomes during a person's lifetime has grown stronger over the last decade. Genomics and related fields of study have made significant contributions to our understanding of the cellular and molecular mechanisms that underpin diet–disease interactions. One of the most important determinants for genome stability is exposure to genotoxins in the environment; nevertheless, it is now known that micronutrients are essential as cofactors for many cellular functions, including DNA replication, uracil incorporation into DNA, methylation of CpG regions, DNA oxidation, and/or repair enzymes.

Many researchers have been drawn to the field of nutrigenomics and nutraceuticals in recent years because it allows them to apply genomics knowledge to nutrition research, resulting in associations between specific nutrients and gene expression that are clinically relevant and can be used in disease prevention. Nutritional genomics (nutrigenomics) is regarded as one of the postgenomic era's most promising study fields. Its core concept is that, while changes in gene expression or epigenetic events can turn a healthy phenotype into chronic disease, this process can be reversed or modulated by the use of particular foods and nutraceuticals.

Nutrigenomics explores the occurrence of links between dietary nutrients and gene expression using cutting-edge genomic and proteomic analyses that track the expression of hundreds of genes in response to food. Individual differences in nutrition response are influenced by genetic differences, which may be at the level of single-nucleotide polymorphisms. To build nutrition research programmer aiming at the prevention and treatment of chronic disease using genomics-based nutritional interventions, it is necessary to integrate and apply genetic and genomics technology into nutrition research.

The knowledge of how nutrition affects metabolic pathways is critical to our understanding of a wide range of human disorders. Because diet has such a broad impact on cellular metabolism and, as a result, many human diseases, exposing the field's acquired knowledge to researchers all over the world can benefit health communities interested in providing dietary advice based on personalized genomic data. The nutraceuticals business is rapidly expanding, which is good news for people's health.

We bring together a diverse group of nutritional specialists from around the world in this book to add to the growing body of knowledge in the fields of nutrigenomics and nutraceuticals. As the title suggests, we have a broad perspective on this subject, and we anticipate that nutrigenomics will quickly become a reference book on the subject.

This new book, titled *Nutrigenomics and Nutraceuticals,* comprises 14 chapters in three sections that provide a fresh viewpoint on illness prevention strategies based on an individual's genomic knowledge and nutraceuticals, as well as the diet he or she consumes.

## PART I: FUNDAMENTALS AND TOOLS OF NUTRIGENOMICS

These breakthroughs offer a chance to lay the groundwork for incorporating biological individuality into dietary recommendations, which could have important deep impacts as discussed in Chapter 1, *Nutrigenetics Nutrigenomics Fundamentals* by Neelam L. Dashputre et al.

Utkarsha P. Patil et al. present the nutrigenomics and nutritional search which are implemented for prevention of diseases, improvement of health, longer and better life in Chapter 2, *Nutrigenomics as a Tool for Prevention of Disease and Better Health.*

## PART II: GENETICS INVOLVEMENT IN NUTRITION

Harshada U. Bagul et al., highlight the study of genetic variation for a useful approach of determining the impact of oxidative stress in diversified disease and human aging in Chapter 3, *Variation in Human Genetics and Oxidative Stress.*

Kishor R. Danao et al. reported that both forms of malnutrition such as undernutrition and overnutrition are harmful to one's health in Chapter 4, *Advances in Food and Nutrition Research.*

Various dietary components are emerging as important biological age moderators, and epigenetic clock models are assisting in unraveling the complex interplay between nutrition and the most common diseases in Chapter 5, *Nutrigenomics Research-Based New Approaches to Food Functions*, which is written by Shubham J. Khairnar et al..

Jayesh D. Kadam et al. reported that cancer is a preventable disease that can be avoided with significant lifestyle modifications in Chapter 6, *Dietary, Genomics, and Cancer-Related Facts and Debates or Controversies.*

Dietary components affect gene expression. The influence of vitamin C, niacin, folate and vitamin $B_{12}$, vitamin D3 and resveratrol, quercetin, fisetin, and myricetin, gallic acid and EGCG, soybeans, fava beans, lupin, kudzu, and *Psoralea*, curcumin, grapes on gene expression is reported by Anjali B. Tajanpure and Pranali P. Deshmukh, in Chapter 7, *The Influence of Dietary Components on Gene Expression*.

## PART III: DIFFERENT APPROACHES OF NUTRIGENOMICS

Diet consists of a mixture of protective, mutagenic, and carcinogenic substances, the majority of which are processed by biotransformation enzymes as discussed in Chapter 8, *The Role of Nutritional Factors in Pathogenesis of Diseases*, written by Rahul R. Sable and his colleagues.

Finally, a case study is explained on how gene expression profiles are different with different dietary patterns which in turn can result in long-term illness is described in Chapter 9, *Gene-Based Dietary Advice and Eating Behavior*, written by Pavan B. Udavant and associates.

Umesh D. Laddha and associates provide overlook on the significance of Nutrigenetics and associated tools in Chapter 10, *The Importance of Nutrigenetics and Nutrigenomics for Functional Foods and Personalized Nutrition*.

The understanding of these gene–nutrient interactions will benefit in the provision of personalized diets based on each person's genotype. As a result, it will be feasible to treat existing ailments or avoid future illnesses as discussed by Vasim T. Pathan et al. in Chapter 11, *Nutrients Are Affected By Genetics*.

Individuals known to possess a higher illness threat through genetic testing may be more encouraged to adhere to a dietary intervention as discussed in Chapter 12, *Nutrigenetics Differences and its Origin*, written by Musab Tanzeel and his colleagues.

Role of diet in human genome adaptation is reported by Surabhi B. Patil et al. in Chapter 13, *Human Genome Adaptation to Dietary Changes*.

Rani Kankate et al. talk about the various approaches to epigenetic epidemiology, including the concept of "epigenetic drift" during adulthood in Chapter 14, *Nutritional Epigenomics*.

CHAPTER 1

# FUNDAMENTALS OF NUTRIGENETICS AND NUTRIGENOMICS

NEELAM L. DASHPUTRE, RAHUL R. SABLE,
SHUBHAM J. KHAIRNAR, SURABHI B. PATIL, and JAYESH D. KADAM

*Department of Pharmacology, MET's Institute of Pharmacy, Adgaon, Nashik, India*

## ABSTRACT

There is a hereditary tendency in all diseases. Large worldwide consortia are uncovering genetic differences that relate to complex disorders using GWAS. However, nutrient data are lacking, which is crucial for developing dietary (Keservani et al., 2010a;b; Keservani et al., 2020) recommendations for disease prevention and management. Nutrigenetics and nutrigenomics research focuses primarily on nutritional and gene connections in health and disease, which is critical for individualized nutrition. When gene-nutrient-disease connections are ready to be analyzed as possible instruments to enhance public health, a process will be required. The innovative and promising topic of nutrigenomics and nutrigenetics is the focus of this edition of "Development in Translational Science and Molecular Biology." The first chapter entails the foundational concepts required for understanding the content presented in the remaining chapters of the book. The idea of individualized nutrition is highlighted, as is its potential function in public health, illness prevention, and therapies. Nutrigenetics is the scientific study of the importance of nutrients in gene expression, while nutrigenomics is the study of the role of nutrients in the expression of genes. Diet has influenced gene expression throughout human history, resulting in phenotypes that can successfully adapt to environmental challenges and allow improved

Nutrigenomics and Nutraceuticals. Raj K. Keservani, Eknath D. Ahire, Shubham J. Khairnar, Sanjay J. Kshirsagar, & Rajesh K. Kesharwani (Eds.)

exploitation of food resources. Human development and growth have been aided by these adaptations. Due to technological innovations, it is possible to study not only a single gene but also the entire genome-wide complement of DNA sequence variations, or transcriptomes, in unbiased designs. These breakthroughs offer a chance to lay the groundwork for incorporating biological individuality into dietary recommendations, which could have important and deep impacts.

## 1.1 INTRODUCTION

The framework for all well-being and illness is the connection between genes and environmental factors, or nature and nurture. Susceptibility to a medical problem is determined by genetic factors, and environmental elements, such as nutrition and exercise, affect who will get the condition or illness among the susceptible. The significance of nutrition as an environmental component cannot be emphasized. The DNA-level study of genetic defects and the study of nutrients at the cellular level have been made easier due to innovations in biology at the molecular level and genetics (Simopoulos, 2010). With the formation of the International Society of Nutrigenetics and Nutrigenomics in 2006 and the publication of the Journal of Nutrigenetics and Nutrigenomics in 2008, these ideas have become more commonly utilized by the scientific community. Despite the fact that not all conceptual concerns have been overcome, considerable progress has been made to allow both terms to be used in ways that promote communication clarity. Nutrigenetics examines the physical and genetic responses to micronutrients, nutritional content, and calories based on single-nucleotide polymorphisms, multiple genetic changes, epigenomic marks, and other genetic markers. Over the years, a considerable body of data has been collected, with a focus on coronary heart disease (CVD), malignancies, and other diseases or risk factors. The majority of this research was observational and concentrated on a single or a few specific genes in a candidate genome (Bouchard et al., 2012). The concept of "nutrigenomics" has given rise to the study of nutrition's influence on gene regulation and biochemical, molecular, and biological mechanisms downstream. In the search to understand the unique impacts of nutrients at the molecular, tissue, and organ levels, nutrigenomics will progressively include transcriptomics, proteomics, and metabolomics, as well as more active techniques brought up by fluxomic technologies. Nutrigenomics studies should eventually progress to include strategies that take advantage of newly discovered gene–nutrient interactions (Bouchard et al., 2012; Surana et al., 2021).

**TABLE 1.1** Food Ingredients That are Shown to Alter Genomic and Epigenomic Events.

| Nutrient group | Nutrient group example |
|---|---|
| **Phytochemicals** | Carotenoids, flavonoids, indoles, isothiocyanates, and allyl sulfur |
| **Zoochemicals** | Omega-3 fatty acids, conjugated linoleic acid, and small molecular weight peptides |
| **Fungochemicals** | Lentinan, schizophyllan, β-glucans, and other compounds found in mushrooms |
| **Bacteriochemicals** | Butyrate, equol, and other compounds formed by gastrointestinal tract flora fermentation |

*Source*: Adapted from Milner (2006).

## 1.2 NUTRIGENETICS

Brennan coined the phrase "nutrigenetics" in his 1975 article "Nutrigenetics: New Concepts for Relieving Hypoglycemia." Nutrigenetics and nutrigenomics may provide a foundation for the design of genotype-dependent food products for promoting good health and long-term illness prevention and management. Despite such findings, National General Dietary Guidelines to Protect Against Chronic Diseases have been published without taking into consideration dietetic responses and genetic differences.

### *1.2.1 GENETIC VARIATION AND DIETETIC RESPONSE IN NUTRIGENETICS*

The knowledge of human variation, such as its scope and nature, origins and maintenance, dispersion in communities, relationship with the environment, particularly diet and practice, as well as the implications for normal growth and equilibrium, is a crucial part of genetically-based approaches to diseases. Few of the oldest studies on human biochemical genetics revealed significant diversity within both populations with nutrition concerns. Individual dietary needs differ because of variations in nutritional requirements and interactions of certain minerals with genetic factors and biochemical and metabolic parameters. This is an inborn parameter that must be separated from differences generated by the life cycle (Simopoulos, 2010). Phenylketonuria (PKU) was the first "inborn metabolic error" induced by a single gene defect and treated with a diet low in phenylalanine for nutrigenetic treatment. There are approximately 6000 single-gene diseases, of which 2000 have been recognized.

CHD, hypertension, hyperglycemia, obesity, and malignancies are all multigenetic and multifactorial disorders. Different heterogeneous genes may cause the same symptom. The interaction of genes and environmental variables is responsible for producing the phenotype in polygenic disorders. Single-gene disorders are uncommon, with an occurrence rate of less than 1 in 1000 births or people; they are often under negative selection pressure. GWASs have recognized genetic changes (alleles) that predispose to hyperglycemia, prostate cancer, lupus, age-related vision problems, Crohn's disease, and other disorders in the last few years. On chromosome 9p21, a significant CHD locus has been discovered.

#### *1.2.1.1 THE ROLE OF GENE–NUTRIENT RELATIONS IN DETERMINING NUTRITIONAL DESIRES: FOLATE*

Nutritional wellness is due to the interaction among genetically determined components of digestion, absorption, distribution, conversion, storage, and elimination by bacteria; transformation, storage, and excretion by enzymes in the form of receptors, transporters, enzymes, and hormones are all examples of this; and the environmental aspects of distribution, bioavailability, utilization, and congestion of dietary components. Individuals with familial hypercholesterolemia, related hypertriglyceridemia, and hereditary combined hyperlipidemia, for example, have distinct requirements as a consequence of these connections. In 5–15% of normal populations, a thermolabile variation of the enzyme related to folate, MTHFR (5,10-methylenetetrahydrofolate reductase), is detected. Mild hyperhomocysteinemia is related to the mutation 677CT, which results in a partial enzyme deficit and is related to CHD and a greater occurrence of neural tube anomalies. Some, but not all, cases of people who are homozygous for the heat-sensitive mutation have reported low serum folate concentrations. The red cell folate concentration can be used to estimate the recognized index of tissue folate reserves. A study of 318 nonpregnant and 242 pregnant healthy women discovered the thermolabile mutation of MTHFR (C677T genotype) and examined red cell folate concentrations. Nonpregnant females who have the TT genotype had considerably lower red cell and plasma folate levels, and pregnant females with the TT genotype had even lower levels. If gene mutations producing changes in nutrient status are frequent, as this study indicates, there may not be such a thing as an ordinary population in terms of micronutrient need, as previously thought whenever the reference values for dietetics were created.

#### *1.2.1.2 NUTRITIONAL CHOLESTEROL, GENETIC VARIANCE*

Investigators have known for over 2 decades that the sensitivity of plasma total cholesterol to cholesterol feeding varies greatly across populations.

Individuals with the apolipoprotein E4/4 genotype react to a low-fat and high-cholesterol diet with a rise in blood cholesterol, whereas those who have Apo E2/2 or Apo E3/2 do not. All variations indicate a lowering of serum cholesterol when administered a low-fat or low-cholesterol diet. As a result, the response of blood cholesterol to dietary cholesterol is genotype-dependent. On a low-fat or low-cholesterol diet, Apo E3/4 male individuals had a 23% drop in low-density lipoprotein (LDL) cholesterol, compared to 14% in Apo E3/3 and 13% in Apo E3/2. Furthermore, the level of LDL cholesterol reduction in males was double that of females. Hypercholesterolemia is connected to Apo E4, while Apo E2 guards against elevated cholesterol levels in the context of obesity, hypothyroidism, and hyperlipoproteinemia (Type III). Apo E2 is connected to the buildup of very low-density lipoprotein (VLDL) and chylomicron in the plasma, or diabetes. In hypothyroidism, obesity, or diabetes, only one out of every 50 people with the Apo E2 mutation develops hypertriglyceridemia. Hypertriglyceridemia occurs due to a rise in calorie consumption, trans-fatty acid diet, or carbohydrate intake (especially in women) because triglyceride elimination is genetically set. According to additional research, women with the phenotype Apo E3/2 will benefit less from a rich polyunsaturated fat:saturated (P:S) diet due to a decrease in the more safeguarding HDL (high-density lipoprotein) cholesterol, whereas males with the phenotype Apo E4/3 will benefit the most from the LDL:HDL ratio improvement. As a result, for women with the Apo E3/2 phenotype, a general suggestion to increase the polyunsaturated fatty acid (PUFA) content of the nutrition to reduce cholesterol levels and the danger of CHD is not acceptable. Oat bran has been proven in certain studies to lower serum cholesterol levels, but not in others. Only those with the phenotype Apo E3/3 had a hypocholesterolemic reaction to oat bran after 4 weeks; those with the Apo E4/3 or E4/4 types showed no significant difference. As a result, particular genetic information is required to find a person's healthiest diet. In most cases, general recommendations result in inconclusive data or a lack of benefit (Surana et al., 2021).

#### *1.2.1.3 THE ROLE OF OMEGA-3 AND OMEGA-6 FATTY ACIDS IN CHD (CORONARY HEART DISEASE): GENETIC VARIATIONS IN THE 5-LIPOXYGENASE*

Most of the PUFAs in the US food supply are arachidonic acid (AA), omega-6 fatty acids, and linoleic acid (LA). They are the most abundant PUFAs in all diets, mainly in western diets. Eicosapentaenoic acid (EPA),

Alpha-linolenic acid (ALA), and docosahexaenoic acid (DHA) are omega-3 fatty acids (PUFAs). Dietary intake and genetic differences in the FADS1 and FADS2 gene clusters affect their blood levels. In the generation of prostaglandins, competition between omega-3 and omega-6 fatty acids occurs. At the lipoxygenase and cyclooxygenase levels, an omega-3 fatty acid (EPA) competes with an omega-6 fatty acid (AA) for leukotriene and prostaglandin leukotriene creation. During inflammatory activation, proinflammatory eicosanoids from AA metabolism are produced from membrane phospholipids. To compete with AA, EPA is secreted for enzymatic degradation, leading to fewer inflammatory and chemotactic compounds being created. Dwyer predicted that a mutation within the 5-lipoxygenase gene developer may be related to atherosclerosis, and this impact could combine with the nutritional patterns of competing 5-lipoxygenase substrates because atherosclerosis requires arterial inflammation. A total of 470 healthy middle-aged females and males were randomly selected from the Los Angeles Atherosclerosis Study. A total of 470 healthy middle-aged females and males were randomly selected from the Los Angeles for Atherosclerosis Study in order to assess inflammatory indicators, the researchers employed six meal recalls over the course of a 24-hour period, CRP, carotid artery intima-media thickness, 5-lipoxygenase (5-LO) genotypes, interleukin-6 (IL-6), dietary LA, ALA, AA, 5- LO, EPA, and DHA variant gene was discovered in 6% of the group, according to the data. When comparing carriers of two variable alleles to carriers of two copies of the typical (wild type) allele, mean intima-media thickness improved by 80 ± 19 μm when adjusted for age, sex, height, and racial or ethnic group. The increase in intima-media thickness among transporters of two variable alleles (62 m, $p = 0.001$) in this cohort was similar to that related to diabetes (64 m, $p = 0.001$), the strongest shared cardiovascular risk factor. Higher dietary AA considerably increased the presumed genotype's atherogenic impact, whereas higher dietary omega-3 FAs, DHA, and EPA considerably lowered this effect. Furthermore, when compared to carriers of the normal allele, the level of CRP in the plasma with two variable alleles was enhanced by two factors. As a result, the 5-LO mutated gene identifies a subgroup with an increased risk of atherosclerosis. The interaction between diet and gene also shows that leukotriene-mediated inflammation is increased by dietary omega-6 fatty acids that contribute to the hardening of the arteries in this subpopulation, but EPA and DHA (marine omega-3 fatty acids) reduce it. The prevalence of variant genotypes varied by racial and ethnic groups, with blacks (24.0%), Asians or Pacific Islanders (19.4%), and other ethnic or racial groupings (18.2%) having a substantially higher prevalence than Hispanics (3.6%) and non-Hispanic whites (18.2%). (3.1%). The findings indicate that genetic variations in an inflammatory process, in this case, the leukotriene pathway, can cause atherosclerosis in people (Ahire et al., 2020).

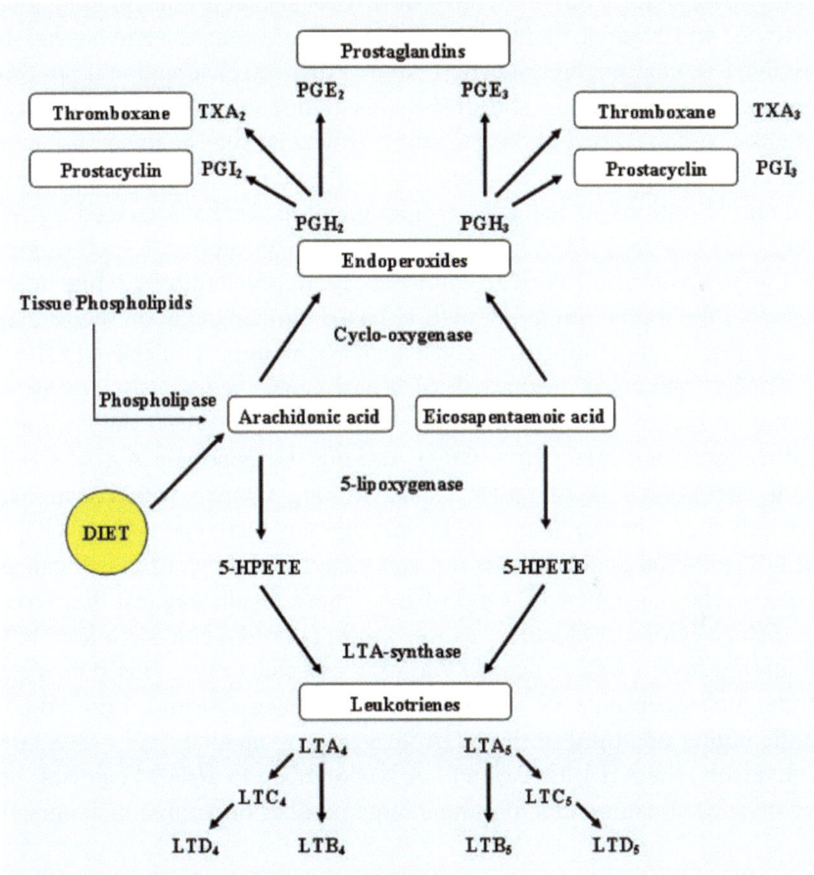


**FIGURE 1.1** Oxidative metabolism of arachidonic acid and eicosapentaenoic acid by the cyclooxygenase and 5-lipoxygenase pathways.

*Source*: Modified from Simopoulos (2010).

These findings may open the path for innovative nutritional and targeted molecular approaches to cardiovascular disease prevention and therapy based on genotype, particularly in non-European countries.

#### 1.2.1.4 BREAST CANCER AND OMEGA-6 FATTY ACIDS: GENETIC VARIANTS IN 5-LIPOXYGENASE

As per several epidemiological data and animal tests, omega-6 fatty acids raise the risk of some malignancies, whereas omega-3 fatty acids lower the risk. Not all studies, however, have shown consistent results. In lung, kidney, prostate, and bladder malignancies, the 5-lipoxygenase pathway has been involved in carcinogenesis progression. Previous observational studies on dietetic fat consumption and breast cancer threats found no link between omega-6 and the threat of breast cancer. However, such analyses did not take into consideration the genetic tendency of omega-6 fatty acid metabolism. Genetic variations in the gene 5-lipoxygenase were discovered by Wang (ALOX5) and the ALOX5AP in association with dietetic LA consumption in Latino, American, African, and San Francisco region's white females based on the multicultural population-based clinical study on breast cancer. The scientists determined no consistent major impact of the ALOX5 and ALOX5AP genomes on the risk of breast cancer across ethnicity or race. However, dietary LA ingestion and ALOX5AP-4900 A>G polymorphisms had a significant effect ($p = 0.03$). Having the genotype AA was related to an increased risk of breast cancer among women who consumed an LA-rich diet, compared to those carrying the genotype AG or GG. The ALOX5AP4900 genotype was not connected to the risk of breast cancer in women who consumed 17.4 g/d of LA. These results suggest that research on dietary calorie content and cancer might take into consideration the type of fat ingested as well as genetic variation and cancer risk. Based on a recent study, males consume 19.3 g/d of LA and women consume 13.4 g/d of LA in the higher one-third of the NHANES (follow-up study 1971–1982 data). Conversely, the AHA advises an LA intake of up to 10% of calories, or 22 g/d on a 2000-calorie diet, putting a large number of women in danger.

#### 1.2.1.5 COX-2 GENETIC VARIANTS AND PROSTATE CANCER

The most common malignancy in men is prostate cancer. Chronic inflammation is gradually being recognized as a part of the elements contributing to the development of this malignancy. Intraprostatic urine reflux, bacterial or viral diseases, or a poor diet can all cause inflammation. Omega-3 PUFAs are dietary elements that are effective anti-inflammatory agents. In studies, COX-2 gene variables have been found to impact prostate inflammation via the COX-2 enzymatic process. COX-2 is an important enzyme in inflammation and fatty acid metabolism. Fradet genotyped nine COX-2 tag SNPs in 466 males diagnosed with advanced prostate cancer and 478 ethnicity- and age-matched controls in a case-control study. A food frequency questionnaire was used to evaluate dietary history in a semiquantitative way. Increasing omega-3 consumption was linked to a reduced aggressive prostate cancer risk, which was particularly strong in males with the rs 4648310 (+ 8897 A/G) gene polymorphisms surrounding COX-2's 3-region. Patients with low omega-3 intake and genetic variables had the most severe cancers, but omega-3 PUFAs protected them. Diet altered the largest dietary effect due to a genetic variation in this gene–nutrient (omega-3) interaction: males with the variable genotype GG or AG and lower consumption of omega-3s had a significantly higher risk than males with the variant gene and high intake of omega-3s. Dark fish consumption was linked to a considerably lower threat of prostate cancer. In comparison to males who never consumed dark fish, males who consumed dark fish one to three times monthly had a 36% decline in prostate cancer risk (OR, 0.64; 95% CI, 0.48–0.86). Furthermore, individuals who consumed dark fish several times a week showed an even lower risk (OR, 0.43; 95% CI, 0.29–0.63) compared to individuals who have never eaten dark fish. Only in males bearing the variant allele rs5275 (+ 6364 A>G) SNP in COX-2 was a frequent intake of fatty fish (rich in omega-3) indirectly correlated with prostate cancer risk, as per a study of 36 Swedish men. Differences in association studies could be addressed by dietary variables and genetic interactions. The most appropriate suggestion for improving public health is to increase omega-3 intake while minimizing omega-6 consumption, concluding with a proper omega-6/omega-3 ratio, as our genes were designed to adapt to the balanced ratio during development.

#### *1.2.1.6 RESPONSE OF BLOOD PRESSURE TO GENOMIC VARIATION AND DIETARY SODIUM*

Essential hypertension is quite prevalent. It's been determined that there is still a link between parental blood pressure and their children's high-tracking profile. Sodium chloride, obesity, alcohol, low calcium, low potassium, stress, an omega-3 fatty acid diet, and physical inactivity are all factors that play a role in the progression of high blood pressure. Blood pressure variations are created by the interaction of many DNA molecules. As a response, various combinations of genes can cause hypertension in different people, even among familial members. Salt retention and pulse pressure are governed by angiotensin II. Humans with angiotensinogen gene variants

(AGT) related to hypertension have significant variations in angiotensinogen serum concentration based on the genetic variation. The angiotensinogen gene is connected to an increase in blood pressure in sibling pairs in England, France, and Utah, as per genetic linkage research. A functional mutation appears to be an A for G gene sequence, a nucleotide stream from the start codon in the region of the promoter angiotensinogen gene-6 start site. The A alteration modifies the binding of a nuclear protein, causing increased gene transcription and angiotensinogen levels, both of which raise blood pressure and modify blood pressure salt sensitivity. After more than 3 years of aftercare in the standard care group, humans with the angiotensinogen gene's AA genotype had higher systolic and diastolic pulse rates and a higher rate of hypertension than those that have the GG genotype, but the variations were minor; whereas, in comparison with the GG genotype, recessive traits were in the middle level of pulse pressure and an increased rate of hypertension. Individuals with the AG and AA genotypes reduced their blood pressure when they decreased their sodium intake; those with the GG gene, on the contrary, were salt intolerant. As a result, encouraging people who have the GG genome to reduce their salt intake would be futile and unnecessary. At 36 months, there was a substantial difference trend in salt restriction lowering diastolic blood pressure compared to normal care ($p = 0.01$), with a greater net lowering of blood pressure with the AA genome (2.2 mm Hg) than people who have the GG genome (+1.1 mm Hg). When it comes to systolic pressure, a similar trend was observed across the three genotypes, but it was not significant. With weight loss, similar blood pressure results were obtained. At P4 107%, a recent genome-wide association study on heart rate and high blood pressure revealed 13 SNPs for systolic pressure, 20 for diastolic pressure, and 10 for hypertension. Interactions with nutrition or exercise, like in other GWA research, have still not been explored. Additionally, neither of the genes shown to have common mutations are part of the well-known blood pressure regulation system. The genes identified are not the same ones that are targeted by current antihypertensive medications. These discoveries could change our way of thinking about the physiology of blood pressure management and provide new targets for treatment or prevention if novel processes emerge from these investigations.

## 1.3 NUTRIGENOMICS

Nutrigenomics is a branch of biology that studies how bioactive dietary factors interact with individual genes at different levels and vice versa in a dynamic, yet regulated, manner. It's only fair that when more evidence about nutrigenomics becomes available, it will provide more relevant details regarding the health consequences of altering dietary habits at the individual level. The overall concept of nutrigenomics is founded on the idea that dietary habits or specific dietary elements can influence cancer risk by changing multiple cellular processes involved in the onset, presence, development, or severity of these diseases. Bioactive food elements can influence the human genome directly or indirectly and hence influence the expression of target genes and gene products. The health implications of a diet are dependent on the proper balance of nutrients. The study of nutrigenomics and its related changes in proteomics and metabolomics events (frequently referred to as "omics") should provide vital clues about bioactive food molecular target components and ultimately provide crucial information that can be applied to a personalized or precautionary approach to nutrition based on accurate data about the probability that an individual would then react to an intervention strategy. A variety of dietary factors are believed to influence genetic and epigenetic events as well as the danger of achieving cancer and tumor behavior. Multiple cancer-related processes have been discovered to be influenced by both essential and nonessential micronutrients. Changes in several cellular systems, including those related to carcinogen metabolism, hormone balance, cell signaling modulation, cell cycle controls, apoptosis, and angiogenesis, may provide this protection. These biologically active components commonly alter multiple processes, and identifying them is the most essential area of active scientific research. The amount and length of exposure to these bioactive components are the most essential determinants in deciding the level and possibly the direction of process modification. According to some data, taking too little of some important nutrients might have negative effects on overall health. Whether this is valid or not for nonessential nutrients is still being researched. When assessing the health repercussions of a change in the intake of food or its active component, the intended use or desired biological response must be analyzed. It may be possible that the eventual phenotypic response is determined by a shift in several systems (Milner, 2006).

### *1.3.1 TECHNOLOGY FOR GENOMIC ANALYSIS*

The HGP (Human Genome Project) provided a lot of knowledge regarding the structure and function of the genome as well as assisted in the annotation of a great number of genes and their end products, that is, proteins. Eukaryotic transcription is now recognized to be variable due to genetic polymorphisms and to be affected by a combination of stimuli like nutrition and infection. As a result, knowing the variety of mRNA types and their abundance aids in predicting contextual regulation processes. However, as the translated mRNA, the protein invariably performs the functions of the produced mRNA, which may not always follow the expected pattern. Multiple proteins (often with very different functions) with variable abundance are generated by alternate sequencing of transcribed RNA and translational regulation, and finally, protein function and half-life are heavily restricted, manifesting in metabolite changes as a consequence of specific genomic information. To gain a greater knowledge of the physiological or pathological status, it's essential to understand the complement of proteins and metabolites in a specific situation. Several disciplines of advancements are been in place for the collaborative characterization of biomolecules. Microarrays 15 (DNA and RNA base pairing mechanisms) and next-generation sequencing methods, for example, assess comparative changes in the global expression of genes, which when cast into pathways aid in adapting to the changes in possible biological pathways modulated by test conditions compared to normal conditions. Soft ionization mass spectrometry techniques like matrix-assisted laser desorption ionization (MALDI) and electrospray ionization were developed, which made biomolecule analysis easier. The ability to detect and analyze a few hundred to thousands of proteins or metabolites in a single process in a few hours is a major benefit of these technologies. Furthermore, the high sensitivity of these techniques overcomes the sometimes limited availability of biological samples. Metabolic pathways are strongly connected, generating a complex network. The real appearance of gene regulation in any physiological or regulatory process is the end product of any metabolism. Metabolomics is the study of all metabolites, including cells, plasma, and tissues, in both qualitative and quantitative terms. NMR and mass spectroscopy (MS) are the two technologies used in metabolomics. Gas and liquid chromatography and capillary electrophoresis are used in MS-based techniques. As a result, under the umbrella name nutrigenomics or systems nutrition, a number of these approaches are used in nutritional research. The term "nutrigenomics" refers to the use of nutrigenetics (the study of DNA, including SNPs), transcriptomics (mRNA), proteomics (protein complement), and metabolomics (metabolite complement) technologies in nutrition (Reddy et al., 2018).

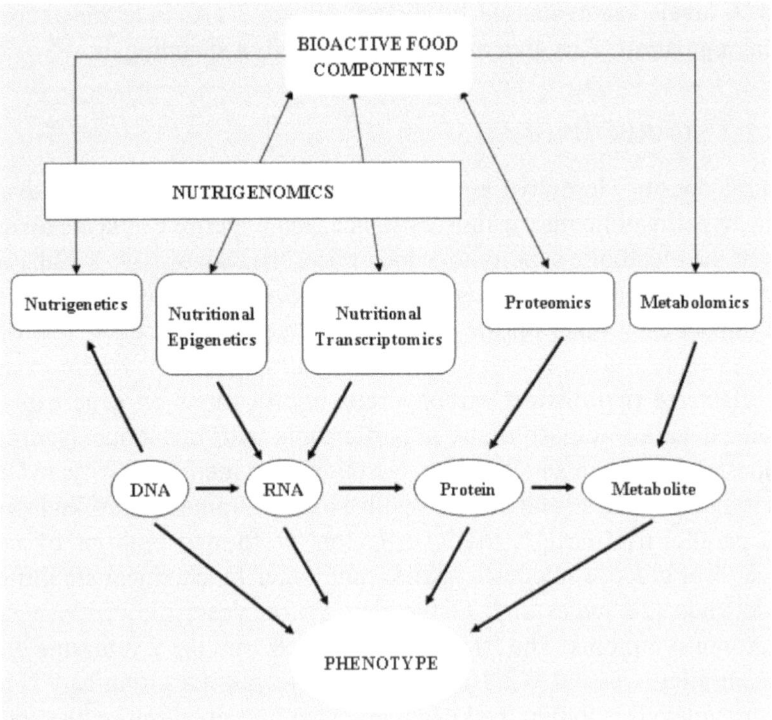


**FIGURE 1.2** Using the "omics" of nutrition to identify how dietary factors contribute to establishing a phenotype.

*Source*: Adapted from Simopoulos (2010).

## 1.3.2 NUTRIENTS, EFFECTS, AND GENE EXPRESSION

### 1.3.2.1 FATTY ACIDS: OMEGA-6 AND OMEGA-3

There seems to have been an increase in the number of applications of molecular biology principles to food ingredient research and important micronutrients as contributors to gene expression control over the last 25 years. The effects of dietary cholesterol and fatty acids on chronic illnesses are particularly important. Dietary cholesterol has a potent effect on the transcription of the hydroxymethylglutaryl (HMG)-CoA reductase gene. In weanling rats and adults, dietary PUFAs reduce fatty acid synthase through hepatic mRNA synthesis and lipoproteinemia. The percentage of unsaturated fatty acids affects the ability to reduce the quantity of mRNA lipogenic proteins. As a result, fish oils containing DHA and EPA are more beneficial than AA. Platelet-derived growth factor (PDGF) and interleukin-1 (IL-1) mRNA levels are decreased by dietary omega-3 FA, indicating transcriptional regulation. This area of research is rapidly expanding.

#### *1.3.2.2 CARBOHYDRATES*

Central obesity (Jarouliya and Keservani, 2019a), hypertension, dyslipidemia, hyperinsulinemia, insulin resistance, and hyperglycemia are all indicators of the metabolic syndrome, which raises the risk of type-2 diabetes and heart disease. Diet, exercise, and genetic susceptibility both seem to interact and impact each other on the onset and course of the disease. The issue is whether or not these interactions affect gene expression. Kallio examined the effects of two distinct carbohydrate modifications on gene expression in subcutaneous adipose tissue in participants with metabolic syndrome: a rye pasta diet with a small insulin secretion after meal reactivity and a high diet with oats, wheat, and potatoes with a high postprandial insulin response. In a parallel trial design, the investigators performed biopsies of adipose tissue, oral glucose tolerance studies, and other biochemical studies on 47 participants (24 males and 23 females, 55–60 years old) with metabolic syndrome symptoms. The BMI was 32.1 ± 3.8, and the 2-h plasma glucose concentration was 8.0 ± 2.3 mmol/L. The rye-plasma group had 71 genes downregulated, including cell death and insulin regulation; on the contrary, 12-week oat-wheat-potato intake had 62 genes upregulated associated with stress, cytokine-chemokine-mediated immunity, and the interleukin cascade.

#### *1.3.2.3 RESTRICTION OF CALORIES*

Crujeiras investigated the interleukin cascade in males on a low-calorie diet for 8 weeks with myeloid cells in the circulatory system. The hypocaloric diet caused all of the subjects to lose a minimum of 5% of their overall body mass. After evaluating microarray data prior to and following the dietary intervention, 385 genes were found to be differently expressed, with 158 genes being overexpressed or upregulated and 227 genes being downregulated. Most of the genes encoding factors involved in nucleotide, DNA, and chromatin metabolism, as well as cellular biosynthetic pathway and regulation of metabolic processes and proteins (Jarouliya and Keservani, 2019b) and lipid pathways, were upregulated (p 0.05), whereas those involved in signal transduction, cell communication, and lipid metabolism were found to be insignificant. Signal transduction, interaction among cells, travel, the response of the immune system, and carbohydrate metabolism genes were mostly upregulated (p 0.05), whereas those involved in nucleotide, DNA, and chromatin degradation, as well as cellular biosynthesis and regulation of metabolic processes and protein and lipid pathways, were mostly downregulated. Intriguingly, IL-8 was linked to a larger decrease in fat mass following the LCD. Bodyweight reduction was linked to a reduction in gene expression in many specific oxidative stress and inflammatory genes. The TNF-alpha/NF-kB signal cascade pathways of caloric restriction reduced the expression of genes involved in signaling pathways, insulin sensitivity, immune response, cell-cell signaling, and protein-DNA assembly, which is complex and involves upregulation of biosynthetic and physiological regulatory processes. After 28 days, extreme LCDs and similar outcomes have been documented in adipose tissue. The regulation of the number of genes implicated in oxidative stress, inflammation, and swelling was altered not just by weight reduction but also by a negative energy balance. It has been proven that calorie restriction affects gene regulation in adipose tissue in the absence of weight reduction. Dahlman examined how energy-restricted diets affected adipose tissue gene regulation in 40 obese women who were assigned at random to a moderate-fat or moderate diet high in carbs or a low-fat/high-carb low-energy (600 kcal/d) diet for 10 weeks. Subcutaneous adipose tissue specimens were tested before and during the diet regime. High-quality RNA samples were collected at both time points from 23 women and hybridized into microarrays containing probes for 8500 human genes. Weight loss of 7.5% of baseline body weight was accomplished on both diets. There were no diet-specific effects on 52 genes that were considerably elevated and 44 genes that were significantly downregulated. The majority of the changes were minor; however, all genes implicated in PUFA synthesis from malonyl-CoA and acetyl-CoA were significantly downregulated (between 25% and 60%). The researchers concluded that macronutrients play a subsidiary role in the process of transformation in adipocyte gene regulation following calorie restriction diets. The most noticeable change, after calorie restriction, was a deliberate attempt to decrease the genetic makeup that controls PUFA synthesis.

#### *1.3.2.4 NUTRITION AND LIFESTYLE*

Carter led the Gene Expression Modulation by Involvement with Nutrition and Lifestyle (GEMINAL) study, a probable single-arm pilot clinical intervention study in 30 men with indolent low-risk prostate cancers, as defined by strict clinical and pathologic criteria, to decrease the threat of metastatic disease. The participants modified their lifestyles totally (low-fat, whole-food, plant-based nutrition; stress management strategies; moderate exercise; psychological group support). For mRNA analysis, prostate needle samples were taken at baseline and 3 months following the lifestyle intervention. After the intervention, microarrays identified 48 transcripts that were upregulated and 453 transcripts that were downregulated. Protein metabolism and modification, intracellular protein traffic, and protein phosphorylation, all of which play crucial roles in cancer, were proven to be affected in a major way by pathway analysis (p 0.05). As a result, intensive dietary and lifestyle changes may influence gene expression in the prostate. Future research may be able to differentiate which components or mixtures of components create specific molecular effects in this intervention due to its complexity. This study, unfortunately, did not look at genetic change.

#### *1.3.2.5 EXERCISE, OBESITY, AND DIABETES*

Frayling discovered a common FTO gene variation that tends to lead to high sugar levels through an influence on BMI while conducting a wide genome search for type-2 diabetes susceptibility genes in 2007. In comparison to those who didn't transfer the risk gene, the 16% of people who were recessive for the FTO risk allele weighed 3 kg more and had a 1.67-fold greater risk of obesity. This link was discovered at the age of seven and reflects a specific rise in fat mass. The FTO gene does not seem to affect fetal weight or infancy body weight. Although the FTO variation (rs9939609) was shown to be linked to type-2 diabetes, this link was lost if BMI was taken into account. The link between obesity and overweight was found in seven white European populations, with 10,172 white European children and 19,424 adults. The rise in BMI was induced by an overall increase in fat, as estimated by waist circumference and fat mass estimates, which included skin folds. Even though the role of the FTO gene product and the biological processes involved is unknown, gene expression profiles suggest that FTO is expressed in a number of ways in tissues, including the brain and muscle. This link has been reported in several other studies that analyzed data from over 80,000 adults and children. In the statistical data, the statistical power is substantially larger than any previous candidate-gene relationship. Andreasen examined the impact of variations of FTO on diabetes, obesity, and other quantitative characteristics of a group of Danes and discovered that low exercise seemed to increase the impact of the FTO rs9939609 genotype on the accumulation of body fat. The FTO rs9939609 has been linked to lower whole body insulin sensitivity in middle-aged people. Physical inactivity is related to a substantial rise in BMI among homozygous carriers of the FTO A allele compared to noncarriers and those heterozygous for the A allele. Cecil investigated the effect of rs9939609 in the FTO gene genetic variable on children's behavior. Energy usage and a preference for high-calorie foods appear to be the key influences. The increase in calorie consumption associated with the SNP A allele is sufficient to explain part or all of the changes in obesity. The rs9939609 genotype did not influence the utilization of resting energy (adjusted for body size), but it increased estimated physical activity. High intake is the most frequent cause of obesity in teenagers (Simopoulos, 2010).

## 1.4 FUTURE PROSPECTS AND RECOMMENDATIONS

The GEI (Genes and Environment Initiative), which contains two primary components, was formed by the US government in 2007. The genetics program for starters is a framework for analyzing genetic diversity in groups of people suffering from various disorders. Second, the exposure biology program is an environmental technology development initiative aimed at developing and validating new tools for monitoring environmental contamination that interacts with genetic variation to cause human disease development. The NIEHS is developing an "exposome." GEI invests in developing new technologies for measuring environmental pollutants, nutritional consumption, and physical activity, as well as determining an individual's biological reaction to these factors using genomics, proteomics, and metabolic engineering. What will the future of genomics and nutrigenetics/nutrigenomics look like? Regional counseling centers and genetic testing will be the sole source of genetic information. Instead, every general practitioner will need to learn how to analyze and explain the genetic information of the patients, as well as how to connect it with a proper nutritional program, the amount and type of exercise, and if necessary, medications. Genetic data will be used by public health specialists to determine groupings of the population that have dramatically varied responses to nutrition, infection, and environmental triggers. Over the next 5–10 years, genotype will become a part of the normal management of a growing number of human illnesses, and pharmacogenetics will be supplemented by nutrigenetics. Knowing who is at risk would be beneficial if it meant minimizing the environmental factors that cause susceptibility to disease. The Pharmacogenomics Knowledgebase will be updated with new information. Nutrigenetics is just getting started in terms of fulfilling its promise. One might see beverages and foods being developed as illness prevention or treatment agents for individuals, families, or subgroups who are prone to a certain condition. The utilization of ketogenic diets in the therapy of children with persistent epilepsy has precedent in pediatrics. Patients with chronic inflammatory disorders such as asthma, arthritis, lupus, atherosclerotic CHD, and ulcerative colitis should eat a balanced diet that is appropriate in fatty acids (omega-3 and omega-6). Novel foods, often referred to as functional foods, have the potential to prevent or treat disease; for example, the addition of omega-3 FAs to products.

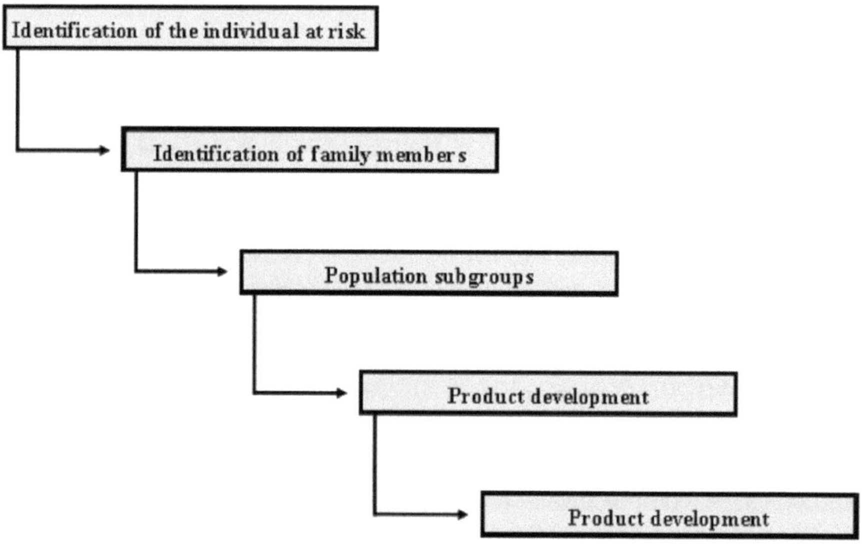


**FIGURE 1.3** Flow of information from genetic screening to marketing of the product.
*Source*: Adapted from Simopoulos (2010).

We can predict the following accomplishments in nutrigenetics and nutrigenomics in the next 15–25 years of the 21st century:

1. Many diseases will have gene and nutrient correlations revealed.
2. Pharmacogenomics and nutrigenomics will be combined and used in the therapeutic process and control of several disorders, including tailored prescriptions and lifestyle changes.
3. To anticipate susceptibility to appropriate nutrition and pharmacological therapies, populations will be classified into subgroups based on genotypes and dietary histories.

4. Gene-based designer diets for coronary artery disease, diabetes, high blood pressure, arthritis, asthma, mental diseases, and other conditions will be developed.
5. Nutrigenomic studies incorporating calorie restriction, specific carbohydrate and fat characteristics, and exercise interventions have identified pathways that help regulate and downregulate gene expression to promote good health, particularly in individuals with metabolic syndrome, obesity, and type-2 diabetes. Increased physical activity and omega-3/omega-6-balanced diets can work as cornerstones for preventative medicine in a variety of chronic diseases.
6. There are hereditary predispositions to all diseases. Human genome variations that affect the risk of complicated diseases are being discovered due to GWA research. Nutrient information is missing from these studies, which is crucial for establishing dietary advice for disease prevention and management as well as individualized nutrition. Age, sex, health condition, ethnicity, obesity, and other factors influence nutritional reactions in addition to genetic variants.
7. As our understanding of gene–nutrient interactions grows, the community will need to build or use suitable educational, legal, social, ethical, and economic frameworks to realize the rewards.
8. Public health and regulatory systems must be established to determine when genomic discoveries, such as gene and nutrient disease links, are ready to be examined as possible tools for improving health screening and dietary ingredients.

Currently, certain dietary guidelines can be given at the community level in terms of public health.

1. Calorie intake requirements must always be evaluated and weighed against energy expenditure. For example, genomic-nutritional research of FTO gene variations in children indicates that early obesity is caused by overeating rather than a decrease in energy expenditure.
2. To be appropriate with an evolutionary understanding of the human diet, diets must be balanced in omega-6 and omega-3 important fatty acids. This balance is best achieved by reducing omega-6-rich oils (sunflower, cottonseed, soybean, safflower, and corn oil) and increasing omega-3-rich oils (canola, flaxseed, perils, chia, and olive oil, which are very low in omega-6).

3. Consume fish a minimum of twice a week.
4. Increase fruit and vegetable consumption (seven or more portions), as well as vitamin D (1500–2000 U) on a daily basis.
5. Consume only lean meats and poultry.
6. Limit salt intake, particularly for individuals with a family history of hypertension, which can be fine-tuned using genotyping. Family history is still an essential part of illness prevention at the family level in public health practice.

## 1.5 CONCLUSION

Having one's genome determined is now relatively affordable, giving information on a lot of differences in essential genes in metabolic pathways that require micronutrients as cofactors. The main question is whether these data can be used to provide reliable and predictable early indicators of undernutrition, micronutrient deficiencies, METs, its consequences, and dietary advice for better health outcomes. Nutrigenomics can also aid in determining interindividual variation in nutrient absorption and utilization, allowing for more individualized dietary recommendations for certain clinical outcomes. As a result, nutrigenomics will have to be a necessary part of nutrition research in the future. We should never ignore the public health messages that are used to promote health; it is noticeable that extremely simplistic messages are not acceptable to all people. Although there are unprecedented opportunities to expand the use of foods and bioactive food components to achieve genetic potential, improve efficiency, and reduce disease risk, the science to make such decisions still has not reached a level of confidence that personalized nutrition suggestions can be made. The ability to find and validate nutrigenetic, nutritional epigenetic, proteomic, and metabolomic biomarkers to establish the origin, effect, and susceptibility to disease will be critical to the success of dietary preemption approaches.

## ACKNOWLEDGMENTS

We thank the MET's Institute of Pharmacy, Bhujbal Knowledge City, Adgaon, Nashik, Principal Dr. Sanjay J. Kshirsagar Sir for their constant support and providing all facilities to complete this work. Also many thanks to all the known and unknowns who helped us right from the writing till the publishing of this chapter.

## KEYWORDS


- **exposome**
- **lipoxygenase**
- **novel foods**
- **nutrigenetics**
- **nutrigenomics**
- **omega fatty acids**
- **PUFA**


## REFERENCES


Ahire, E. D.; Surana, K. R.; Patil, C. D.; Shah, H. S.; Sonwane, G. B.; Talele, S. G. Role of Omega-3 Fatty Acids in Different Neurodegenerative Disorders. In *Applied Pharmaceutical Science and Microbiology: Novel Green Chemistry Methods and Natural Products*, 2020; p 173.

Anjana, R. M.; Deepa, M.; Pradeepa, R.; Mahanta, J.; Narain, K.; Das, H. K.; Adhikari, P.; Rao, P. V.; Saboo, B.; Kumar, A.; Bhansali, A. Prevalence of Diabetes and Prediabetes in 15 States of India: Results from the ICMR–INDIAN Population-Based Cross-Sectional Study. *Lancet Diabetes Endocrinol.* **2017,** *5* (8), 585–596.

Ashrafi, K.; Chang, F. Y.; Watts, J. L.; Fraser, A. G.; Kamath, R. S.; Ahringer, J.; Ruvkun, G. Genome-Wide RNAi Analysis of Caenorhabditis Elegans Fat Regulatory Genes. *Nature* **2003,** *421* (6920), 268–272.

Avis, I. M.; Jett, M.; Boyle, T.; Vos, M. D.; Moody, T.; Treston, A. M.; Martínez, A.; Mulshine, J. L. Growth Control of Lung Cancer by Interruption of 5-lipoxygenase-Mediated Growth Factor Signaling. *J. Clin. Invest.* **1996,** *97* (3), 806–813.

Bailey, R. L.; West Jr, K. P.; Black, R. E. The Epidemiology of Global Micronutrient Deficiencies. *Ann. Nutr. Metab.* **2015,** *66* (Suppl. 2), 22–33.

Berdanier, C. D.; Hargrove, J. L. Eds.; In *Nutrition and Gene Expression*; CRC Press: Boca Raton, FL, 1993; p 579.

Bonne-Tamir, B.; Adam, A. Genetic Diversity among Jews. Oxford University Press: New York, NY, USA, 1992.

Bouchard, C.; Ordovas, J. M. Fundamentals of Nutrigenetics and Nutrigenomics. In *Progress in Molecular Biology and Translational Science*, 2012; vol *108*, pp 1–15.

Bowman, J. E. In *Genetic Variation and Disorders in Peoples of African Origin*; Johns Hopkins University Press, 1990.

Braunwald, E.; Zipes, D. P.; Libby, P. A Textbook of Cardiovascular Medicine. Examination of the Patient, 1992.

Brennan, R. O.; Mulligan, W. C. In *Nutrigenetics: New Concepts for Relieving Hypoglycemia*; M Evans & Company, 1975.

Castro, C. E.; Towle, H. C. In *Nutrient-genome Interaction: Introduction*, Federation Proceedings, 1986; vol *45* (9), 2392–2393.

Caulfield, M.; Lavender, P.; Newell-Price, J.; Farrall, M.; Kamdar, S.; Daniel, H.; Lawson, M.; De Freitas, P.; Fogarty, P.; Clark, A. J. Linkage of the Angiotensinogen Gene Locus to Human Essential Hypertension in African Caribbeans. *J. Clin. Invest.* **1995,** *96* (2), 687–692.

Chadwick, A. C.; Musunuru, K. Genome Editing for the Study of Cardiovascular Diseases. *Curr. Cardiol. Rep.* **2017,** *19* (3), 1–8.

Clark, L. C.; Combs, G. F.; Turnbull, B. W.; Slate, E. H.; Chalker, D. K.; Chow, J.; Davis, L. S.; Glover, R. A.; Graham, G. F.; Gross, E. G.; Krongrad, A. Effects of Selenium Supplementation for Cancer Prevention in Patients with Carcinoma of the Skin: A Randomized Controlled Trial. *JAMA* **1996,** *276* (24), 1957–1963.

Dhirar, N.; Dudeja, S.; Khandekar, J.; Bachani, D. Childhood Morbidity and Mortality in India—Analysis of National Family Health Survey 4 (NFHS-4) Findings. *Indian Pediatr.* **2018,** *55* (4), 335–338.

Dwyer, J. H.; Allayee, H.; Dwyer, K. M.; Fan, J.; Wu, H.; Mar, R.; Lusis, A. J.; Mehrabian, M. Arachidonate 5-lipoxygenase Promoter Genotype, Dietary Arachidonic Acid, and Atherosclerosis. *New Engl. J. Med.* **2004,** *350* (1), 29–37.

Elias, J. E.; Haas, W.; Faherty, B. K.; Gygi, S. P. Comparative Evaluation of Mass Spectrometry Platforms used in Large-Scale Proteomics Investigations. *Nat. Methods* **2005,** *2* (9), 667–675.

Engbersen, A. M.; Franken, D. G.; Boers, G. H.; Stevens, E. M.; Trijbels, F. J.; Blom, H. J. Thermolabile 5, 10-methylenetetrahydrofolate Reductase as a Cause of Mild Hyperhomocysteinemia. *Am. J. Hum. Genet.* **1995,** *56* (1), 142.

Faronato, M.; Muzzonigro, G.; Milanese, G.; Menna, C.; Bonfigli, A. R.; Catalano, A.; Procopio, A. Increased Expression of 5-lipoxygenase is Common in Clear Cell Renal Cell carcinoma. *Histol. Histopathol.* **2007,** *22* (10), 1109–1018.

Fradet, V.; Cheng, I.; Casey, G.; Witte, J. S. Dietary Omega-3 Fatty Acids, Cyclooxygenase-2 Genetic Variation, and Aggressive Prostate Cancer Risk. *Clin. Cancer Res.* **2009,** *15* (7), 2559–2566.

Frosst, P.; Blom, H. J.; Milos, R.; Goyette, P.; Sheppard, C. A.; Matthews, R. G.; Boers, G. J. H.; Den Heijer, M.; Kluijtmans, L. A. J.; Van Den Heuve, L. P.; Rozen, R. A Candidate Genetic Risk Factor for Vascular Disease: A Common Mutation in Methylenetetrahydrofolate Reductase. *Nat. Genet.* **1995,** *10* (1), 111–113.

Geisterfer-Lowrance, A. A.; Christe, M.; Conner, D. A.; Ingwall, J. S.; Schoen, F. J.; Seidman, C. E.; Seidman, J. G. A Mouse Model of Familial Hypertrophic Cardiomyopathy. *Science* **1996,** *272* (5262), 731–734.

Ghosh, J.; Myers, C. E. Arachidonic Acid Stimulates Prostate Cancer Cell Growth: Critical Role of 5-Lipoxygenase. *Biochem. Biophys. Res. Commun.* **1997,** *235* (2), 418–423.

Hennekens, C. H.; Buring, J. E.; Manson, J. E.; Stampfer, M.; Rosner, B.; Cook, N. R.; Belanger, C.; LaMotte, F.; Gaziano, J. M.; Ridker, P. M.; Willett, W. Lack of Effect of Long-term Supplementation with Beta Carotene on the Incidence of Malignant Neoplasms and Cardiovascular Disease. *N Engl. J. Med.* **1996,** *334* (18), 1145–1149.

Hunt, S. C.; Cook, N. R.; Oberman, A.; Cutler, J. A.; Hennekens, C. H.; Allender, P. S.; Walker, W. G.; Whelton, P. K.; Williams, R. R. Angiotensinogen Genotype, Sodium Reduction, Weight Loss, and Prevention of Hypertension: Trials of Hypertension Prevention, Phase II. *Hypertension* **1998,** *32* (3), 393–401.

Inoue, I.; Nakajima, T.; Williams, C. S.; Quackenbush, J.; Puryear, R.; Powers, M.; Cheng, T.; Ludwig, E. H.; Sharma, A. M.; Hata, A.; Jeunemaitre, X. A Nucleotide Substitution in the Promoter of Human Angiotensinogen is Associated with Essential Hypertension and Affects Basal Transcription in vitro. *J Clin. Invest.* **1997,** *99* (7), 1786–1797.

Jarouliya, U.; Keservani, R. K. Pathways Leading to Child Obesity, Chapter 12. In *Global Perspectives on Childhood Obesity: Current Status, Consequences and Prevention*, Second Ed.; Bagchi, D., Ed.; Academic Press, Elsevier, 2019a; pp 137–146, ISBN: 9780128128404.

Jarouliya, U.; Keservani, R. K. Protein Function as Cell Surface and Nuclear Receptor in Human Diseases, Chapter 1. In *Medicinal Chemistry with Pharmaceutical Product Development*; Bharti, S.; Mahapatra, D., Apple Academic Press, CRC Press, Taylor & Francis Group, 2019b; pp 1–32, ISBN: 9781771887106.

Jeunemaitre, X.; Inoue, I.; Williams, C.; Charru, A.; Tichet, J.; Powers, M.; Sharma, A. M.; Gimenez-Roqueplo, A. P.; Hata, A.; Corvol, P.; Lalouel, J. M. Haplotypes of Angiotensinogen in Essential Hypertension. *Am. J. Hum. Genet.* **1997,** *60* (6), 1448–1460.

Jeunemaitre, X.; Soubrier, F.; Kotelevtsev, Y. V.; Lifton, R. P.; Williams, C. S.; Charru, A.; Hunt, S. C.; Hopkins, P. N.; Williams, R. R.; Lalouel, J. M.; Corvol, P. Molecular Basis of Human Hypertension: Role of Angiotensinogen. *Cell* **1992,** *71* (1), 169–180.

Kang, S. S.; Passen, E. L.; Ruggie, N.; Wong, P. W.; Sora, H. Thermolabile Defect of Methylenetetrahydrofolate Reductase in Coronary Artery Disease. *Circulation* **1993,** *88* (4), 1463–1469.

Kaput, J.; Rodriguez, R. L. Nutritional Genomics: The Next Frontier in the Postgenomic Era. *Physiol. Genomics* **2004,** *16* (2), 166–177.

Keservani, R. K.; Kesharwani, R. K.; Sharma, A. K.; Vyas, N.; Chadoker, A. Nutritional Supplements: An Overview. *Int. J. Curr. Pharm. Rev. Res.* **2010b,** *1* (1), 59–75.

Keservani, R. K.; Kesharwani, R. K.; Vyas, N.; Jain, S.; Raghuvanshi, R.; Sharma, A. K. Nutraceutical and Functional Food as Future Food: A Review. *Der Pharmacia Lettre* **2010a,** *2* (1), 106–116.

Keservani, R. K.; Sharma, A. K.; Kesharwani, R. K. Eds.; In *Nutraceuticals and Dietary Supplements: Applications in Health Improvement and Disease Management*, CRC Press, 2020; ISBN: 9781771888738.

Kirke, P. N.; Mills, J. L.; Whitehead, A. S.; Molloy, A.; Scott, J. M. Methylenetetrahydrofolate Reductase Mutation and Neural Tube Defects. *Lancet* **1996,** *348* (9033), 1037–1038.

Klampfer, L.; Huang, J.; Swaby, L. A.; Augenlicht, L. Requirement of Histone Deacetylase Activity for Signaling by STAT1. *J. Biol. Chem.* **2004,** *279* (29), 30358–30368.

Kluijtmans, L. A.; Van den Heuvel, L. P.; Boers, G. H.; Frosst, P.; Stevens, E. M.; van Oost, B. A.; den Heijer, M.; Trijbels, F. J.; Rozen, R.; Blom, H. J. Molecular Genetic Analysis in Mild Hyperhomocysteinemia: A Common Mutation in the Methylenetetrahydrofolate Reductase Gene is a Genetic Risk Factor for Cardiovascular Disease. *Am. J. Hum. Genet.* **1996,** *58* (1), 35.

Langley-Evans, S. Nutrigenetics and the Early Life Origins of Health and Disease: Effects of Protein Restriction. In *Principles of Nutrigenetics and Nutrigenomics*; Academic Press, 2020; pp 113–119.

Laplante, M.; Sabatini, D. M. mTOR Signaling in Growth Control and Disease. *Cell* **2012,** *149* (2), 274–293.

Marian, A. J.; Roberts, R. The Molecular Genetic Basis for Hypertrophic Cardiomyopathy. *J. Mol. Cell. Cardiol.* **2001,** *33* (4), 655–670.

Molloy, A. M.; Daly, S.; Mills, J. L.; Kirke, P. N.; Whitehead, A. S.; Ramsbottom, D.; Conley, M. R.; Weir, D. G.; Scott, J. M. Thermolabile Variant of 5, 10-methylenetetrahydrofolate Reductaseassociated with Low Red-Cell Folates: Implications for Folate Intake Recommendations. *Lancet* **1997,** *349* (9065), 1591–1593.

Reddy, V. S.; Palika, R.; Ismail, A.; Pullakhandam, R.; Reddy, G. B. Nutrigenomics: Opportunities & Challenges for Public Health Nutrition. *Indian J. Med. Res.* **2018,** *148* (5), 632.

Rucker, R.; Tinker, D. The Role of Nutrition in Gene Expression: A Fertile Field for the Application of Molecular Biology. *J. Nutr.* **1986,** *116* (2), 177–189.

Schaeffer, L.; Gohlke, H.; Müller, M.; Heid, I. M.; Palmer, L. J.; Kompauer, I.; Demmelmair, H.; Illig, T.; Koletzko, B.; Heinrich, J. Common Genetic Variants of the FADS1 FADS2 Gene Vluster and their Reconstructed Haplotypes are Associated with the Fatty Acid Composition in Phospholipids. *Hum. Mol. Genet.* **2006,** *15* (11), 1745–1756.

Simopoulos, A. P. Omega-6/Omega-3 Essential Fatty Acids: Biological Effects. *World Rev. Nutr. Diet* **2009,** *99* (1), 1–16.

Simopoulos, A. P. Nutrigenetics/Nutrigenomics. *Ann. Rev. Public Health* **2010,** *31*, 53–68.

Sundaram, S.; Ghosh, J. Expression of 5-oxoETE Receptor in Prostate Cancer Cells: Critical Role in Survival. *Biochem. Biophys. Res. Commun.* **2006,** *339* (1), 93–98.

Surana, K. R.; Ahire, E. D.; Sonawane, V. N.; Talele, S. G. Biomolecular and Molecular Docking: A Modern Tool in Drug Discovery and Virtual Screening of Natural Products. In *Applied Pharmaceutical Practice and Nutraceuticals*; Apple Academic Press, 2021; pp. 209–223.

Surana, K. R.; Ahire, E. D.; Sonawane, V. N.; Talele, S. G.; Talele, G. S. Molecular Modeling: Novel Techniques in Food and Nutrition Development. In *Natural Food Products and Waste Recovery: Healthy Foods, Nutrition Design, and Extraction of Valuable Compounds*, 2021; p 17.

Thissen, J. P.; Ketelslegers, J. M.; Underwood, L. E. Nutritional Regulation of the Insulin-like Growth Factors. *Endocr. Rev.* **1994,** *15* (1), 80–101.

Trujillo, E.; Davis, C.; Milner, J. Nutrigenomics, proteomics, metabolomics, and the Practice of Dietetics. *J. Am. Diet. Assoc.* **2006,** *106* (3), 403–413.

Van Ommen, B.; Stierum, R. Nutrigenomics: Exploiting Systems Biology in the Nutrition and Health arena. *Curr. Opin. Biotechnol.* **2002,** *13* (5), 517–521.

Wolfe, A. R.; Ogbonna, E. M.; Lim, S.; Li, Y.; Zhang, J. Dietary Linoleic and Oleic Fatty Acids in Relation to Severely Depressed Mood: 10 Years Follow-up of a National Cohort. *Prog. Neuro-Psychopharmacol. Biol. Psychiatry*, **2009,** *33* (6), 972–977.

Ye, Y. N.; Wu, W. K. K.; Shin, V. Y.; Bruce, I. C.; Wong, B. C. Y.; Cho, C. H. Dual Inhibition of 5-LOX and COX-2 Suppresses Colon Cancer Formation Promoted by Cigarette Smoke. *Carcinogenesis* **2005,** *26* (4), 827–834.

Zangani, D.; Darcy, K. M.; Shoemaker, S.; Ip, M. M. Adipocyte–Epithelial Interactions Regulate the in vitro development of Normal Mammary Epithelial Cells. *Exp. Cell Res.* **1999,** *247* (2), 399–409.

Zhang, H. F.; Qiu, L. X.; Chen, Y.; Zhu, W. L.; Mao, C.; Zhu, L. G.; Zheng, M. H.; Wang, Y.; Lei, L.; Shi, J. ATG16L1 T300A Polymorphism and Crohn's Disease Susceptibility: Evidence from 13,022 Cases and 17,532 Controls. *Hum. Genet.* **2009,** *125* (5), 627–631.

# CHAPTER 2

# NUTRIGENOMICS AS A TOOL FOR PREVENTION OF DISEASE AND BETTER HEALTH

AKASH M. BHOITE[1], UTKARSHA P. PATIL[1], ATUL R. BENDALE,[1] VAISHALI NAPHADE,[2] and ANIL G. JADHAV[1]

[1]*Sandip Institute of Pharmaceutical Sciences, Nashik, Maharashtra, India*

[2]*School of Pharmaceutical Sciences, Sandip University, Nashik, Maharashtra, India*

## ABSTRACT

The association between nutrition and health depends on the favorable range of genes and functions based on the diet consumed. Dietary changes can improve well-being and diminish the dangers of improper diet-related illness, resulting in a growing interest in nutrigenomics, which directly focuses on nutrient and gene interactions. Nutrigenomics serve as a comprehension tool related to the effect of nutrition on metabolism and homeostatic regulation, which is disturbed in diet-related diseases such as cardiovascular disease, diabetes, eye disease, inflammation, cancer, obesity, etc. To suppress the expansion of such diseases, nutrition researchers are working with nutritional optimization to provide better health to patients. This chapter presents nutrigenomics and nutritional search, which are the tools for prevention of diseases, improving health, and providing a longer and better life.

Nutrigenomics and Nutraceuticals. Raj K. Keservani, Eknath D. Ahire, Shubham J. Khairnar, Sanjay J. Kshirsagar, & Rajesh K. Kesharwani (Eds.)

## 2.1 INTRODUCTION

Molecular biologics and genetics have become increasingly important areas in nutrition research over the last 10 years, replacing epidemiology and physiology. This is mainly a result of three facts that have revealed the inference of nutrition on disease and health, which cannot be recognized without understanding the activity of nutrients at the molecular level (Raju et al., 2016). First, various genome projects get completed and markedly change the research agenda by focusing attention on the significant relevance of genes in human nutrition, which gives new genetic data to be explored. Second, micronutrients and macronutrients are becoming increasingly recognized as powerful dietary signals that can control the metabolic programming of cells and regulate the homeostasis of the body (Francis et al., 2002). Third, nutrition researchers have started to think that genetic disposition may be a major contributor to the mortality associated with diets like cardiovascular disease, cancer, and type-2 diabetes. The connection between diet and health is well established (Willett, 2002; Surana et al., 2021). In 1996, a patent was filed by Ghai et al., which showed that human diseases developed due to an alteration in gene expression because of food development. This was the start of the study of the correlation of nutrition with genes (Elliott and Ong, 2002).

### *2.1.1 WHAT IS NUTRIGENOMICS?*

In 1999, Della Pena introduced the first concept of nutrigenomics as an elementary approach to the discovery of genes that are presently most suitable for substances that contain nutritional importance and are produced or amassed by plants and organisms. Chavez et al. (2003) define nutrigenomics as the study of the molecular link between dietary factors and gene expression. According to Muller and Kersten (2003), nutrigenomics provide knowledge about how nutrition may influence the pathways of metabolism and homeostasis, how regulation interferes with diet-related diseases, and the degree to which genotypes are responsible for the development of diseases. Nutrigenomics examine the nutrients actions on the genome, proteome, and metabolome and also focuses on the consequences of genetic dissimilarity on the diet and diseases interactions with nutrient requirements. Nutrigenomics is the main connection between diet, health, and genomics (Surana et al., 2021).

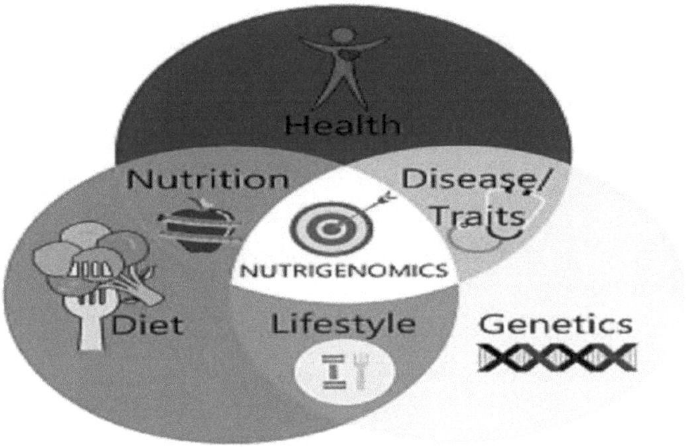


**FIGURE 2.1** Interaction among gene, diet, and human health.

Nutrigenomics is based on five principles. First, the human genome is altered by dietary (Keservani et al., 2010a; Keservani et al., 2010b; Keservani et al., 2020) ingredients, either directly or indirectly, altering the gene expression or structure of genes. Second, diet can play a crucial role in causing several diseases, under certain circumstances for some people. Third, the onset, progression, and severity of chronic disease are caused by certain diet-controlled genes. Fourth, a person's genetic makeup is influenced by the health and disease state balance, and fifth, long-term diseases may be prevented, minimized, or cured through diet based on nutritional needs, nutritional status, and genotype (De Busk et al., 2005).

## Advantages of nutrigenomics

- Increased focus on a healthy diet and lifestyle.
- Increased awareness about the risk of certain conditions.
- Improved health quantity of life.
- Focus on the prevention of diseases.
- Decreased the morbidity and premature morbidity.
- Better understanding of the mechanisms involved in disease susceptibility.

## Disadvantages of nutrigenomics

- Attention is drawn away from other modifiable risk factors.
- Focus only specific nutrients/foods.

- Misleading claims.
- Increased costs associated with personalized diets and designer foods.

**FIGURE 2.2** Impact of diet on genes.

## 2.2 NUTRIGENOMICS AND OMIC TECHNOLOGIES

Genomic tools can be utilized as two different but complementary approaches to molecular nutrition research. The first approach uses molecular tools such as genetics (including DNA), transcriptomics (RNA), proteomics (proteins), and metabolites (metabolites) to hypothesize for the analysis of specific genes and proteins whose expression is determined by nutrition. This existing strategy eventually led to the identification of the regulatory pathways through which diet influences homeostasis (Müller and Kersten, 2003). In the future, this model may provide a solution for understanding the relationship between metabolic and inflammatory signaling pathways. The second, currently mostly theoretical, approach is that of systems biology. Genes, proteins, and metabolic markers associated with certain nutrients or diets can be cataloged to provide molecular "early warning" biomarkers for nutrient-induced homeostasis changes (Müller and Kersten, 2003). While the first strategy provides detailed molecular data on the connection between nutrition and the genome, the second strategy could be of great importance for human nutrition if the problem of taking tissue samples from "healthy" people is solved. With these two general strategies in mind, the following objectives of nutritional genomics research can be defined: Identify genotypes that are harmful factors for diet-related diseases (such as diabetes, high blood pressure, or atherosclerosis) and measure their impact (Raju et al., 2016).

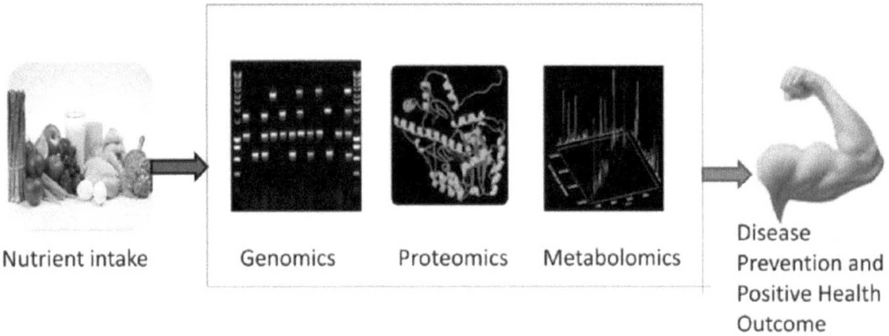


**FIGURE 2.3** Nutrigenomics approach in disease prevention.

## 2.3 NUTRIENTS-GENE INTERACTIONS

The chief agents through which gene expression is influenced by nutrients are transcription factors. In the total human genome, 48 members belong to the nuclear receptor class of transcription factors. Nuclear receptors have a very important role in the regulation of nutrient metabolism, embryonic development, and cellular differentiation. There are three types of nutrient-gene interactions: direct interaction, epigenic interaction, and gene variations. Direct interaction occurs when, sometimes after the interaction with receptors, the nutrients work as transcription factors which bind to DNA and induce gene expression on acute levels. Second, epigenic interactions, in which the composition of DNA is changed or altered by nutrients or sometimes histone proteins, so as to change gene expression, and third, gene variations, in which commonly known gene variations, for example single nucleotide polymorphisms (SNPs), can also lead to alteration in gene function and expression.

### *2.3.1 EFFECT OF NUTRIENTS ON GENE EXPRESSIONS*

#### *2.3.1.1 CARBOHYDRATES-RICH DIET MODIFIES HEPATIC GENE EXPRESSION*

In the pancreatic beta cells, glucose acts as a primary stimulus for insulin synthesis and secretion. In the liver, glucose induces gene expression through insulin, which encodes glucose transporters and glycolytic and lipogenic enzymes such as L-type pyruvate kinase, acetyl CoA carboxylate, and fatty acid synthase. It is observed that genes such as glucokinase can be induced by an insulin concentration apart from the presence of glucose. Glucose-6-phosphate dehydrogenase (G6PD) is a group of enzymes that participate in several metabolic pathways. Its activities are enhanced by the presence of a carbohydrate-rich diet. These changes in the activity of G6PD are responsible for causing an expression change of the G6PD gene. The more lipogenic enzymes, the larger the changes to the rate of gene transcription as a response to dietary changes are seen (Siddique et al., 2009). There are carbohydrates response elements known as ChRE, which are short stretch of DNA, and refer to glucose sensitivity to the gene promoter and have been identified in several glucose responsive genes.

#### *2.3.1.2 FATTY ACID MODULATED TRANSCRIPTION FACTORS*

Either or both high-fat diets and periods of starvation increase the oxidation of fatty acids in several tissues. Among those, the effect of fatty acids on the transcription factor PPAR alpha (peroxisome proliferator-activated receptor alpha) is the most important one. Other factors such as hepatocyte nuclear factor 4 alpha (HNF4 alpha), NF-κB (nuclear factor kappa B), and SREBP1C (sterol regulatory element binding protein 1C) are also responsible (Lee et al.). Researchers have found that polyunsaturated fatty acids (PUFA) suppress SREBP1 expression, thus inducing changes in it at mRNA levels. It can also reduce the SRE transcription-regulated gene, the mature SREBP1 protein, which may lead to increased cholesterol in the stool.

PUFAs show their consequences on gene expression quite rapidly, due to which changes in lipogenic gene expression take place. This has been observed in animals within hours of feeding diets rich in PUFAs. PUFAs are also involved in the increasing expression of genes, which are responsible for encoding enzymes that are critical in hepatic and skeletal muscle fatty acid beta-oxidation and also repress the genes that encode glycolytic, lipogenic, and cholesterol enzymes.

#### *2.3.1.3 EFFECT OF VITAMINS ON GENE EXPRESSION*

Almost all vitamins are involved in gene expression. Vitamin C and vitamin E are useful in preventing DNA and lipid oxidation, but their deficiency is correlated with increased DNA backbone breaks, chromosome breaks, oxidative DNA damage, and the addition of DNA lipid peroxides. Folic acid and cobalamin act as important factors in maintaining and methylating TMP DNA synthesis and efficient recycling of dUMP folic acid, but when deficient, uracil does not properly bind to DNA, increases chromosomal abnormalities, and leads to DNA hypomethylation. Niacin is required as a substrate for poly(ADP-ribose) polymerase (PARP), which is involved in DNA cleavage, recombination, and the maintenance of telomere length. Its deficiency causes an increased rate of unmodified DNA cleavage, an increased number of chromosome breakdowns and rearrangements, and susceptibility to mutagens. Vitamin A is involved in the phosphoenolpyruvate carboxykinase (PEPCK) and insulin-like growth factor (IGF) genes expression. PEPCK is a vitamin A-dependent enzyme that plays a role in the conversion of oxaloacetate to phosphoenolpyruvate, an important step in gluconeogenesis. The deficiency of vitamin A causes changes in the chromosomal structure of retinoic acid reactive elements (RARE), which further leads to changes in binding and coregulatory activities. The complex interaction of PEPCK-RARE and preinitiation reduces the binding of RNA polymerase II to the PEPCK promoter and also increases gluconeogenesis in the absence or presence of PEPCK. In mice with adequate amounts of vitamin A, PEPCK gene expression was strongly induced under fasting conditions, resulting in a decrease in blood sugar levels. However, PEPCK gene expression was negligible or absent in vitamin A-deficient individuals. Dietary vitamin A levels modulated serum IGF-I concentrations and the expression of IGF-I, IGF-IR, and IR genes in various tissues. Vitamin E has many biological functions, including being an important fat-soluble antioxidant, and unlike many other vitamins, it is not an enzyme cofactor. In some cases, vitamin E plays an important role in gene expression regulation. Several studies reported a reverse relationship between L-α-tocopherol (the main biologically active vitamin E) and some types of cancer. However, not all studies come to the same conclusions, and others have found neutral or harmful effects on the risk of cancer and heart disease. This lack of consensus may be due to the significant differences in doses tested. A 2005 metaanalysis showed that a study used a dose range of 16.5–2000 IU/day. Biotin is directly involved in important metabolic processes such as gluconeogenesis, fatty acid synthesis, and amino acid catabolism (through pyruvate carboxylase) and is a component of acetyl coenzyme carboxylase, propionyl-CoA carboxylase, and b-methylcrotonyl-CoA carboxylase. It also has an important function as a component of ornithine transcarbamylase (OTC) gene expression. Deficiency in biotin utilization causes hyperammonemia and acidosis in humans, in addition to clinical symptoms such as lethargy, seizures, rashes, and alopecia, as well as the synthesis of glutamine, which is catabolized by the urea cycle enzyme and glutamine synthase. Biotin can alter the activity and expression of urea cyclase genes. The mechanism is that over-the-counter drugs convert ornithine to citrulline, which is a key step in the urea cycle. Biotin is an OTC prosthetic group, so biotin deficiency is OTC.

#### *2.3.1.3.1 Effect of Protein on Gene Expression*

Proteins are very essential for the growth, development of immunity, the normal functions of the entire body, and the structure, besides processes like reproduction and production.

The purpose of proteins in the body does not only follow the functions mentioned above at macro levels but also at gene levels. Both quantitatively and qualitatively, proteins affect gene expression.

A low-protein diet was fed to rats, which in turn caused a reduction in insulin secretion as there was depletion in the pancreatic beta-cell mass, a low response of the remaining beta cells to nutrients, and a lowering of the protein kinase activity (PKA). The genes that encode IGF are highly sensitive to the nutritional status of the body.

The diet containing low protein that was fed to rats reoriented the gene expression, which is responsible for proteins related to insulin biosynthesis, secretion, and cellular remodeling.

Insulin is the primary regulator of IGFBP-1 expression in these tissues, and the primary effect of insulin is rapid inhibition of transcription (Lee et al.). Some important protein factors affecting their functions through genetic levels can be stated as: Soy protein decreases cholesterol levels by acting at the gene level, whereas calcium binding protein (CaBP) is an important protein that is mainly responsible for the absorption of calcium from the intestine. Calcium absorption is influenced by the level of vitamin D, which acts at the nuclear level and is involved in transcription and translation. Inhibition of protein and RNA synthesis also prevented downregulation of AS mRNA levels by high concentrations of asparagines. The secretions of insulin are affected by levels of protein kinase C (PKC), K+ channel protein, calcium ion ($Ca^{2+}$), and PKA. The increase in the ATP to ADP ratio levels is caused by glucose metabolism, which in turn closes the K+ ATP channels, which leads to the depolarization of beta cells. The depolarized beta cells open the voltage-gated $Ca^2$+ channels, which lead to the influx of $Ca^2$+ ions and the exocytosis of insulin granules. Feeding a low-protein diet not only decreases the insulin levels but also decreases the movement of intracellular calcium.

#### 2.3.1.3.2 *Effect of Minerals on Gene Expression*

Zinc is essential for many structural, catalytic, and regulatory functions. Its deficiency is associated with stunted growth, dermatitis, taste disturbance, delayed puberty, hematological abnormalities, and immune dysfunction. As a result of the analysis of the oligonucleotide sequence of the transcriptome of HT-29 cells under conditions of low or normal zinc levels for 16 h, changes in the mRNA levels in 309 genes out of 10,000 analyzed genes were detected. When intracellular zinc was depleted, 231 genes were downregulated and 78 genes were activated. The Zn-dependent transcription factor KLF4 (Kruppel-like factor 4) is involved in the production of the cellular protein HT-29. Other proteins containing Zn as a component are ATP synthetase, cytochrome c, and Zn-regulated NADP dehydrogenases I and II.

Iron and aluminum: It has been shown to increase miR-125b and -146a expression through an NF-κB-dependent mechanism. These miRNAs are involved in astrocyte proliferation and inflammatory responses, respectively, and are significantly upregulated in Alzheimer's disease. Specific regulation of microRNAs by Fe alone has not been demonstrated. However, an important role of heme iron in the regulation of miRNA processing has been demonstrated. DGCR8 forms a very stable and active complex with heme iron. When reduced to iron, the DGCR8 complex loses its ability to process basal (pri)-miRNA transcripts.

Magnesium: Mg has been shown to interact with the microRNA processing mechanism along with the $Mg^{2+}$-dependent protein RISC (RNA-induced silencing complex). Mg ions are also found in the small RNA-binding domain of argonaute proteins and are important for sequence-specific miRNA-target interactions, facilitating miRNA binding to argonaute proteins and thus playing a role in miRNA-target cleavage. Mg ions have also been shown to regulate the overall stabilization of the argonaute protein.

## 2.4 NUTRIGENOMICS: GENES AND HEALTH AFFECTED BY NUTRITION

Nutrigenomics is the study of the interaction between genes and nutrients. The flowchart gives a cavernous understanding of the effect of gene variation on health. The DNA sequence is responsible for health outcomes and

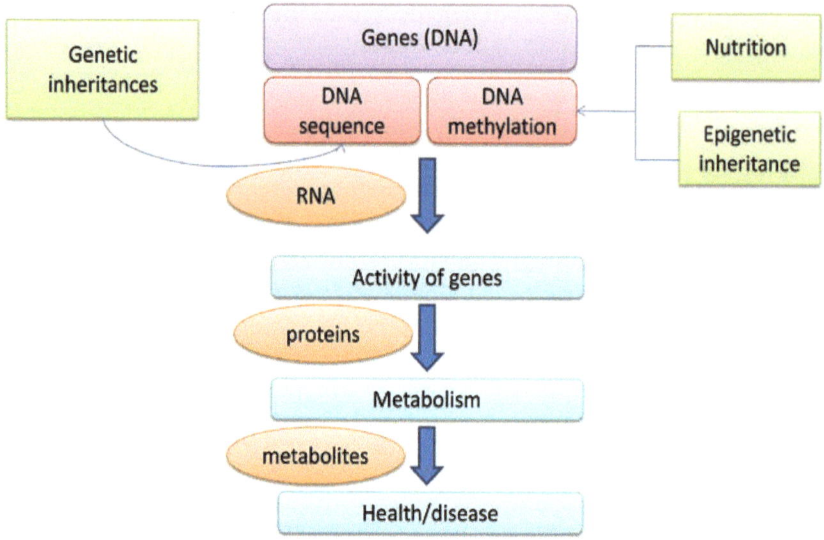


is an exponent of "epigenetics," which facilitates the idea that traits may be inherited from one generation to the next. DNA methylation is a process in which DNA may be chemically altered without change in DNA sequence, that is, methyl groups get attached to DNA nucleotides. Some genes affected by DNA methylation may be inherited and changed by dietary components. Nutrition is a complex result of both genetics and DNA methylation, which results in nutrition and epigenetics.

## 2.5 APPLICATIONS OF NUTRIGENOMICS IN PREVENTION OF DISEASES

### *2.5.1 CARDIOVASCULAR DISEASE*

The very most common origin of morbidity and mortality is multiplex disease, caused by a combination of genetic factors and lifestyle factors such as diet. Cardiovascular disease (CVD) includes a number of related conditions such as coronary heart disease (CHD), peripheral arterial disease, cerebrovascular disease, congenital and rheumatic heart disease, and venous thromboembolism. Research in nutrigenomics shows that genotype affects response to nutrients that may help reduce CVD disease (Corella and Ordovas, 2009). In dietary personalization, cardiovascular medicine, and CVD prevention and treatment, nutrigenomics has become a reality. Atherosclerosis and CHD are usually linked to hyperlipidemia. The therapy in the treatment involves changes in the patient's diet, pharmaceutical therapy, and physical exercise, but each individual responds individually to the therapy. Women with a SNP (-75G/A) in the apolipoprotein gene have a higher level of high-density lipoprotein cholesterol with an intake of dietary unsaturated fatty acids. A variant individual manifests growing protective HDL levels by following an elevation of dietary unsaturated fatty acids consumption when compared with the G variant, which shows similar quantities of dietary unsaturated fatty acids (Corthésy, 2005).

### 2.5.2 CANCER

Diet is a significant, mutable environmental factor that affects tumor and cancer risk. The etiology and prevention of cancer are closely linked to diet. Plant-based medicines are prescribed in Ayurveda to treat cancer (Sinha et al., 2003). Interactions between diet and genes have also made a pronounced contribution to various cancer risks. Diet influences about 30–40% of cancer cases, but the precise amount is unknown and depends on the type of cancer and a particular component of the diet. Research has proposed that prostate, colon, breast, lung, and liver cancers are related to dietary intake (Davis and Milner, 2004). Vegetables and fruits contain bioactive ingredients that can protect against carcinogenesis through several mechanisms, like inhibiting metabolism by extending detoxification. Several constituents of plant foods are known to modulate detoxified enzymes in vivo and in vitro, for example, flavonoids like quercetin, rutin, and genistein; phenols like curcumin, epigallocatechin-3-gallate, and resveratrol; allyl sulfur compounds, indoles, isothiocyanates, and selenium (Keum et al., 2004; Milner, 2001). In 2005, Kamra et al. studied dietary bioactive components, their physiological actions in organisms, and the role of diet in the treatment of cancer.

#### 2.5.2.1 OPPORTUNITIES FOR IDENTIFYING TARGETS BY USING RODENT MODELS

Researchers have developed mouse strains that contain cancer-linked genes that are inactivated or overexpressed to study carcinogenesis and evaluate prevention strategies. It is estimated that approximately 50% of overall human tumors have a mutation in the p53 tumor suppressor gene. The energy restriction in p532 knockout mice accelerated spontaneous tumor expansion by 75%, reduced leptin levels, and reduced IGF-1, which also decreased the growth of thrombocyte cells. The tumor restrainer gene p21 has been evaluated in other studies for its role in determining how long people will live on a diet rich in fat and low in vitamin D and calcium, as well as the effect of alleles from each parent on the size of the response with regard to longevity. Both animal and human findings suggest that genetics has a substantial role in influencing the absorption, metabolism, and molecular targeting of bioactive dietary substances (Raju et al., 2016).

### *2.5.3 DIABETES*

An impaired metabolism of carbohydrates, proteins, and lipids is a sign of type-2 diabetes. This is caused by a sedentary lifestyle and poor nutritional habits. In studies on gene-diet interactions, the adipopectin gene polymorphism was implicated in the evolution of resistance to insulin and type-2 diabetes, which was exacerbated in individuals consuming diets with high glycemic loads (Mohan et al., 2007). Patients with diabetes may have individual responses to certain foods due to their genetic differences. According to a study by Kaput et al. (2007), physicians suggest modifying the diet and increasing exercise for patients with type-2 diabetes, yet only 20% of patients literally control their symptoms (Kaput et al., 2007). This might be where nutrigenomics help control type-2 diabetes. Nutritional genomics can inform clinical diagnosis of phenotypes like glucose tolerance and insulin level, as can metabolomics diagnostics, where diabetes biomarkers can be evaluated. Research has shown that higher levels of SREB-1a and SREBP-1c (t-RNAs that activate genes that elaborate in cholesterol synthesis and triglyceride uptake) contribute to diabetes progression. Some fibers also play an important role in determining nutrient bioavailability by modulating cholesterol absorption in the digestive tract. Researchers may use this to understand the composite relationship between diabetes and genes and come up with more effective dietary strategies for diabetes.

The effect of some vitamins on type-2 diabetes can be attributed to their actions on nuclear receptors. The derivatives of vitamin A, mostly retinoic acid, control gene expression by triggering retinoic acid receptors and RXRs. In several insulin-sensitive tissues, such as the liver and adipose tissues, retinoids have been shown to regulate gene expression (Kaput et al., 2007).

### 2.5.4 *EYE DISEASES*

Researchers have discovered a connection between nutrition and gene-related factors in age-related eye diseases such as age-related macular degeneration (AMD) and cataract (Resnikoff et al., 2004). The eye is particularly responsive to oxidative stress because of its direct exposure to light. In AMD and probably in cataracts, antioxidants like vitamins C, E, and zinc are protective. Furthermore, lutein and zeaxanthin are two carotenoids that play important roles in the eye: they form macular pigment in the retina and accumulate in the lens. Docosahexaenoic acid (DHA) plays an important role in the photoreceptor, which serves structural and protective functions. Protective functions involve anti-inflammatory, antiapoptotic, and antiangiogenic actions, as well as particular actions like leading to growing lipid degeneration in the retinal pigment epithelium and improving lysosomal acid lipase. The result of two cross-sectional studies suggests a 50–60% reduction in AMD risk after multivariate adjustment for fish consumption, which is an important source of DHA. According to a case-control study of eye disease, elevated intake of omega-6-PUFA was related to a growing risk of exudative AMD by multivariate adjustment. Consumption of fish and omega-3-PUFA was significantly associated with a decrease in risk for AMD by 31% after gender and age adjustment, but not after multivariate adjustment. The consumption of large quantities of DHA was linked to a reduced risk of AMD, while the consumption of large amounts of alpha-linolenic acid was linked to a growing risk. Fish consumption was associated with a decrease in the risk of AMD, only for those with low intakes of linolenic acid. Recently, researchers discovered that omega-3 and omega-6 PUFA may also be used in other eye diseases like glaucoma or dry eye syndrome. AMD and cataracts are strongly affected by genetics, as familial and twin studies show (Cécile et al., 2007).

In the case of cortical cataract, a recent study identified more loci on chromosome 6p12q12 for cortical cataract. The interaction between nutritional factors and genetic variability poses a new challenge in explaining health differences between individuals. Several nutritional factors have direct effects, including the antioxidant properties or functions of DHA; however, numerous elements influence cellular effects and interrelate with genes. Nutrigenomics is potentially applicable to all genes involved in metabolism or activities of nutritional components related to eye disease, as well as nutrients implicated in gene-related activities with eye disease, thus opening a rich research field.

**TABLE 2.1** Nutrient Deficiency Diseases and Preventive Food Sources.

| Nutrients | Gene alteration | Diet deficiency disease | Potential food intake |
|---|---|---|---|
| | **Vitamins** | | |
| Vitamin A (Retinol) | Repression of PEPCK gene | Termination of pregnancy and fetal death | Carrots, spinach, turnip, kale, apricots, cantaloupe, bell pepper, papaya, mango, peach, beef liver, and chicken liver |
| Vitamin B3 (Niacin) | DNA hampering | Nerve complications and memory loss | Tuna, kidney, liver, poultry, carrots, turnips, pork, prawns, celery, mushrooms, beans, almonds, wheat products, rice bran, and milk |
| Vitamin B6 (Pyridoxine) | Breaking of chromosome and trammel DNA repair or methylation | Cancer, coronary heart disease, brain problem, male infertility, and leukemia | Spinach, potato, bell peppers, turnip, mushroom, garlic, cauliflower, banana, chicken, pork, beef, salmon, tuna, and turkey |
| Vitamin B9 (Folic acid) | Breaking of chromosome and trammel DNA repairing/methylation | Heart disease, brain problems, male infertility, and leukemia | Liver, egg yolk, beetroot, kidney, orange, spinach, asparagus peas, and cowpeas |
| Vitamin B12 (Cobalamin) | Breaks chromosome and trammel DNA repair/methylation | Memory loss, cancer, heart disease, male infertility, and leukemia | Liver, sardines, salmon, clam, beef, milk, cheese, and yogurt |
| Vitamin D (Calciferol) | Prevention of gene variation | Cancer | Cod liver oils, beef liver, mackerel, tuna, egg, orange juice, cow milk, yogurt, and cheese |
| Vitamin E (Tocopherols) | Mimics radiation damage | Colon cancer, heart disease, and immune dysfunction | Tomato, spinach, broccoli, blueberries, mangoes, kiwi, papaya, and almonds |
| | **Other nutrients** | | |
| Flavonoids | Alters gene expression | Cancer | Onion, green bean, broccoli, curly kale, endive, celery, cranberry, orange juice, grape fruits, lemons; red, blue, and purple berries, peppers, tomatoes, and eggplants |

**TABLE 2.1** *(Continued)*

| Nutrients | Gene alteration | Diet deficiency disease | Potential food intake |
|---|---|---|---|
| Fatty acids | Alters gene expression | Obesity, CVD, and diabetes | Salmon, sardines, herring, mackerel, soy oil, sunflower oil, palm oil, flaxseeds, rapeseeds, peanuts, walnuts, almonds, mustard seeds, cloves, oregano, cauliflower, and broccoli |
| Protein | Changes gene expression | Kwashiorkor and marasmus | Milk, soya milk, tofu, yogurt, egg, cheese, almonds, peanuts, cashew, poultry, and broccoli |
| Zinc | Chromosome breaks | Brain and immune dysfunction | Oysters, beef, crab, pork, lobster, chicken, spinach, broccoli, cashewnuts, almond, milk, cheese, and yogurt |

## 2.6 CONCLUSION

Nutrigenomics is a hastily rising science. The improvements in vitamin studies with the use of numerous molecular technologies will truly replace the expertise of nutrient, gene, and fitness interactions. By focusing on the precise gene through dietary manipulation, it may be viable to get the favored overall performance in terms of health in addition to production. The microarrays or different advances in DNA technology in nutrigenomics studies allow screening of massive numbers of genes simultaneously, giving a complete picture of the variants of gene expression styles and offering factors for complex regulatory interactions, such as those among diet, vitamins, and genes. These strategies will decorate the researcher's skills to hold an individual's health, optimize overall performance, and enhance unerringly analogous to the declaration of Hippocrates, the father of drugs, who said, "Leave your drug within the chemist's pot if you could heal the affected person with food."

Nutrigenomics methods promise to set new standards for nutritional research and will revolutionize the perspective of nutrient necessities and interactions. The gene expression studies will outline the unknown outcomes of precise vitamin and nutrient interactions. With knowledge of those new nutrient outcomes, it will likely be capable of defining and taking a look at new nutritional concepts. Nutrigenomics is a tool that has been seeded for programmed nutrition in both animals and humans. More importantly, the extra-precise measures of nutrient effects afforded through biomarkers and molecular profiling techniques will provide new gear for comparing nutrient necessities and food regimen methods that ultimately result in customized vitamins in keeping with the individual's genotype.

## KEYWORDS

- **nutrigenomics**
- **diet supplement**
- **health**
- **nutrients-gene interaction**
- **gene alteration**
- **nutrition research**

## REFERENCES

Afman, L.; Muller, M. Nutrigenomics: From Molecular Nutrition to Prevention of Disease. *J. Am. Diet. Assoc.* **2006,** *106* (4), 569–576.

Berná, G.; Oliveras-López, Mj.; Jurado-Ruíz, E.; Tejedo, J.; Bedoya, F.; Soria, B. Nutrigenetics and Nutrigenomics Insights into Diabetes Etiopathogenesis. *Nutrients* **2014,** *6*, 5338–5369.

Calerina, R. De; Martinez, J. A.; Kohlmeier, M. In *Principles of Nutrigenetics and Nutrigenomics: Fundamental for Individualized Nutrition*; Elsevier: 2020.

Chavez, A.; De Chávez, M. M. Nutrigenomics in Public Health Nutrition: Short Term Perspectives. *Euro. J. Clin. Nutr.* **2003,** *57*, S97–S100.

Cho, E.; Hung, S.; Willett, W. C.; Spiegelman, D.; Rimm E. B.; Seddon, J. M.; Colditz, G. A.; Hankinson, S. E. Prospective Study of Dietary Fat and the Risk of Age-Related Macular Degeneration. *Am. J. Clin. Nutr.* **2001,** *73*, 209–218.

Congdon, N.; Broman, K. W.; Lai, H.; Munoz, B.; Bowie, H.; Gilbert, D.; Wojciechowski, R.; West, S. K. Cortical, but not Posterior Subcapsular, Cataract Shows Significant Familial Aggregation in an Older Population after Adjustment for Possible Shared Environmental Factors. *Ophthalmology* **2005,** *112*, 73–77.

Corella, D.; Ordovas, J. M. Nutrigenomics in Cardiovascular Medicine. Circ. Cardiovasc. Genet. **2009,** *2* (6), 637–651.

Corthésy, T. I. Nutrigenomics: The Impact of Biomics Technology on Nutrition Research. *Ann. Nutr. Metab.* **2005,** *49*, 355–365.

Davis, C. D.; Milner, J. Frontiers in Nutrigenomics, Proteomics, Metabolomics and Cancer Prevention. *Mutat. Res.* **2004,** *551* (1-2), 51–64.

Debusk, R. M.; Fogarty, C. P.; Ordovas, J. M.; Kornman, K. S. Nutritional Genomics in Practice: Where do we Begin? *J. Am. Diet. Assoc.* **2005,** *105*, 589–598.

Delcourt, C. Application of Nutrigenomics in Eye Health. In *Nutrigenomics Opportunities in Asia*, 2007; *60*, pp 168–175.

Delcourt, C.; Cristol, J. P.; Tessier, F.; Leger, C. L.; Descomps, B.; Papoz, L. Age-related Macular Degeneration and Antioxidant Status in the POLA Study. POLA Study Group. Pathologies Oculaires Liees a l'Age. *Arch. Ophthalmol.* **1999,** *117*, 1384–1390.

Elliott, R.; Ong, T. J. Nutritional Genomics. BMJ **2002,** *324* (7351), 1438–1442.

Fisher, S. A.; Abecasis, G. R.; Yashar, B. M.; Zareparsi, S.; Swaroop, A.; Iyengar, S. K.

Francis, G. A.; Fayard, E.; Picard, F.; Auwerx, J. Nuclear Receptors and the Control of Metabolism. *Annu. Rev. Physiol.* **2003,** *65*, 261–311.

Ghosh, D. Future Perspectives of Nutrigenomics Foods: Benefits vs. Risks. Indian. *J. Biochem. Biophy.* **2009,** *46*, 31–36.

Gomase, V. S.; Tripathi, A. K.; Tagore, S. Genomics: New Aspect of Cancer Research. *Int. J. Syst. Biol.* 2009, *1* (1), 01–19.

Gorin, M. B.; Breitner, J. C.; De Jong, P. T.; Hageman, G. S.; Klaver, C. C.; Kuehn, M. H.; Seddon, J. M. The Genetics of Age-Related Macular Degeneration. *Mol. Vis.* **1999,** 5–29.

Heuberger, R. A.; Mares Perlman, J. A.; Klein, R.; Klein, B. E.; Millen, A. E.; Palta, M. Relationship of Dietary Fat to Age-Related Maculopathy in the Third National Health and Nutrition Examination Survey. *Arch. Ophthalmol.* **2001,** *119*, 1833–1838.

Iyengar, S. K.; Klein, B. E.; Klein, R.; Jun, G.; Schick, J. H.; Millard, C.; Liptak, R.; Russo, K.; Lee, K. E.; Elston, R. C. Identification of a Major Locus for Age-Related Cortical Cataract on Chromosome 6p12-q12 in the Beaver Dam Eye Study. *Proc. Natl. Acad. Sci. U. S. A.* **2004,** *101*, 14485–14490.

Kamra, A.; Kalavapudi, S.; Sudhakaran, T. R. Exfoliated Colonic Epithelial Cells: Surrogate Targets for Evaluation of Bioactive Food Components in Cancer Prevention. *J. Nutr.* **2005,** *135* (11), 2719–2722.

Kang, J. H.; Pasquale, L. R.; Willett, W. C.; Rosner, B. A.; Egan, K. M.; Faberowski, N. Hankinson, S. E. Dietary fat consumption and primary open-angle glaucoma. *Am. J. Clin. Nutr.* **2004,** *79*, 755–764.

Kaput, J.; Noble, J.; Hatipoglu, B.; Kohrs, K.; Dawson, K.; Bartholomew, A. Application of Nutrigenomic Concepts to Type 2 Diabetes Mellitus, Nutrition, Metabolism & Cardiovascular Diseases. *Nutr. Metab. Cardiovasc. Dis.* **2007,** *17*, 89–103.

Keservani, R. K.; Kesharwani, R. K.; Sharma, A. K.; Vyas, N.; Chadoker, A. Nutritional Supplements: An Overview. *Int. J. Curr. Pharm. Rev. Res.* **2010b,** *1* (1), 59–75.

Keservani, R. K.; Kesharwani, R. K.; Vyas, N.; Jain, S.; Raghuvanshi, R.; Sharma, A. K. Nutraceutical and Functional Food as Future Food: A Review. *Der Pharmacia Lettre* **2010a,** *2* (1), 106–116.

Keservani, R. K.; Sharma, A. K.; Kesharwani, R. K. Eds.; In *Nutraceuticals and Dietary Supplements: Applications in Health Improvement and Disease Management*; CRC Press, 2020; ISBN: 9781771888738.

Klein, B. E.; Klein, R.; Lee, K. E.; Majewski, J.; Schultz, D. W.; Klein, M. l., Seddon, J. M.

Malek, G.; Johnson, L. V.; Mace, B. E.; Saloupis, P.; Schmechel, D. E.; Rickman, D. W.; Toth, C. A.; Sullivan, P, M.; Rickman, C. B. Apolipoprotein E Allele-dependent Pathogenesis: A Model for Age-Related Retinal Degeneration. *Proc. Natl. Acad. Sci. U. S. A.* **2005,** *102*, 11900–11905.

Mohan, V.; Sandeep, S.; Deepa, R.; Shah, B.; Varghese, C. Epidemiology of Type 2 Diabetes: Indian Scenario. *Indian J. Med. Res.* **2007,** *125*, 217–230.

Mohan, V.; Sudha, V.; Radhika, G.; Radha, V,; Rema, M.; Deepa, R. Gene-environment Interactions and the Diabetes Epidemic in India. *Forum Nutr.* **2007,** *60*, 118–126.

Molloy, A. M.; Daly, S.; Mills, J. L.; Kirke, P. N.; Whitehead, A. S.; Ramsbottom, D.; Conley, M. R.; Weir, D. G.; Scott, J. M. Thermolabile Variant of 5, 10-methylenetetrahydrofolate Reductaseassociated with Low Red-cell Folates: Implications for Folate Intake Recommendations. *Lancet* **1997,** *349* (9065), 1591–1593.

Muller, M.; Kersten, S. Nutrigenomics: Goals and Strategies. *Nat. Rev.* **2003,** *4*, 315–322.

Munshi, A.; Duvvuri, V. S. Nutrigenomics: Looking to DNA for Nutrition Advice. *Indian J. Biotechnol.* **2008,** *7*, 32–40.

Ordovas, J. M.; Corella, D. Nutritional Genomics. *Annu. Rev. Genomic Hum. Genet.* **2004,** *5*, 71–118

Ordovs, J. M.; Mooser, V.; Nutrigenomics and Nutrigenetics. *Curr. Opin. Lipidol.* **2004,** 101–108.

Pericak-Vance, M. A.; Gorin, M. B.; Schulz, H. L.; Pardi, F.; Lewis, C. M.; Weber, B. H. Meta-Analysis of Genome Scans of Age-Related Macular Degeneration. *Hum. Mol. Genet.* **2005,** *14*, 2257–2264.

Raju, J.; Reddy, P. R. K.; Priyanka, B.; Narasimha, J.; Guru Vishnu, P. Nutrigenomics – Emerging Tool for the Prevention of Diseases. *Int. J. Sci. Environ. Technol.* **2016,** *5* (6), 3724–3730.

Ramachandran, A. The Indian Diabetes Prevention Programme Shows that Lifestyle Modification and Metformin Prevent Type 2 Diabetes in Asian Indian Subjects with Impaired Glucose Tolerance (IDPP1). *Diabetologia* **2006,** *49*, 289–297.

Rastogi, T.; Reddy, K. S.; Vaz, M.; et al. Diet and Risk of Ischemic Heart Disease in India. *Am. J. Clin. Nutr.* **2004,** *79*, 582–592.

Resnikoff, S.; Pascolini, D.; Etya'ale, D.; Kocur, I.; Pararajasegaram, R.; Pokharel, G. P.; Mariotti, S. P. Global Data on Visual Impairment in the Year 2002. *Bull. World Health Organ.* **2004,** *82*, 844–851.

Sangiovanni, J. P.; Chew, E. Y. The Role of Omega-3 Long-chain Polyunsaturated Fatty Acids in Health and Disease of the Retina. *Prog. Retin. Eye Res.* **2005,** *24*, 87–138.

Seddon, J. M.; Rosner, B.; Sperduto, R. D.; Yannuzzi, L.; Haller, J. A.; Blair, N. P.; Willett, W. Dietary Fat and Risk for Advanced Age-related Macular Degeneration. *Arch. Ophthalmol.* **2001,** *119*, 1191–1199.

Selvam, R. Nutrigenomics- an Interaction of Nutrition, Health and Genomics, A Stepping Stone to Personalized Nutrition, 2019; pp 327–330.

Sharma, R.; Singh, R. B. Bioactive Foods and Nutraceutical Supplementation Criteria in Cardiovascular Protection. *Open Nutraceuticals J.* 2010, *3*, 141–153.

Siddique, R. A.; Tandon, M.; Ambwani, T.; Rai, S. N.; Atreja, S. K. Nutrigenomics: Nutrient-gene Interactions. *Food Rev. Int.* **2009,** *25* (4), 326–345.

Sinha, R. Cancer Risk and Diet in India. *J. Postgrad. Med.* **2003,** *49*, 222–228.

Sivasankaran, S. The Cardio-protective Diet. *Indian J. Med. Res.* **2010,** *132* (5), 608–616.

Smith, W.; Mitchell, P.; Leeder, S. R. Dietary Fat and Fish Intake and Age-Related Maculopathy. *Arch. Ophthalmol.* **2000,** *118*, 401–404.

Surana, K. R.; Ahire, E. D.; Sonawane, V. N.; Talele, S. G. Biomolecular and Molecular Docking: A Modern Tool in Drug Discovery and Virtual Screening of Natural Products. In *Applied Pharmaceutical Practice and Nutraceuticals*; Apple Academic Press, 2021; pp 209–223.

Surana, K. R.; Ahire, E. D.; Sonawane, V. N.; Talele, S. G.; Talele, G. S. Molecular Modeling: Novel Techniques in Food and Nutrition Development. In *Natural Food Products and Waste Recovery: Healthy Foods, Nutrition Design, and Extraction of Valuable Compounds*, 2021 ; p 17.

Thompkinson, D. K.; Bhavana, V.; Kanika, P. Dietary Approaches for Management of Cardio-Vascular Health- A Review. *J. Food Sci. Technol.* 2012.

Sudhakar Reddy, V.; Palika, R.; Ismail, A.; Pullakhandam, R.; Bhanuprakash Reddy, G. Nutrigenomics: Opportunities & Challenges for Public Health Nutrition. Indian J. Med. Res. **2018,** *148* (5), 632–641.

Willett, W. C. Balancing Life-Style and Genomics Research for Disease Prevention. *Science* **2002,** *296*, 695–698.

Zduńczyk, Z.; Pareek, C. S. Application of Nutrigenomics Tools in Animal Feeding and Nutritional Research. *J. Anim. Feed Sci.* **2009,** *18* (1), 3–16.

Zeisel, S. H. Nutritional Genomics: Defining the Dietary Requirement and Effects of Choline. *J. Nutr.* **2011,** *141*, 531–534.

## CHAPTER 3

# VARIATION IN HUMAN GENETICS AND OXIDATIVE STRESS

PRADNYA H. VAIDYA,[1] HARSHADA U. BAGUL,[2] SANDHYA BORSE,[3] LAXMIKANT B. BORSE,[2] and ANIL G. JADHAV[2]

*[1]University of Bourgogne Franche Comte Master Innovative Drugs, Dijon, France*

*[2]Sandip Institute of Pharmaceutical Sciences, Nashik, Maharashtra, India*

*[3]School of Pharmaceutical Sciences, Sandip University, Nashik, Maharashtra, India*

## ABSTRACT

Oxidative stress is experienced because of a variance between the manufacturing of reactive species and the body's potential to overcome them by utilizing antioxidants. Individual genetic adaptation impacts the digestion of exogenous antioxidants that are consumed from regular meals and eventually participates in the expansion of oxidative stress. While alteration in the quality of gene coding might affect the endogenous cell reinforcement protein's enzymatic action, and accordingly, increase the level of reactive species, oxidative stress, and the probability of disease evolution, oxidative stress has been reported to increase the progression of various diseases like type-2 diabetes, cardiovascular disorders, malignant tumors, Alzheimer's disease, and rheumatoid arthritis. Maximum oxidative stress can create impaired functional homeostasis, depicting human aging. Aging seems to have an advanced conservative stress response framework, presumably obtained from an amalgam of a diet containing natural antioxidants,

Nutrigenomics and Nutraceuticals. Raj K. Keservani, Eknath D. Ahire, Shubham J. Khairnar, Sanjay J. Kshirsagar, & Rajesh K. Kesharwani (Eds.)

a dynamic lifestyle, and a positive genetic establishment, particularly rich in hereditary variations ready to neutralize the stress burden. The blend of these variables could permit aging to keep up with moderate degrees of free radicals that produce profitable signaling and modulator effects on cellular metabolism. Herein, we attempt to focus on the current knowledge of genetic variance in human genes related to oxidative stress. Highlighting the study of genetic variation will be a useful approach for determining the impact of oxidative stress in diversified diseases and human aging.

## 3.1 INTRODUCTION

The production of reactive species, such as free radicals and reactive oxygen species (ROS), leads to normal physiological activity and plays a vital role in cell communication, transcription of genes, and immunologic response (Valko et al., 2007). In the aerobic respiration process, electron leakage in the electron transport chain is responsible for the formation of superoxide anion. Apart from these oxidative bursts, cytochrome $P_{450}$ and xanthine oxidase enzymes participate in the production of highly reactive species (Serafini, 2006). When a redox reaction occurs, the formation of reactive species in excessive amounts can have deleterious activity on macromolecules, DNA, and the cell membrane (Valko et al., 2007; McCord, 2000). To deal with oxidative stress, an antioxidant strategy has evolved, made up of gene-recovery enzymes and antioxidants (Singh et al., 2017a) that restrict or suppress oxidation in cells. As a result, these cause cells' death when the biological functions of the cell membrane, enzyme receptor activity, and cell function are altered (Dalle-Donne et al., 2006). For this purpose, a complex system of defenses exists in the human body to protect itself against excess production and affliction caused by reactive species in order to maintain "redox homeostasis" (Da Costa et al., 2012). Apart from destroying biological molecules or tissues, oxidative stress can stimulate transcription factors, including nuclear factor kappa B (NF-κB), which activate a biochemical cascade that leads to an inflammation response and cytokine release syndrome (García-Bailo et al., 2011). Individual oxidative stress vulnerability exploits the estimation of associated damage, free radical development, antioxidant capability, or transcription factors. These measures offer an overall assessment of complex interaction between diet, physiological state, and individual genetics; additionally, factors related to oxidative stress are also estimated. The human genetic makeup is being decoded, and gene variation is being documented with maximum effort, making it practical to learn genetic factors corresponding to oxidative stress (Forsberg et al., 2001). The chapter is increasingly devoted to efforts targeted at cataloguing human genetic variation in some genes as well as correlation studies of variation and diseases. The prolongation of life span in model microorganisms such as *Caenorhabditis elegans* treated with antioxidant superoxide dismutase or catalase mimic suggests that antioxidant activity may delay aging and ward off diseases. There may be multiple longevity-promoting medications that activate mitochondrial oxygen consumption that increases the production of ROS. ROS can cause oxidation and destruction of important constituents of plasma membrane. Due to the incessant formation of free radicals across cell life, mitochondria act as the main target for their attack. These act as molecular indications and may ultimately lead to the activation of the endogenous defense process, resulting in a longer life span and increased stress resistance. This response is commonly known as mitohormesis (Ristow et al., 2011). The disease associated with age can be developed with an elevation of the DNA oxidation level in both the nucleus and the mitochondria (Dato et al., 2013; Surana, et al., 2021).

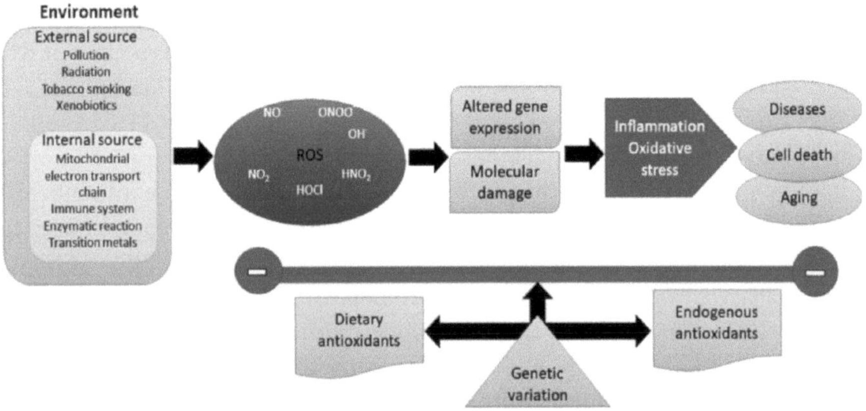


**FIGURE 3.1** Outline of the interrelation between the formation of reactive oxygen species (ROS), oxidative stress, evolution of disease, action of antioxidants, and genetic variation.

## 3.2 CAUSES OF ROS

The endogenous source is further divided into two types: intracellular and extracellular sources of ROS. It consists of mitochondria, peroxisomes, and inflammatory cell activation. Mitochondria are the significant origin of intracellular ROS through the creation of superoxide radicals during typical oxidative phosphorylation (Tangvarasittichai, 2015). Mitochondrial ROS creation is tissue and cell-cycle explicit and increments as the cell ages (Ku et al., 1993). ROS delivered with mitochondrial respiration has been seen to be higher in neoplastic cells when contrasted with noncancerous cells (Trachootham et al., 2006). Mitochondrial ROS is the aftereffect of mitochondrial oxidative phosphorylation, where 1–2% of the electrons set free from the electron transfer chain (ETC) bring about superoxide arrangement (St-Pierre et al., 2002). Mitochondria complex I and complex III are responsible for most of the superoxide creation. Mitochondrial complex III ROS compeers are engaged with the ubiquinone cycle (Betteridge et al., 2000), which manage the initiation of hypoxia-inducible factors (HIFs) transcription factors. The peroxisome is another intracellular organelle, which seems to be a source of ROS production inside the cell (Bell et al., 2007). Both hydrogen peroxide and superoxide are generated in peroxisomes by means of peroxisomal oxidases, including acyl-CoA oxidase and xanthine oxidase (Corton et al., 2018; Klaunig et al., 2003; Corton et al., 2014). However, the immediate reason and relationship aren't yet clear in the literature, but the enlistment of peroxisomal ROS remains a contributing conceivable path or modifier for PPAR alpha-enacted liver carcinogenesis. Inflammatory cells, including neutrophils and macrophages, are additionally an endogenous source of ROS and add to the tissue ROS load. Utilizing NADPH oxidase, these phagocytic cells produce ROS. The NADPH oxidase pathway comprises two subunits, gp91phox and p22phox, which are membrane bound like a film. Moreover, p47phox, p67phox, and Rac are found in the cytoplasm of the inflammatory cells (Babior et al., 1999). Whenever the neutrophils and macrophages are enacted, the phagocytes go through an expansion in oxygen uptake, bringing about an expansion in ROS species through NADPH oxidase. The produced superoxide can be dismutated by superoxide dismutase into hydrogen peroxide. Physiologically, the manufacturing of hydrogen peroxide by the phagocytic cells is a significant tissue protection for killing microorganisms. Other than acting as a protection system, the phagocyte-produced ROS have been shown to have an association with all phases of the carcinogenesis cycle. On account of synthetic cancer-causing agents, that is, chemical carcinogens, the Kupffer cell (liver macrophage) is actuated by exposure to PPAR alpha, initiating hepatic malignant growth (Klaunig et al., 2004). Researchers propose that the initiated Kupffer cell takes part in the advancement phase of carcinogenesis, as proven by an expansion in hepatic preneoplastic central volume and DNA combination that is suppressed by inactivators of macrophage actuation (Rose et al., 1997). ROS may likewise give rise to exogenous sources, including ionizing radiation and xenobiotics. Ionizing radiation destructs cells, including DNA damage and gene mutation, cell demise, and malignant growth that are mediated by ROS development. ROS created from radiation are quickly delivered through the radiolysis of water particles. This expansion of ROS in the cell can deliver intense outcomes or endure and diffuse inside the cell, bringing about toxic effects. Natural and synthetic drug substances can produce ROS in cells either in an indirect way through the digestion of the compound or simply by initiating an endogenous source of ROS. Oxidative pressure and oxidative damage have been accounted by exposing to an assortment of synthetically different cancer-causing xenobiotics. These include chlorinated compounds, radiation, metal particles, barbiturates, phorbol esters, and some peroxisome-proliferating compounds, among the classes of mixtures that have been shown to incite oxidative pressure and oxidative damage in vitro and in vivo (Klaunig et al., 1997; Klaunig et al., 1998).

## 3.3 IMPACT OF OXIDATIVE STRESS DUE TO GENETIC VARIATION ON CHRONIC DISEASES

### *3.3.1 CANCER*

Numerous investigations have affirmed a robust connection between oxidative pressure and the development of a few human pathologies, including cancer (Agarwal et al., 2012; Dalle-Donne et al., 2006). Broad investigational testimony holds that ROS is involved in tumor initiation, promotion, and progression (Guyton et al., 1993; Kumar et al., 2008). An overabundance of ROS can harm genomic and mitochondrial DNA (mtDNA), causing DNA damage, molecule transformation, and distorted signaling pathways. The DNA molecule is susceptible to being hit by hydroxyl radicals, resulting in damage and alteration of purine and pyrimidine bases (Hwang et al., 2007). The subsequent adducts, 8-OH deoxyguanosine (8-OHdG), can be produced during ROS production (particularly through the hydroxyl radical), which are directly associated with an expansion in mutagenesis (Forcados et al., 2017). Expanded oxidative stress can change the redox capability of the cell following the induction of gene expression (Benhar et al., 2002). Both endogenous and exogenous mediators have been shown to have consequences for ROS induction in cancer (Gottlieb et al., 2005). If ROS from endogenous and exogenous sources are not detoxified by antioxidants, they can provoke an increase in oxidative stress in the cell. The resulting oxidative stress is found to be severe enough to modify the structure and function of critical cellular macromolecules, including DNA, following cell growth, mutation, and chromosome variability rising in neoplasm formation.

#### 3.3.1.1 EFFECT OF OXIDATIVE STRESS ON DNA AND RNA

Genomic transformation of DNA is a fundamental part of the advancement of cancer. Oxidative damage and the subsequent changes in DNA bases can prompt mutations, deletions, insertions, or chromosomal movements, which may create oncogene initiation or tumor suppressor gene quality inactivation (Toyokuni, 2006). The development of 8-OHdG is exceptionally mutagenic due to the mispairing of the guanine with the adenine during DNA replication. 8-OHdG can change GC pairs to TA pairs on top of DNA replication. Raised levels of 8-OHdG in genomic DNA have been noted in both human tumors and rodent neoplasms (Valavanidis et al., 2000). The oxidative DNA base, 8-OHdG, has often been utilized as a biomarker of oxidative DNA damage, with its estimation as a sign of oxidative stress. The utilization of oxidative DNA adducts incorporating 8-OHdG in blood (WBC) and urine has the potential to observe both the creation of oxidative DNA damage and its maintenance (Kawanishi et al., 2006). This has potentially significant applications in chemoprevention and dietary (Keservani et al., 2010a; Keservani et al., 2010b; Keservani et al., 2020) intercession studies. MtDNA has been demonstrated to be more susceptible to oxidative damage than nuclear DNA. MtDNA changes detailed in neoplasms can be characterized into two categories: tumorigenic transformations and adaptive mutations. Tumorigenic mtDNA transformations bring about the hindrance of oxidative phosphorylation (OXPHOS) combined with the restraint of typical electron movement through the mitochondrial electrotransfer chain (St-pierre et al., 2002). This results in expanded ROS creation and adds to both malignant growth and progression. Oxidative RNA damage includes alterations of bases and ribose, base extraction, and strand breaks (Frenkel, 1992).

#### 3.3.1.2 EFFECT OF OXIDATIVE STRESS ON GENE EXPRESSION

Vulnerability of cells to ROS or ROS-creating compounds has been shown to bring about apoptosis or necrosis, as well as adjusting quality gene articulation, especially the quality genes associated with development factors and protooncogenes articulation. The impacts of ROS on the regulation of cell development have shown that the induction of cell multiplication happens at low vulnerable concentrations of ROS or transient vulnerabilities to ROS and is cell type dependent (Klaunig, 2011). At higher dosages of ROS, the cells go through necrosis or apoptosis; thus, the impact of ROS on malignant growth is very dose dependent. ROS as a signaling messenger can initiate basic targeted particles engaged with quality gene articulation pathways associated with neoplasm development, including PKC, Nrf2, MAP kinase, AP-1, NF-κB, and HIF-1α. As mentioned above, the quantity of ROS in the cell seems to impact which transcription factor is activated (Wallace, 2005; Kensler et al., 2007).

### 3.3.2 RHEUMATOID ARTHRITIS

This disease is a chronic inflammatory condition affecting the synovium. Various immune cells are found in the synovial tissue and synovial liquid, including lymphocytes, macrophages, and neutrophils. It is possible that reactive oxidants are partly responsible for the damage in rheumatoid arthritis due to the presence of neutrophils and macrophages in the inflamed joints. Among protein (Singh et al., 2017b) molecules, carbonyl groups are known to reflect impairment caused by reactive oxidants, and proteins from rheumatoid arthritis patients contain more carbonyl groups than those from healthy individuals (Chapman et al., 1989). Another marker of oxidation is tyrosine nitration, which indicates the release of NO and $O_2$, respectively, around nitrated proteins. Those patients with RA have nitrated proteins in their synovial liquid (Kaur et al., 1994).

#### 3.3.2.1 EVIDENCES OF GENOTOXIC EFFECT ON OXIDATIVE STRESS IN RA

The evidence has shown a reverse correlation between antioxidant intake and RA prevalence and a reverse correlation between antioxidant levels and inflammation (Cerhan et al., 2003; Heliovaara et al., 1994). In RA synovial tissue, iron acts as a catalyst for the production of hydroxyl radicals from hydrogen peroxide and is linked to a poor prognosis (Dabbagh et al., 1993). A number of studies have shown that RA serum and synovial fluid exhibit increased levels of oxidative enzymes along with depleted antioxidants (De Leo et al., 2002; Hajizadeh et al., 2002). As a result, evidence for oxidative stress in RA has been produced by detecting variations in molecules caused by oxidants (Taysi et al., 2002; Henrotin et al., 2003; Halliwell, 1995). When reactive oxygen is generated, it directly affects genes and inhibits gene formation mechanisms. In this process, DNA strands can be broken or nucleotide bases may be damaged (Tak et al., 2000; Bashir et al., 1993). RA tissue has a microsatellite variability that indicates constant mutagenesis. The mismatch repair system (MMR) normally corrects this kind of mutagenesis caused by DNA damage. However, MMR is inoperative in RA, which is probably related to oxidative stress. It is found that oxidative stress decreases the expression of hMSH6, a constituent of the MutSβ complex that is crucial for repairing single base mismatches that are affected by oxidative stress, as well as increases expression of hMSH3, a constituent of MutSβ that is significant for repairing insertion and deletion loops. According to Lee et al.'s (2003) suggestion, the summary of MMR expression was reproduced, and oxidative stress plays a role in the expansion of microsatellite instability. This pattern may contribute to a temporary cell life span by avoiding damage to DNA, although allowing minor DNA damage or the possibility that it could create a mutated phenotype that is more beneficial to survival. MtDNA is predominantly vulnerable to oxidative stress, and extended contact leads to tenacious mtDNA impairment without effectual improvement, mitochondrial dysfunction, cell cycle arrest, and programmed cell death (Yakes et al., 1997). The significance of mtDNA in inflammatory arthritis is found in studies signifying that in the plasma and synovial fluid of RA, extracellular mtDNA is elevated (Hajizadeh et al., 2002), and oxidative impairment of mtDNA also leads to murine arthritis (Collins et al., 2004).

### *3.3.3 CARDIOVASCULAR DISEASE (CVD)*

Heart attacks and strokes (CVD) are multifactorial diseases, and their etiology is complicated by a large number of risk factors. In particular, hypercholesterolemia, high blood pressure, smoking habits, diabetes, overweightness, and an inactive lifestyle are the most widely recognized risk factors (Bahorun et al., 2006; Chatterjee et al., 2007). In recent years, researchers have shown that oxidative stress is either the main or subordinate cause of CVD. Oxidative stress mainly damages the arteries. Plaques of atheromatic derivation are usually formed as a result of endothelial tenderness, which produces ROS by immune cells in situ. ROS oxidize LDL in the circulation, causing foam cells and lipid accretion (Pacher et al., 2007).

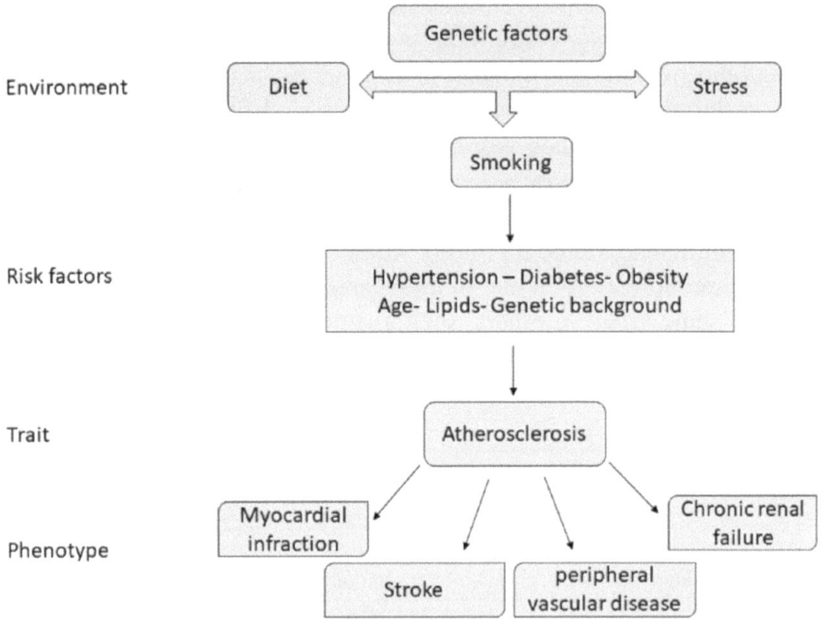


**FIGURE 3.2** Several relationships among genomic and environmental influences leading to cardiac target organ impairment.

Heart disease is a condition that is greatly influenced by oxidative stress due to the cardiovascular system's vulnerability to prooxidant damage. During oxidative stress, heart blood vessels and myocytes are irreparably damaged, which further aggravates the condition and results in disease progression (Montezano et al., 2015). Oxidative stress causes changes in cell metabolism, signaling, transference, and existence processes that result in tissue impairment. This effect is caused by ROS, which can act directly on biomolecules during the oxidative stress process. Often, oxidative stress-induced heart damage is caused by polymorphisms in enzymes that regulate redox metabolism and are genetically influenced (Maiolino et al., 2015).

## 3.3.4 EFFECT OF OXIDATIVE STRESS ON AGING AND LONGEVITY DUE TO GENETIC VARIABILITY

### 3.3.4.1 ROLE OF AGING AND LONGEVITY IN ANIMAL MODELS

The development of strategies to manage oxidative stress at an average level depends on discovering the factors that contribute to the oxidative stress reaction in aging organisms. Evolutionary conserved pathways, such as organism response to stress, are associated with aging and longevity in animal and mammalian models, in conjugation with nutrient sensing and DNA restoration and improvement. For the study of aging models, organisms like mice, nematode worms (*Caenorhabditis elegans*), and fruit flies (Drosophila) with genetic alteration can expand life expectancy approximately six folds, seemingly by reducing aging activity (Lin et al., 1997; Lin et al., 1998; Guarente et al., 2000). Additionally, several genes responsible for increased longevity in one of these species, for example, fruit flies, also have the same effect in others, such as mice and nematodes. The study of these genes may help to better understand how they have been conserved during evolution, particularly since the genes involved in these pathways

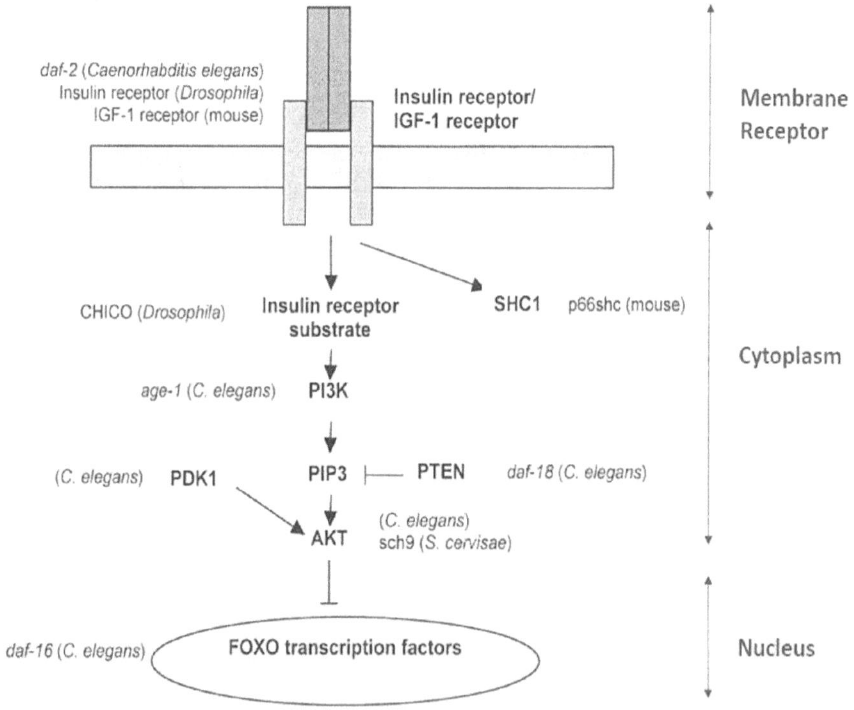


**FIGURE 3.3** Framework indicating how the IGF-1 signaling cascade affects mammals' life span. The IGF-1 receptor has a similar structure to insulin. Not all constituents are mentioned, but those that influence life span in the model are titled (e.g., CHICO increases longevity in fruit flies. The expanded life span of p66shc eliminates mice due to class O factor. Arrow demonstrates stimulus phase, T-shape line demonstrates inhibitory phase).

have human homologs (Browner et al., 2004). Initially, it appears that invertebrates evolved mechanisms to recognize adverse environmental conditions, like food scarcity and adverse climates, thereby favoring variation into prolonged stress-resistant forms instead of susceptible breeding forms. In *C. elegans*, alteration occurs in various genetic materials that manage to approach the "Clock" and "Dauer" stages, where clk-1, clk-2, daf-2, and daf-16 formation perform characteristics of expansion of life span in organisms (Lakowski, 1996). In fruit flies, changes in heredity in the methuselah gene expand longevity up to 35% and enhance resistance to free radical damage, malnutrition, and elevated temperatures (Lin et al., 1998).

**TABLE 3.1** List of Genetic Materials in Signaling Cascade.

| | |
|---|---|
| AKT | AKT oncogene/protein kinase B |
| PDK1 | 3-phosphoinositide-dependent protein kinase 1 |
| PIP3 | Phosphatidylinositol 3,4,5-trisphosphate |
| PI3K | Phosphatidylinositol-3-kinase |
| PTEN | Phosphatase and tension homolog |

### 3.3.5 DIABETES MELLITUS

Since ages, type-2 diabetes mellitus (T2DM) has been known as a multiplex and multifactorial metabolic disorder that is described by the unusual metabolism of carbohydrates, sugars, fats, and proteins, which prompts elevated levels of glucose and lipids inside the blood (Akash et al., 2013b). Persistent exposure to prominent levels of glucose and lipids triggers different pathways that are considered to instigate impaired insulin secretion from the β-cells of pancreatic islets, insulin resistance in peripheral tissues, diminished glucose usage in fringe tissues, and atypical hepatic glucose production.

An adequate proof from the consequences of clinical investigations demonstrates that oxidative stress is firmly connected with T2DM, and the degrees of supportive oxidants and oxidative stress-induced tissue destruction biomarkers (e.g., oxidation of DNA bases, 4-hydroxy-2-nonenal (HNE) proteins, hydroperoxides, 8-hydroxy-deoxyguanine, and 8-epi-prostaglandin F2α) have been found to be high in pancreas biopsies, platelets, plasma, and serum of type-2 diabetic patients (Sakuraba et al., 2002; Shin et al., 2001).

Elevated levels of autooxidative and nonenzymatic glycosylation are one of the potential ways that, together, trigger the arrangement of free radicals and radical-induced lipid peroxidation (Hunt et al., 1990). Extended formation of

ROS disturbs the equilibrium between the levels of oxidants and antioxidants, which achieves the situation known as the prooxidative condition.

#### *3.3.5.1 BIOMARKERS OF OXIDATIVE STRESS USED IN THE DIAGNOSIS OF T2DM*

Different biomarkers of oxidative pressure are utilized for the identification of the pathogenesis of T2DM. Below, we have momentarily portrayed the different significant oxidative stress biomarkers for the diagnosis of T2DM.

- **Glycated proteins:** Long-term unmasking of hyperglycemia is responsible for inciting protein glycation by creating oxidative stress. The measurement of protein glycation is assessed by estimating the degrees of glycated hemoglobin and fructosamine in serum, and additionally, in blood. Myeloperoxidase changes L-tyrosine to 3,3-dityrosin, which acts as a crosslink between polypeptide chains for similar or potentially unique proteins. This protein has been considered as helpful oxidative pressure biomarker for protein oxidation during the pathogenesis of T2DM (Yla-Herttuala, 1999).
- **Lipid peroxidation:** During the advancement of T2DM, the amount of lipid profiles is diffused, which makes the body cells allow more lipid peroxidation (Perez-Matute et al., 2009). It has been proven from clinical studies that because of the presence of numerous bonds inside polyunsaturated fatty acids in the cell layer, they are more exposed to free radicals (Butterfield et al., 1998). Lipid hyperperoxides (LHPs) are poisonous lipid radicals that produce greasy fatty acids, which are perceived to be exceptionally toxic and can instigate the development of oxidative pressure (Matough et al., 2012). Similarly, lipid peroxidation is another significant oxidative stress biomarker, and malondialdehyde (MDA) is generated because of lipid peroxidation, and it tends to be used to quantify the degree of lipid peroxidation when it responds with thiobarbituric acid (Pasaoglu et al., 2004).
- **Vitamins:** Vitamins A, C, and E, being a significant portion of different natural biological systems, have been thought of as strong antioxidants as they can detoxify free radicals. If there is an occurrence of oxidative stress, the typical levels of these vitamins are changed, and this demonstrates that the degrees of these nutrients can be assessed as a critical biomarker of oxidative stress.

- **Glutathione:** Glutathione reductase and glutathione peroxidase are regarded as two of the major oxidative stress biomarkers during the pathogenesis of T2DM. These enzymes exist inside the cells that metabolize peroxide to water, and any adjustment in the levels of these enzymes makes the cells more vulnerable to oxidative stress (Maritim et al., 2003). In diabetes, glutathionylated hemoglobin has been considered a biomarker of oxidative stress (Chen et al., 2014; Ghezzi, 2013).
- **Catalase (CAT):** CAT is the principal controller of $H_2O_2$ and breaks $H_2O_2$ chemically into water and oxygen. If there is an insufficiency of CAT, β-cells of pancreatic islets are more inclined to exorbitant ROS creation and oxidative pressure, which prompts the dysfunction of β-cells of pancreatic islets and T2DM (Asmat et al., 2016). Lately, it has been considered that persistent exposure to hyperglycemia results in the manufacturing of $H_2O_2$ and decreases modulation of the quality gene articulation of CAT (Patel et al., 2013).
- **4-Hydroxynonenal (HNE) and Malondialdehyde (MDA):** Since long, it has been seen that with age, the degree of serum MDA has been seen to be elevated in T2DM (Holvoet et al., 1996). Also, lately, it has been seen that the oil of *Eruca sativa* seeds helped enhance oxidative pressure in alloxan-induced diabetes mellitus in rodents, and unknowingly, it was tracked down that the degrees of both MDA and 4-HNE were altogether decreased, showing a reasonable decrease in oxidative stress (El-Missiry and El Gindy, 2000).

#### *3.3.5.2 APPROACHES TO TREATMENT*

To resist the oxidative stress-induced evolution of T2DM, acknowledgment of the creation of ROS and the vitally fundamental risks engaged with the advancement of different molecular mechanisms in the pathogenesis of T2DM may lay out a reason for the avoidance and treatment of oxidative stress-induced T2DM.

Many examinations have led to research on the utilization of exogenous antioxidants in safeguarding the β-cells of pancreatic islets from the destructive impacts of ROS and oxidative stress. Nicotinamide and desferrioxamine are the main antioxidants, and it has been concluded that these agents can to some extent safeguard the β-cells of pancreatic islets from oxidative stress during low-portion streptozotocin-initiated insulitis (Mendola et al.,

1989). Metformin and troglitazone, however, are traditional ways of treating diabetes. They have been accounted to have antioxidant properties and are seen to shield β-cells of pancreatic islets from oxidative stress-induced hyperglycemia in ZDF rodents (Harmon et al., 1999).

It has been determined by various scientific investigations that natural therapeutic plants contain enormous phytochemical constituents that have different antioxidant properties against varied diseases. These therapeutic plants have been utilized since ancient times for the treatment of oxidative stress-induced diseases. To the extent that the utilization of therapeutic plants for the treatment of oxidative stress-induced T2DM is concerned, we have proactively revealed that most ordinarily utilized kitchen products, eminently *Nigella sativa* (Akash et al., 2011), espresso (coffee) (Akash et al., 2014a), ginger (Akash et al., 2015), and onion (Akash et al., 2014b), are loaded with naturally occurring antioxidants against oxidative stress-initiated T2DM. Latest investigations have detailed that among different kinds of drug constituents present in therapeutic plants, flavonoids and polyphenols are the most significant ones that have accounted for their antioxidant properties against oxidative stress-induced disorders.

## 3.4 CONCLUSION

Numerous chronic diseases, such as malignant tumors and heart disease, are associated with oxidative stress. Many studies show that genetic variations in genes encoding antioxidant enzymes affect cancer risk significantly. However, there are discrepancies between studies, which make it strenuous to draw ultimate conclusions. The investigation of environmental aspects such as dietary source consumption led to an explanation for these discrepancies. Studying oxidative stress-induced diseases using genetic epidemiology is a relatively new approach. As has been the case for additional connotation studies, it is predictable that optimistic studies will be contradicted by undesirable results. Despite oxidative stress as a phenomenon being a leading cause of harm to individuals' health and well-being, it can also be used as a therapeutic tool if we are capable of modifying it within the human body. Studying the interaction between genes and the environment is not only an excellent way to understand disease pathology at a molecular level, but it also makes it possible to target specific information and treatment to high-risk patients.

## KEYWORDS


- **oxidative stress**
- **reactive species**
- **antioxidants**
- **gene coding**
- **homeostasis**
- **enzymatic action**


## REFERENCES


Agarwal, A.; Aponte-Mellado, A.; Premkumar, B. J.; Shaman, A.; Gupta, S. The Effects of Oxidative Stress on Female Reproduction: A Review. *Reprod. Biol. Endocrinol.* **2012,** *10*, 49.

Akash, M. S.; Rehman, K.; Chen, S. Goto-Kakizaki Rats: Its Suitability as Non-obese Diabetic Animal Model for Spontaneous Type 2 Diabetes Mellitus. *Curr. Diabetes Rev.* **2013a,** *9*, 387–396.

Akash, M. S. H.; Rehman, K.; Chen, S. Role of Inflammatory Mechanisms in Pathogenesis of Type 2 Diabetes Mellitus. *J. Cell Biochem.* **2013b,** *114*, 525–531.

Akash, M. S. H.; Rehman, K.; Chen, S. Effects of Coffee on Type 2 Diabetes Mellitus. *Nutrition* **2014a,** *30*, 755–763.

Akash, M. S. H.; Rehman, K.; Chen, S. Spice Plant Allium Cepa: Dietary Supplement for Treatment of Type 2 Diabetes Mellitus. *Nutrition* **2014b,** *30*, 1128–1137.

Akash, M.; Rehman, K.; Rasool, F.; Sethi, A.; Abrar, M.; Irshad, A.; Abid, A.; Murtaza, G. Alternate Therapy of Type 2 Diabetes Mellitus (T2DM) with Nigella (Ranunculaceae). *J. Med. Plants Res.* **2011,** *5*, 6885–6889.

Akash, M. S. H.; Rehman, K.; Tariq, M.; Chen, S. Zingiber Officinale and Type 2 Diabetes Mellitus: Evidence from Experimental Studies. *Crit. Rev. Eukaryote Gene Expr.* **2015,** *25*, 91–112.

Asmat, U.; Abad, K.; Ismail, K. Diabetes Mellitus and Oxidative Stress-A Concise Review. *Saudi Pharm. J.* **2016,** *24*, 547–553.

Babior, B. M. Nadph Oxidase: An Update. *Blood* **1999,** *93* (5), 1464–1476.

Bahorun, T.; Soobrattee, M. A.;Luximon-Ramma, V.;Aruoma, O. I. Free Radicals and Antioxidants in Cardiovascular Health and Disease. *Int. J. Med. Update* **2006,** *1*, 1–17.

Bashir, S.; Harris, G.; Denman, M. A.; Blake, D. R.; Winyard, P. G. Oxidative DNA Damage and Cellular Sensitivity to Oxidative Stress in Human Autoimmune Diseases. *Ann. Rheum. Dis.* **1993,** *52*, 659– 666.

Bell. E. L.; Klimova, T. A.; Eisenbart, J. The Qo Site of the Mitochondrial Complex III is Required for the Transduction of Hypoxic Signaling Via Reactive Oxygen Species Production. *J. Cell Biol.* **2007,** *177* (6), 1029–1036.

Benhar, M.; Engelberg, D.; Levitzki, A. ROS, Stress-Activated Kinases and Stress Signaling in Cancer. *EMBO Rep.* **2002,** *3* (5), 420–425.

Betteridge, D. J. What is Oxidative Stress? *Metabolism* **2000,** *49* (2) (Suppl. 1), 3–8.

Browner, W. S.; Kahn, A. J.; Ziv, E.; Reiner, A. P.; Oshima, J.; Cawthon, R. M.; Hsueh, W. C.; Cummings, S. R. The Genetics of Human Longevity. *Am. J. Med.***2004,** *117*, 851–860.

Butterfield, D. A.;, Koppal, T.; Howard, B.; Subramaniam, R.; Hall, N.; Hensley, K.; Yatin, S.; Allen, K.; Aksenov, M. Structural and Functional Changes in Proteins Induced by Free Radical- Mediated Oxidative Stress and Protective Action of the Antioxidants N- tert-Butyl- α- phenylnitrone and Vitamin Ea. *Ann. N. Y. Acad. Sci.* **1998,** *854*, 448–462.

Cerhan, J. R.; Saag, K. G.; Merlino, L. A.; Mikuls, T. R.; Criswell, L. A. Antioxidant Micronutrients and Risk of Rheumatoid Arthritis in a Cohort of Older Women. *Am. J. Epidemiol.* **2003,** *157*, 345–354.

Chapman, M. L.; Rubin, B. R.; Gracy, R. W. Increased Carbonyl Content of Proteins in Synovial Fluid from Patients with Rheumatoid Arthritis. *J. Rheumatol.* **1989,** *16*, 15–18.

Chatterjee, M.; Saluja, R.; Kanneganti, S.; Chinta, S.; Dikshit, M. Biochemical and Molecular Evaluation of Neutrophil NOS in Spontaneously Hypertensive Rats. *Cell. Mol. Biol.* **2007,** *53*, 84–93.

Chen, H. J.; Lin, W. P.; Chiu, S. D.; Fan, C. H. Multistage Mass Spectrometric Analysis of Human Haemoglobin Glutathionylation: Correlation with Cigarette Smoking. *Chem. Res. Toxicol.* **2014,** *27*, 864–872.

Collins, L. V.; Hajizadeh, S.; Holme, E.; Jonsson, I. M.; Tarkowski, A. Endogenously Oxidized Mitochondrial DNA Induces in vivo and in vitro Inflammatory Responses. *J. Leukoc Biol.* **2004,** *75*, 995–1000.

Corton, J. C.; Cunningham, M. L.; Hummer, B. T.; Lau, C.; Meek, B.; Peters, J. M.; Klaunig, J. E. Mode of Action Framework Analysis for Receptor-Mediated Toxicity: The Peroxisome Proliferator-Activated Receptor Alpha (PPARα) as a Case Study. *Crit. Rev. Toxicol.* **2014,** *44* (1), 1–49.

Corton, J. C.; Peters, J. M.; Klaunig, J. E. The PPARα-dependent Rodent Liver Tumor Response is not Relevant to Humans: Addressing Misconceptions. *Arch. Toxicol.* **2018,** *92* (1), 83–119.

Dabbagh, A. J.; Trenam, C. W.; Morris, C. J.; Blake, D. R. Iron in Joint Inflammation. *Ann. Rheum. Dis.* **1993,** *52*, 67–73.

DA Costa, L. A.; Badawi, A.; EL-Sohemy, A. Nutrigenetics and Modulation of Oxidative Stress. *Ann. Nutr. Metab.* **2012,** *60* (suppl 3), 27–36.

Dalle-Donne, I.; Rossi, R.; Colombo, R.; Giustarini, D.; Milzani, A. Biomarkers of Oxidative Damage in Human Disease. *Clin. Chem.* **2006,** *52*, 601–623.

Dato, S.; Crocco, P.; D'aquila, P.; De Rango, F.; Bellizzi, D.; Rose, G.; Passarino, G. Exploring the Role of Genetic Variability and Lifestyle in Oxidative Stress Response for Healthy Aging and Longevity: *Int. J. Mol. Sci.* **2013,** *14*, 16443–16472.

De Leo, M. E.; Tranghese, A.; Passantino, M.; Mordente, A.; Lizzio, M. M.; Galeotti, T.; Zoli, A. Manganese Superoxide Dismutase, Glutathione Peroxidase, and Total Radical Trapping Antioxidant Capacity in Active Rheumatoid Arthritis. *J. Rheumatol.* **2002,** *29*, 2245–2246.

El-Missiry, M. A.; El Gindy, A. M. Amelioration of Alloxan Induced Diabetes Mellitus and Oxidative Stress in Rats by Oil of Eruca Sativa Seeds. *Ann. Nutr. Metab.* **2000,** *44*, 97–100.

Forcados, G. E.; James, D. B.; Sallau, A. B.; Muhammad, A.; Mabeta, P. Oxidative Stress and Carcinogenesis: Potential of Phytochemicals in Breast Cancer Therapy. *Nutr. Cancer* **2017,** *69* (3), 365–74.

Forsberg, L.; de Faire, U.; Morgenstern, R. Oxidative Stress, Human Genetic Variation, and Disease. *Arch. Biochem. Biophys.* **2001,** *389* (1), 84–93.

Frenkel, K. Carcinogen-Mediated Oxidant Formation and Oxidative DNA Damage. *Pharmacol. Ther.* 1992, *53* (1), 127–166.

García-Bailo, B.; El-Sohemy, A.; Haddad, P.; Arora, P.; Benzaied, F.; Karmali, M.; Badawi, A. Vitamins D, C, and E in the Prevention of Type 2 Diabetes Mellitus: Modulation of Inflammation and Oxidative Stress. *Biologics* **2011**, *5*, 1–13.

Ghezzi, P. Protein Glutathionylation in Health and Disease. *Biochim. Biopsy's Acta* **2013,** *1830*, 3165–3172.

Gottlieb, E.; Tomlinson, I. P. Mitochondrial Tumour Suppressors: A Genetic and Biochemical Update. *Nat. Rev. Cancer* **2005,** *5* (11), 857–866.

Guarente, L.; Kenyon, C. Genetic Pathways that Regulate Ageing in Model Organisms. *Nature* **2000,** *408*, 255–262.

Guyton, K. Z.; Kensler, T. W. Oxidative Mechanisms in Carcinogenesis. *Br. Med. Bull.* **1993,** *49* (3), 523–544.

Hajizadeh, S.; Degroot, J.; Tekoppele, J. M.; Tarkowski, A.; Collins, L. V. Extracellular Mitochondrial DNA and Oxidatively Damaged DNA in Synovial Fluid of Patients with Rheumatoid Arthritis. *Arthritis Res. Ther.* **2003,** *5*, R234–R240.

Halliwell, B. Oxygen Radicals, Nitric Oxide and Human Inflammatory Joint Disease. *Ann. Rheum. Dis.* **1995,** *54*, 505–510.

Harmon, J. S.; Gleason, C. E.; Tanaka, Y.; Oseid, E. A.; Hunter-Berger, K. K.; Robertson, R. P. In vivo Prevention of Hyperglycemia also Prevents Glucotoxic Effects on PDX-1 and Insulin Gene Expression. *Diabetes* **1999,** *48*, 1995–2000.

Heliovaara, M.; Knekt, P.; Aho, K.; Aaran, R. K.; Alfthan, G.; Aromaa, A. Serum Antioxidants and Risk of Rheumatoid Arthritis. *Ann. Rheum. Dis.* **1994,** *53*, 51–53.

Henrotin, Y. E.; Bruckner, P.; Pujol, J. P. The Role of Reactive Oxygen Species in Homeostasis and Degradation of Cartilage. *Osteoarthr. Cartil.* **2003,** *11*, 747–755.

Holvoet, P.; Donck, J.; Landeloos, M.; Brouwers, E.; Luijtens, K.; Arnout, J.; Lesaffre, E.; Vanrenterghem, Y.; Collen, D. Correlation Between Oxidized Low-Density Lipoproteins and von Willebrand Factor in Chronic Renal Failure. *Thromb. Haemost.* **1996,** *76*, 663–669.

Hunt, J. V.; Smith, C. C.; Wolff, S. P. Autoxidative Glycosylation and Possible Involvement of Peroxides and Free Radicals in LDL Modification by Glucose. *Diabetes* **1990,** *39*, 1420–1424.

Hwang, E. S.; Bowen, P. E. DNA Damage, a Biomarker of Carcinogenesis: its Measurement and Modulation by Diet and Environment. *Crit. Rev. Food Sci. Nutr.* **2007,** *47* (1), 27–50.

Kaur, H.; Halliwell, B. Evidence for Nitric Oxide-Mediated Oxidative Damage in Chronic Inflammation. Nitrotyrosine in Serum and Synovial Fluid from Rheumatoid Patients. *FEBS Lett.* **1994,** *350*, 9–12.

Kensler, T. W.; Wakabayashi, N.; Biswal, S. Cell Survival Responses to Environmental Stresses via the Keap1-Nrf2-ARE Pathway. *Annu. Rev. Pharmacol. Toxicol.* **2007,** *47*, 89–116.

Keservani, R. K.; Kesharwani, R. K.; Vyas, N.; Jain, S.; Raghuvanshi, R.; Sharma, A. K. Nutraceutical and Functional Food as Future Food: A Review. *Der Pharmacia Lettre* **2010a,** *2* (1), 106–116.

Keservani, R. K.; Kesharwani, R. K.; Sharma, A. K.; Vyas, N.; Chadoker, A. Nutritional Supplements: An Overview. *Int. J. Curr. Pharm. Rev. Res.* **2010b,** *1* (1), 59–75.

Keservani, R. K.; Sharma, A. K.; Kesharwani, R. K. Eds.; *Nutraceuticals and Dietary Supplements: Applications in Health Improvement and Disease Management*; CRC Press, 2020; ISBN: 9781771888738.

Klaunig, J. E.; Kamendulis, L. M. The Role of Oxidative Stress in Carcinogenesis. *Annu. Rev. Pharmacol. Toxicol.* **2004,** *44*, 239–67.

Klaunig, J. E.; Wang, Z.; Pu, X.; Zhou, S. Oxidative Stress and Oxidative Damage in Chemical Carcinogenesis. *Toxicol. Appl. Pharmacol.* **2011,** *254* (2), 86–99.

Klaunig, J. E.; Xu, Y.; Bachowski, S.; Jiang, J. Free-Radical Oxygen Induced Changes in Chemical Carcinogenesis. *Free Radical Toxicol.* **1997,** 375–400.

Klaunig, J. E.; Xu, Y.; Isenberg, J. S. The Role of Oxidative Stress in Chemical Carcinogenesis. *Environ. Health Perspect.* **1998,** *106* (Suppl. 1), 289–295.

Ku, H. H.; Brunk, U. T.; Sohal, R. S. Relationship Between Mitochondrial Superoxide and Hydrogen Peroxide Production and Longevity of Mammalian Species. *Free Radic. Biol. Med.* **1993,** *15* (6), 621–627.

Kumar, B.; Koul, S.; Khandrika, L.; Meacham, R. B.; Koul, H. K. Oxidative Stress is Inherent in Prostate Cancer Cells and is Required for Aggressive Phenotype. *Cancer Res.* **2008,** *68* (6), 1777–1785.

Lakowski, B.; Hekimi, S. Determination of Life-Span in Caenorhabditis Elegans by Four Clock Genes. *Science* **1996,** *272*, 1010–1013.

Lee, S. H.; Chang, D. K.; Goel, A.; Boland, C. R.; Bugbee, W.; Boyle, D. L.; Firestein, G. S. Microsatellite Instability and Suppressed DNA Repair Enzyme Expression in Rheumatoid Arthritis. *J. Immunol.* **2003,** *170*, 2214–2220.

Lin, K.; Dorman, J. B.; Rodan, A.; Kenyon, C. DAF-16: An HNF-3/forkhead Family Member that can Function to Double the Life-Span of Caenorhabditis Elegans. *Science* **1997,** *278*, 1319–1322.

Lin, Y. J.; Seroude, L.; Benzer, S. Extended Life-Span and Stress Resistance in the Drosophila Mutant Methuselah. *Science* **1998,** *282*, 943–946.

Matough, F. A.; Budin, S. B.; Hamid, Z. A.; Alwahaibi, N.; Mohamed, J. The Role of Oxidative Stress and Antioxidants in Diabetic Complications. *Sultan Qaboos Univ. Med. J.* **2012,** *12*, 5–18.

Mccord, J. M. The Evolution of Free Radicals and Oxidative Stress. *Am. J. Med.* **2000,** *108*, 652–659.

Mendola, J.; Wright, J. R.; Lacy, P. E. Oxygen Free-Radical Scavengers and Immune Destruction of Murine Islets in Allograft Rejection and Multiple Low-Dose Streptozocin-Induced Insulitis. *Diabetes* **1989,** *38*, 379–385.

Pacher, P.; Beckman, J. S.; Liaudet, L. Nitric Oxide and Peroxynitrite in Health and Disease. *Physiol. Rev.* **2007,** *87*, 315–424.

Pasaoglu, H.; Sancak, B.; Bukan, N. Lipid Peroxidation and Resistance to Oxidation in Patients with Type 2 Diabetes Mellitus. *Tohoku J. Exp. Med.* **2004,** *203*, 211–218.

Patel, H.; Chen, J.; Das, K. C.; Kavdia, M. Hyperglycemia Induces Differential Change in Oxidative Stress at Gene Expression and Functional Levels in HUVEC and HMVEC. *Cardiovasc. Diabetol.* **2013,** *12*, 142–146.

Perez-Matute, P.; Zulet, M. A.; Martinez, J. A. Reactive Species and Diabetes: Counteracting Oxidative Stress to Improve Health. *Curr. Opin. Pharmacol.* **2009,** *9*, 771–779.

Ristow, M.; Schmeisser, S. Extending Life Span by Increasing Oxidative Stress. *Free Radic. Biol. Med.* **2011,** *51*, 327–336.

Rose, M. L.; Rivera, C. A.; Bradford, B. U. Kupffer Cell Oxidant Production is Central to the Mechanism of Peroxisome Proliferators. *Carcinogenesis* **1999,** *20* (1), 27–33.

Sakuraba, H.; Mizukami, H.; Yagihashi, N.; Wada, R.; Hanyu, C.; Yagihashi, S. Reduced Beta-Cell Mass and Expression of Oxidative Stress-Related DNA Damage in the Islet of Japanese Type II Diabetic Patients. *Diabetologia* **2002,** *45*, 85–96.

Serafini, M. The Role of Antioxidants in Disease Prevention. *Medicine* **2006,** *34*, 533–535.

Shin, C. S.; Moon, B. S.; Park, K. S.; Kim, S. Y.; Park, S J.; Chung, M. H.; Lee, H. K. Serum 8-hydroxy-guanine Levels are Increased in Diabetic Patients. *Diabetes Care* **2001,** *24*, 733–737.

Singh, P.; Kesharwani, R. K.; Keservani, R. K. Antioxidants and Vitamins: Roles in Cellular Function and Metabolism. In *Sustained Energy for Enhanced Human Functions and Activity*; Academic Press, 2017a; pp 385–407, ISBN: 9780128054130.

Singh, P.; Kesharwani, R. K.; Keservani, R. K. Protein, Carbohydrates, and Fats: Energy Metabolism. In *Sustained Energy for Enhanced Human Functions and Activity*; Academic Press, 2017b; pp 103–115, ISBN: 9780128054130.

St-Pierre, J.; Buckingham, J. A.; Roebuck, S. J.; Brand, M. D. Topology of Superoxide Production from Different Sites in the Mitochondrial Electron Transport Chain. *J. Biol. Chem.* **2002,** *277* (47), 44784–44790.

Surana, K. R.; Ahire, E. D.; Sonawane, V. N.; Talele, S. G. Biomolecular and Molecular Docking: A Modern Tool in Drug Discovery and Virtual Screening of Natural Products. In *Applied Pharmaceutical Practice and Nutraceuticals*; Apple Academic Press, 2021; pp 209–223.

Tak, P. P.; Zvaifler, N. J.; Green, D. R.; Firestein, G. S. Rheumatoid Arthritis and p53: How Oxidative Stress Might Alter the Course of Inflammatory Diseases. *Immunol. Today* **2000,** *21*, 78–82.

Tangvarasittichai, S. Oxidative Stress, Insulin Resistance, Dyslipidemia and Type 2 Diabetes Mellitus. *World J. Diabetes* **2015,** *6* (3), 456–480.

Taysi, S.; Polat, F.; Gul, M.; Sari, R. A.; Bakan, E. Lipid Peroxidation, Some Extracellular Antioxidants, and Antioxidant Enzymes in Serum of Patients with Rheumatoid Arthritis. *Rheumatol. Int.* **2002,** *21*, 200–204.

Toyokuni, S. Novel Aspects of Oxidative Stress-Associated Carcinogenesis. *Antioxid. Redox Signal* **2006,** *8* (7-8), 1373–1377.

Trachootham, D.; Zhou, Y.; Zhang, H. Selective Killing of Oncogenically Transformed Cells through a ROS-Mediated Mechanism by Beta-phenylethyl Isothiocyanate. *Cancer Cell* **2006,** *10* (3), 241–252.

Valavanidis, A.; Vlachogianni, T.; Fiotakis, C. 8-hydroxy-2′ - deoxyguanosine (8-OHdG): A Critical Biomarker of Oxidative Stress and Carcinogenesis. *J. Environ. Sci. Health C Environ. Carcinog. Ecotoxicol. Rev.* **2009,** *27* (2), 120–139.

Valko, M.; Leibfritz, D.; Moncol, J.; Cronin, M. T. D.; Mazur, M.; Telser, J. Free Radicals and Antioxidants in Normal Physiological Functions and Human Disease. *Int. J. Biochem. Cell Biol.* **2007,** *39*, 44–84.

Wallace, D. C. A Mitochondrial Paradigm of Metabolic and Degenerative Diseases, Aging, and Cancer: A Dawn for Evolutionary Medicine. *Annu. Rev. Genet.* **2005,** 39.

Yakes, F. M.; Van Houten, B. Mitochondrial DNA Damage is More Extensive and Persists Longer than Nuclear DNA Damage in Human Cells Following Oxidative Stress. *Proc. Natl. Acad. Sci. U. S. A.* **1997,** *94*, 514–519.

Yla-Herttuala, S. Oxidized LDL and Atherogenesis. *Ann. N. Y. Acad. Sci.* **1999,** *874*, 134–137.

## CHAPTER 4

# ADVANCES IN FOOD AND NUTRITION RESEARCH

KISHOR DANAO,[1] VIJAYSHRI ROKDE,[2] AKHIL NAGAR,[3] ATUL R. BENDALE,[3] and NILESH KARANDE[4]

[1]*DadasahebBalpande College of Pharmacy, Nagpur, Maharashtra, India*

[2]*R. C. Patel Institute of Pharmaceutical Education and Research, Shirpur, Maharashtra*

[3]*Sandip Institute of Pharmaceutical Sciences, Nashik, Maharashtra, India*

[4]*Institute of Pharmaceutical Education and Research, Wardha, Maharashtra, India*

### ABSTRACT

Food and nutritional research are one among the most popular, efficient, and cost-effective methods for decreasing the problem of noncommunicable diseases (NCD) and their risk factors, such as nutritional deficits. The dietary research is crucial to learning more about the reasons of obesity and its comorbidities, and so it has the potential to have a significant impact on world health and the economy. Antioxidant (Singh et al., 2017a) characteristics in many functional foods, often recognized as common foods with health advantages beyond their nutritional value, can moderate the degenerative processes of aging. Most countries suffer from malnutrition and noncommunicable diseases, but there are no globally accepted standards for determining dietary superiority in populations and subgroups. Different nutritional assessment methods are used, such as domestic food production (Level 1), total food availability (Level 2), and household measurements

Nutrigenomics and Nutraceuticals. Raj K. Keservani, Eknath D. Ahire, Shubham J. Khairnar, Sanjay J. Kshirsagar, & Rajesh K. Kesharwani (Eds.)

(Levels 3 and 4). Quantity of food ingesting in individuals (Level 5) is used. Omics, bioinformatics, records, biomarkers, and cost-effectiveness analysis are also key tools for the upgrading of nourishment research, according to the American Society for Nutrition (ASN). Microarray and nanotechnology are two developing technologies that have the potential to promote nutrition research by supporting in the discovery, progress, as well as delivery of a variety of intervention techniques to improve health and minimize the risk and complications of a variety of diseases. The study's key finding is that human health requires enough nourishment daily. Previous studies and review papers have found that both forms of malnutrition such as under nutrition and over nutrition are harmful to one's health.

## Graphical Abstract

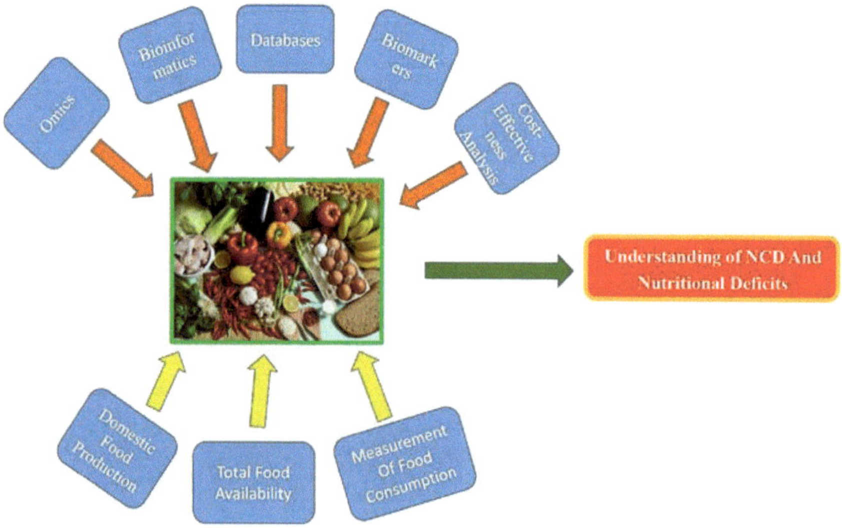


## 4.1 INTRODUCTION

Developments in Food and Nutrition Research identify the essential connection amid the nutrition and dietary sciences (Keservani et al., 2010a; Keservani et al., 2010b; Keservani et al., 2020). Food is the source of nutrition. In the nutrition, body nurtures itself through transforming food into energy in addition to body components. Also, it includes exploiting food for development, metabolism, and restoration of tissues, and encompasses ingestion, digestion, absorption, transport, assimilation into cells, and excretion. Good nutrition is dependent on an integrated supply of food. Foods obtained through plants and animals, and their major elements, are the principal sources of nutrition for humans (Ensminger, 1993). Nutrition is crucial for human and animal growth and development, and also for the prevention and treatment of numerous diseases. Moreover, nutrition is necessary for the conservation of better health and functionality. The foundations for healthy populations and thriving economies include research on interrelationships between nutrition and noncommunicable illnesses, nutritional composition, and nutrition monitoring. As a result, improvements in food and nutrition research are laying the groundwork for answers to bigger health-related challenges, allowing people to live healthier in addition to supplementary productive lives (Lawrence, 1964). Currently, the human populace is increasing at a breakneck speed consequently the Earth's capability to refill its own resources on a constant and dramatic basis. Recently, experts have developed radical procedures to increase food production and discover new ingredients. Also, experts design and construct innovative food structures, as well as integrate technology into the food system (Ehrlich, 1971; Henderson, 2018). The food and nutrition research is one among the most popular, successful, and cost-efficient approaches to minimize the impact of a variety of disorders and their related risk factors, such as obesity. Nutrition study is used to understand the origins of overweightness and its comorbidities, and so has the potential to have a significant impact on world health and economics (Surana et al., 2021; Ahire et al., 2020). Improved study will help us better understand and reduce the harmful consequences of both low and high nutrient intakes on disease progression and overall health. Research will aid in determining the optimal consumption of nutrients that are both necessary and not-necessary on their own and in combination with other nutrients in the diet. According to the American Society for Nutrition (ASN), omics, bioinformatics, databases, biomarkers, and cost-effectiveness analysis are key tools for advancing nutrition research (Skinner, 2012). New technologies such as microarray technology and nanotechnology are emerging to advance nutrition research and its ability to promote nutrition research by supporting in the research and development, and implementation of a variety of intervention techniques to improve health and minimize the risk and complications of a variety of diseases. Emerging technologies can help in several areas, including identifying the site of action for bioactive food components, discovering biomarkers that represent exposure, reaction, and allergy to food and its components, and identifying new target delivery systems for improving health (Page, 2003; Muller, 2003). World Health Organization (WHO) had revealed that for the wellbeing and development, nutrition is the central aspect. Proper nutrition is attributed to healthier neonatal, kid, sturdier immunity, safer gravidity, and labor, a lesser risk of noncommunicable illnesses, and higher life expectancy (WHO, 2009; WHO, 2014). The consequences of nutritional deficiencies lead to causing malnutrition and under nutrition which are related to nutritional diseases. It can include obesity and eating disorders that are all caused by nutritional deficiencies or excesses, as well as chronic diseases like hypertension, cardiovascular disease, cancer, and diabetes mellitus. Malnourish condition is produced by inadequate food to encounter the energy requirements; weight loss, body fat waste, muscle wasting, and letdown to flourish are all signs. The majority of children resulted in the death from their mothers' low and unhealthy nutritional status, as well as lack of possibilities enacted by poverty. Under nutrition is a global problem that affects over 925 million people. Nutrient toxicity is another parameter that may affect human health (Elia, 2000). It entails consuming an extreme quantity of a nutrient such as iron, vitamin A, and may lead to occurrence of poisoning, acute toxicity, and even death in severe condition. Excessive intake of most nutrients can result in negative health effects or chronic toxicity. Overconsumption of calorie-producing foods (fat, carbohydrate, and protein) as well as alcohol raises the risk of obesity and chronic diseases (including the health hazards connected with obesity). The nutrient's need is in a safe range, and there is no risk of negative consequences if it is exceeded (Jorgensen, 2018; and Surana et al., 2021).

## 4.2 NUTRITION AND HEALTH

### *4.2.1 BASIC OF NUTRITION*

The term "Food" can be considered any edible substance which is essential for the nourishment of body or maintaining the physiological and anatomical functions of cells, tissues, or organ and which is directly reflected in system. The term nutrition may be defined as the method of taking and utilizing the food for proper physiological and anatomical functions of body. Plants are autrotrophs as they can produce their own foods by photosynthesis reactions whereas human and other animals are heterotrophs in respect to nutrition as they depend on plants or another animal.

Food and utilization of food, that is, nutrition can fulfill the following functions in living multi-cellular living system:

i. Growth and multiplication cells
ii. Energy production of cells
iii. Differentiation tissues and organ
iv. Adaptability to external environment
v. Maintenance of homoeostasis
vi. Synthesis of biochemical for growth, development, and communication with external environment
vii. In this book, we will only discuss the role of food and nutrition related to human subject not to other animals and plants.

### 4.2.2 *ROLE OF DIET IN NUTRITION AND HEALTH*

Food includes **nutrients** and other ingredients which are essential for maintaining the physiological and anatomical functions of the body. **Diet** or different varieties of food consumed by individual human being play an important role in human health and physiological and anatomical function (Bendale, 2022). The human nutrition process involves the following stages:

**TABLE 4.1** Human Nutrition Process.

| **Ingestion** | **Process of food taking in** |
|---|---|
| **Digestion** | Process of breaking down of complex food material into simpler with the help of enzymes |
| **Absorption** | Process of entering and diffusion of simpler components of food in blood stream and required part |
| **Assimilation** | Incorporation of nonliving components of food in living protoplasm and utilization for energy production and other physiological purposes |
| **Metabolism** | Process of utilization of food for energy production |
| **Egestion** | Process of removal of residue left in the process of nutrition |

As we discussed, the role of nutrients or the components of food is essential for normal physiological and anatomical function of body, so a proper balanced diet is a fundamental requirement for every individual human life.

Living cell consists of mainly **protein (or amino acids), carbohydrate, and** fat (Singh et al., 2017b; Keservani et al., 2015), so for the structure and function of cells all 3 components or nutrients are required in large quantity in human diet and called **macronutrients. Vitamins and minerals** are compulsory in less quantity for the functioning of cells so they are called **micronutrients.** Along with this sufficient quantity of water is essential for the role of cells. Nonstarch polysaccharide (NSP) or dietary fibers are important part of human diet for roughage production and should be present in Human diet.

### *4.2.3 NUTRITIONAL REQUIREMENT OF HEALTHY SUBJECT-MACRO AND MICRO NUTRIENTS*

For a healthy subject, healthy diet consumption is essential to prevent malnutrition and protect the body from different lifestyle-mediated diseases; macro- and micronutrients constituents of healthy diet are discussed as follows:

### *4.2.4 DIETARY FIBERS*

Dietary fibers are also known as roughage. Dietary fibers are components in plant meals that cannot be entirely broken down by human digestive enzymes, according to the British Nutrition Federation. In his article "Dietary 'Fibre' and Pregnancy Toxaemia," Dr. Eben Hipsley developed the term "dietary fiber," referring to dietary fiber (Keservani et al., 2020) as a nondigestible element of the plant cell wall. Dietary fibers are defined by the American Association of Cereal Chemists as the edible fraction of a plant or its equivalent (carbohydrates) that are resistant to absorption and digestion in the small intestine but can be thoroughly fermented in the large intestine.

According to European commission, carbohydrate polymers having three or more monomeric units are known as dietary fibers. In the small intestine, they are unable to be absorbed and digested.

Dietary fiber consists of:

- Carbohydrate polymers obtained from food raw material using a physical, chemical, or enzymatic approach;
- Carbohydrate polymers obtained from food raw material using a physical, chemical, or enzymatic technique;
- Synthetic carbohydrate polymers that have been shown in scientific studies to have a favorable physiological effect.

Dietary fibers include waxes, lignin, and nonstarch polysaccharides (NSP) such as cellulose and pectin. Some of the dietary fibers can be fermented

**TABLE 4.2** Macro Nutrients.

| Name of nutrient | Chemical nature | Dietary sources | Major role |
|---|---|---|---|
| Carbohydrate<br>(energy value:<br>**4 Kcal/g)** | Polyhydroxy aldehyde /ketone;<br>monosaccharide (glucose, fructose)<br>Disaccharide (sucrose)<br>Polysaccharide (starch) | Monosaccharide: - fruits, vegetables, and honey<br>Disaccharide: sugar and jaggery<br>Polysaccharide: root vegetables, cereals, pulses | • Main source of energy<br>• Spare proteins for body construction purpose<br>• Ingredient of cell |
| Protein<br>(energy value:<br>**4 Kcal/g)** | Complex polymer of amino acids | Milk, fish, meat, and eggs; legumes and pulses | • Primary organizational and purposeful components of cell<br>• Development, upkeep, and reparation of tissues<br>• Essential for the creation of hormones, enzymes, antibodies, hemoglobin, etc.<br>• Source of energy<br>• Source of essential amino acids (not synthesized by human body) |
| Fats and Lipids<br>(energy value:<br>**9 Kcal/g)** | Triglycerides of fatty acids<br>Compound phospholipid<br>Derived lipid (Cholesterol) | Oils, butter, ghee,<br>oil-seeds, nuts<br>Meat, poultry, fish, eggs, and milk products | • Concentrated source of energy<br>• Means of transportation for fat-soluble vitamins A, D, E, and K and carotenes, they also encourage their absorption<br>• Sources of essential polyunsaturated fatty acids<br>• Maintain body temperature<br>• Reduce the use of proteins for energy<br>• Protects certain vital organs |

**TABLE 4.3** Micronutrients.

## a. Minerals

| Micronutrient | Major food source | Function |
|---|---|---|
| **Zinc** | Breast milk; meat; egg; seafood; cereals, whole-grain breads, and other grain products; legumes | Components of various enzyme systems |
| | | Helps in activity of enzymes |
| | | Stimulates insulin activity |
| | | Assists antioxidant enzymes |
| | | Boosts immunity |
| **Sodium** | Table salt | Regulates acid–base equilibrium |
| | | Upholds liquid and electrolyte equilibrium |
| | | Essential for appropriate nerve transmission and muscle retrenchment |
| **Calcium** | Breast milk, yogurt, cheese, paneer, tofu, fishes (sardines, salmon) | Bones and teeth; development and maintenance, essential in blood clotting; controls ions transmission across cell membranes; necessary in nerve transmission |
| **Iron** | Meat; legumes and pulses; whole-grain breads, cereals | Necessary for the establishment of hemoglobin and oxygen transportation; boosts resistance to infection |
| **Fluoride** | Fluorinated drinking water | Helps protect teeth against decay; helps in reducing bone loss |
| **Chloride** | Breast milk, sodium chloride (table salt) | Helps to adjust acid–base balance of body fluids |
| **Selenium** | Sea foods, whole-grain breads, cereals, grain products, meats, onions | Necessary to tissue respiration; connected with fat metabolism and vitamin E; used as an antioxidant |
| **Chromium** | Meat, whole-grain breads, cereals, and other fortified or enriched grain products | Essential for normal glucose metabolism |
| **Potassium** | Soy products; vegetables, fruits; meat; yogurt; potatoes; fish; poultry | Regulates acid–base equilibrium and osmotic pressure of body fluids; influences muscle activity, especially heart muscle |

**TABLE 4.3** *(Continued)*

| | | |
|---|---|---|
| **Copper** | Liver; kidney; poultry; legumes and pulses; whole-grain breads, cereals, and other grain products | Catalyzes the function of numerous enzymes and iron; important part of RNA, DNA |
| **Iodine** | Breast milk, seafood, iodized salt | Control thyroid hormones |
| **Phosphorus** | Breast milk; fishes; cheese; egg yolk; meat; poultry; fish; whole-grain breads, cereals, and other grain products; legumes | Constituents of bone and teeth; component of nucleic acids, phospholipids; as coenzyme functions in energy metabolism; buffers intra cellular fluid |

## b. Vitamins

| Micronutrient | Major food source | Function |
|---|---|---|
| | **Fat-soluble vitamins** | |
| **Vitamin** A | Carrots, cod liver oil, tomatoes, pumpkin, green leafy vegetables, milk, butter, and egg | Maintains epithelial cell integrity; produces rhodopsin for low-light vision; and is required for wound healing, growth, and appropriate immunological function |
| **Vitamin** D | Sunlight, milk, cheese, butter, fish liver oil, green vegetables | It is required for appropriate bone development and enhances calcium and phosphorus absorption in the intestine |
| **Vitamin** E | Whole grains, nuts, seeds, leafy vegetables, vegetable oils, whole grains, butter, milk, nuts, seeds, spinach | Acts as an antioxidant in the tissues; acts as a coenzyme; neuromuscular function |
| **Vitamin K** | Soyabeans, cabbage, green leafy vegetables, spinach and cereals, cauliflower | Prothrombin production is catalyzed by this enzyme; it is also essential for the synthesis of other blood-clotting components |
| | **Water-soluble vitamins** | |
| **Vitamin C** | Citrus fruits, peppers, and tomatoes. | It is necessary for collagen production, iron absorption, and transport, and it is a water-soluble antioxidant |

**TABLE 4.3** *(Continued)*

| Micronutrient | Major food source | Function |
|---|---|---|
| | | **Vitamin B complex** |
| **VitaminB1 (Thiamine)** | Bread, nuts, cereals, and seeds, peas and beans | It combines with phosphorus to generate thiamin pyrophosphate (TPP), which is required for protein, carbohydrate, and fat metabolism; it is also vital for growth, appropriate appetite, digestion, and nerve function |
| **VitaminB2 (Riboflavin)** | Milk products, eggs, cereals, mushrooms, and almonds | Is essential for growth; has an enzymatic role in tissue respiration and operates as a hydrogen ion transporter; FMN and FAD production |
| **Vitamin B3 (niacin)** | Meat; fish; breast milk; poultry; whole-grain breads, cereals, and egg yolk | Aids in the release of energy from food and the maintenance of normal skin. It also aids in the correct functioning of the nervous system and the reduction of fatigue |
| **Vitamin B6 (Pyridoxine)** | Meat, fish, yeast extract, poultry, egg yolk, sesame seeds, soyabeans, some fruit, and vegetables | Helps to make red blood cells, which carry oxygen around the body. In the synthesis and break-down of amino acids and unsaturated fatty acids from essential fatty acids; essential for conversion of tryptophan to niacin; essential for normal growth |
| **VitaminB12 (Cobalamin, Cyanocobalamin)** | Breast milk, poultry, meat, fish, cheese, egg | Required for nucleic acid and nucleoprotein production, as well as RBC maturation and folate metabolism |
| **Folacin (Folate)** | Peas and beans, green leafy vegetables, oranges, berries | It is required for the appropriate maturation of RBCs and for the production of nucleic acids |
| **Pantothenic acid** | Meat; poultry; fish; egg yolk; legumes; vegetables | It is involved in the intermediate metabolism of glucose, fat, and protein |
| **Biotin** | Meat, yeast, egg yolk, bananas, vegetables, water melon, strawberries, grape fruit | Coenzyme carrier of carbon dioxide; plays a vital function in the metabolism of fatty acids and amino acids; essential component of enzymes; involved in reactions requiring the lengthening of carbon chains |

in the large intestine by gut bacteria, producing short chain fatty acids and gases like $CO_2$, methane, and hydrogen. Produced fatty acids are absorbed into the blood stream and provide a small amount of energy. The amount of gas produced depends on the type of fiber eaten and the gut bacteria present.

Dietary fibers are broadly categorized into the following 2 groups:

a. Water-insoluble/less fermented fibers
b. Water-soluble/well-fermented fibers

a. **Water-insoluble/less fermented fibers:** Water-insoluble fibers do not dissolve in water and are not digested by digestive enzymes. These mainly help in bulking of stool and quick easy passage of digestive tract. These are hemicellulose, cellulose, and lignins.

- **Cellulose:** Polysaccharides composed of up to 10,000 glucose units are organized linearly with a 1,4 glycosidic bond. Amylase is unable to break down cellulose. The fundamental structural component of the plant cell wall is cellulose. Concentrated alkali renders them insoluble, while concentrated acid renders them soluble. Grains, vegetables, fruit, nuts, and cereal bran are all good sources of cellulose in the diet.
- **Hemicellulose:** Hemicellulose is a type of polysaccharide found in cell walls that has a backbone of 1,4 glucosidic bonds. In dilute alkali, it dissolves. Cereal grains, vegetables, fruit, legumes (such as peas, beans, chickpeas, and lentils), and nuts are all good sources of iron.
- **Lignins:** Lignins are a non-carbohydrate component of the cell wall. Phenyl propane polymer with a complex cross-linking structure. These are resistant to bacterial degradation. Dietary source of lignins is foods with a woody component, for example, celery and the outer layers of cereal grains.

b. **Water-soluble/well-fermented fibers:** Soluble fibers dissolve in water and form a gel that binds the fecal matter and also prevent non-propulsive colon contraction. Gut bacteria digest soluble fibers successfully in the large intestine, creating short chain fatty acids and gases (methane, hydrogen, and CO2). Fatty acids enter the bloodstream and supply a tiny quantity of energy. The amount of gas produced is determined by the fiber type consumed as well as the presence of bacteria in the colon. Abdominal distension, pain, and wind are all symptoms of a highly soluble fiber diet. Bacteria in the large intestine and intestine respond to increased intake with time, and symptoms normally subside (Bendale, 2022).

Examples of soluble fibers are pectin, gums, and mucilage.

- **Pectin:** D-galacturonic acid is one of the key ingredients and structural components of the walls of vegetable cells, and it also works as an intercellular cementing element in pectin. Pectin is a water-soluble, gel-forming substance. Fruits and vegetables, legumes, nuts, and potatoes are examples of dietary sources.
- **Gums and mucilage:** Nonstarch polysaccharides are thick gelling fibers that help maintain the walls of plant cells together. These are described to be very branched polysaccharides which form gels, bind water and other organic matters. Gums are sticky exudates formed in response to shock (i.e., gum arabic). They are mostly made from guar gum and gum Arabic. Guar gum is a galactomannan that is isolated from Cyamopsis tetragonolobus seed (guar). Partial enzymatic hydrolysis produces a product which can be used as a soluble dietary fiber. Mucilage is secreted in the endosperm of plant seeds and acts to prevent unnecessary dehydration. Leguminous seed plants (guar, locust bean), seaweed extracts (carrageenan, alginates), and microbial gums are all examples of gum (xanthan, gellaren). Examples of mucilages are gum acacia, gum tragacanth, gum karaya, gum tragacanth, and isapgul husk.

### *4.2.5 OTHER DIETARY FIBERS*

1. **Beta-glucans:** These are glucose polymers that have a branched structure obtained from cell wall of oats and barley.
2. **Resistant starch:** Starch and starch breakdown products that are not absorbed by the small intestine. Legumes, potatoes, and cereal grains are used to make this dish.
3. **Oligosaccharides:** 3-9 monomers make up a short chain carbohydrate. Onions, chicory, and Jerusalem artichokes contain fructo-oligosaccharides and galacto-oligosaccharides.

4. **Micro components** (waxes, cutin, and suberin): Dietary sources include cereal grain. The assessment of under nutrition requires various parameters that include nutritional intake, nutritional status, and functional indicators.

    - Nutritional intake: The adequacy of dietary calorie, protein, and micronutrient (vitamins and minerals) intake is determined by estimating the amount of food consumed. Individuals are classified as well-nourished or malnourished based on whether their food intake meets their macronutrient (energy, protein, and fat) and micronutrient requirements.
    - Nutritional status: To evaluate whether a person is well-nourished or undernourished, physiological, biochemical, or anthropometric variables are used. Growth in children and changes in body weight in adults are the most common and trustworthy assessments. Biochemical tests are also a viable option. Biochemical assessments alone, with the exception of particular cases such as iron deficiency anemia, are insufficient to determine deficient nutritional status; in general, they become aberrant only in advanced under nutrition.
    - Functional indicators: Most organ systems experience functional implications as a result of malnutrition, with detectable and occasionally quantitative impacts.

Research on food and nutrition is critical for their long-term security (Jorgensen, 2018). Global food security is achieved by the circular economy which is increasing effective exploitation of natural resources that have limiting food waste and loss and also, mitigating biodiversity loss, the circular economy, when included into the food chain, offers options to ensure universal food sustainability. In a repeated practice, the resources are consumed, lowering the requirement for fresh raw materials in food production. This effective usage of natural resources for food in a circular economy aids biodiversity recovery by limiting further conversion of natural habitats to agricultural land, which is a single leading cause of biodiversity loss. The problem of dropping food loss and waste needs a comprehensive understanding of what constitutes food loss and waste. Waste is theoretically nonexistent in a circular food system because it is employed as a feedstock for another cycle, resulting in a system that replicates natural regeneration (Merli, 2017).

**FIGURE 4.1** Approaches to decrease wastage of food in the food supply chain in a spherical food system (Valoppi, 2021).

*Source*: Adapted from Valoppi, 2021.

Better understanding of the specific nutritional science is still relatively new, despite the fact that food and nutrition have been studied for millennia. In 1926, less than 100 years ago, the first vitamin was extracted and chemically defined, kicking off a half-century of research into single-nutrient insufficiency disorders (Carpenter, 2003). Nutritional research on complex noncommunicable chronic diseases such as cardiovascular disease, diabetes, obesity, and cancer is even more recent, having surged in the previous two or three decades, notably around the year 2000. Many of the known important vitamins and minerals were identified and synthesized during the first two decades of the twentieth century, and they were used to prevent and cure nutritional deficiency-related disorders such as scurvy, beriberi, pellagra, rickets, xerophthalmia, and nutritional anemias (ACS, 2002). Casimir Funk proposed the concept of a "vital amine" in food in 1913, after noticing that the hull of raw rice protected birds against a beriberi-like disease. This "vital amine" or vitamin was initially identified in 1926 and given the name thiamine before being manufactured as vitamin B1 in 1936. Vitamin C was isolated and definitely shown to defend against scurvy for the first time in 1932 (Backstrand, 2002). Animal and human being studies established the nutritive origin of severe nutrient deficiencies, which led to the development of dietary strategies to combat beriberi (Thiamin), pellagra (Niacin), scurvy (Ascorbic acid), pernicious anemia (Cobalamin), rickets (Cholecalciferol), and also additional deficit diseases. The chemical manufacture of vitamins, on the other hand, swiftly displaced food-based techniques in favor of treatment with specific nutritional supplements. This foreshadowed the contemporary use and promotion of individual and bundled multivitamins to prevent deficiency, beginning the vitamin supplement business as a whole (Bishai, 2002).

New research objectives in nutrition science are emerging, based on evidence for the diverse effects of various foods, processing methods, and eating patterns. These include the impact of certain fatty acids, flavonoids (Keservani and Sharma, 2014). To jointly address maternal fitness, child growth, infection hazard, and noncommunicable illnesses, comprehensive inquiry is essential for lower-income countries and populations (Mozaffarian, 2016; Mozaffarian, 2011).

## 4.3 METHODS

Various approaches can be used to check food supply and consumption. From national statistics to personal use, several strategies were used. Individual, home, and institution data can be integrated to create a regional or national level of consumption. Different nutritional assessment methods are used, such as domestic food production (level I), total food availability (level II), and household measurements (levels III and IV). Measurement of food consumption in individuals (level V) is used (Bingham 1987; Biró, 2002; Cameron, 1988; Bingham, 1991; Burke, 1947; Nelson, 1997).

**A. (Level I) Domestic food production**

Agriculturalists and food producers, such as farmers, are subject to the majority of government requirements under this strategy. They have to provide a report on how much food they produce. This is part of a longer-term food experiment that provides insight into food supply accessibility. It also includes information on food security in the countries as well as levels of agricultural self-sufficiency. It is useful for organizing food supplies, but not so much on its own. However, there is nothing like it as a completely self-sufficient country in practice. The quality of reporting varies depending on the country. Developed countries that have big farms and industrial development give extensive and accurate statistics data than developing countries.

**B. (Level II) Total food availability**

The UN's Food and Agriculture Organization (FAO) (FAOSTAT database 2013) is in charge of international hunger relief initiatives and provides data on food balance and consumption. The FAO's major mission is to encourage people to live active and healthy lives that ensure that everyone has access to food and security, as well as that people consume sufficient amounts of high-quality food on a regular basis. As shown in Figure 4.1 which shows balance sheet that is updated annually, it relates to any changes with respect to type of consumable food and the necessities of nutritional components in proper range. The first component of a food balance sheet, which is related to a country's food system, is production, imports, and stock movements in relation to food make up the domestic food supply. The second component is domestic food use, which includes feed, seed processing, waste, and food export. Per capita numbers, which include food commodity supply, are the third factor.

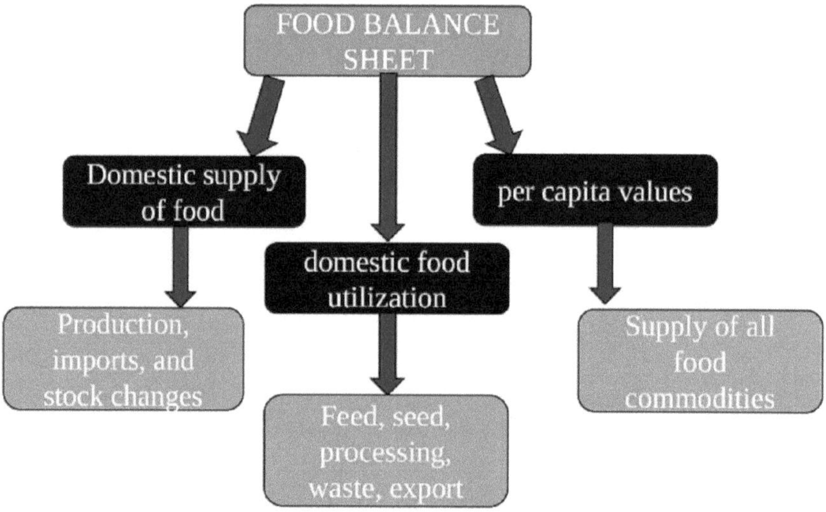


**FIGURE 4.2** Food balance sheet.

As per the data of balance sheet, it is helpful to monitor disease condition and healthy condition. In a recent study, the Global Dietary Database's nationally representative individual-based dietary surveys were compared to estimates published in FAO food balance sheets.

**C. Household measurements (Levels III and IV)**

At the domestic level to estimate food consumption four levels are used that are food accounting, inventories, household recall, and last one is the list recall are the basic methodologies.

- **Food Accounting Method**

  All household food purchases are properly recorded using this approach, either in terms of amount or cost, or both. Typically, no data on changes in storeroom stocks are obtained before or throughout the study period. Menu data can be used to assess the consumed diet percentage away from home. The proportion of consumed diet away from house can be estimated using menu records. Records of foods acquired and eaten away from home are also recorded. It is estimated that 5–10% of waste which is generated is fixed. Using appropriate consumption statistics such as food consumption data, the energy and nutritional composition of the household diet can also be assessed. Although variations in stocks are not occupied into consideration, it is thought that the average alteration across a large enough number of families is zero. The National Food Survey (NFS) (herc.ox.au, 2022) and the Family Expenditure Survey (FES) were combined in 2001 to establish the Expenditure and Food Survey (EFS). When the EFS became part of the Integrated Household Survey (HIS), which is maintained by the Office for National Statistics, it was renamed the Living Costs and Food Survey.

- **Inventory Method**

  The inventory strategy is like the food account method, in the same way that all food records that enter the house are generated by asking respondents. At the opening and the completion of the survey period inventories of stored food are collected. This technique was employed by the National Food Survey (NFS) previous to 1952, and it comprised the first survey of city working-class communities. The capacity of respondents to provide perfect inventory data is usually good. The method gives a precise estimate of how much food and nutrients are accessible for consumption in a given residence.

- **Household Records Method**

  The method is beneficial in countries where a substantial component of the food is prepared at home rather than bought or preprocessed. In this method record of household members food consumption is assessed. The total estimate is reduced by the amount of food consumed by visitors. This method is a fusion of recall and recording methods. The interviewer usually by telephonic contacts in the morning, finds the household composition, and enquires the respondent to remember the amounts of diet used to make breakfast. At the time of launch the foods that will be used are measured in household procedures. An afternoon interview allows for the estimation of lunch waste in addition to the measurement and recording of ingredients for the evening meal. The method's shortcomings are comparable to those who belong to the food account method.

- **List recall Method**
  This method is based on a structured survey, in which particularly for a specific period generally for 1 week the respondent is asked to recollect the amount and charge of food purchased for household consumption. It considers not only food purchases and acquisitions, but also food consumption. Previously, the approach of this method was employed in the US Food Consumption Analysis. The technique is perfectly suitable to communities where the majority of food is bought rather than home-based. Food eaten away from home, food consumed by visitors, and food and nutrition distribution within households continue to be issues.

## D. Measurement of food consumption in individual Method (level V)

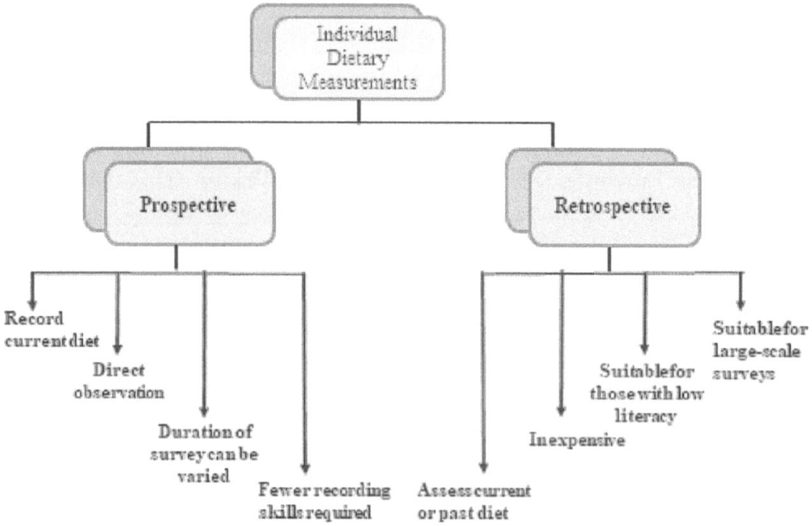


**FIGURE 4.3** Individual dietary measurements.

As shown in Figure 4.3, individual dietary assessment can be carried out by two methods like prospective and retrospective methods. Both are important as per the record of diet.

There are different techniques for prospective records that are used like: (1) weighed record, (2) unweighed record, (3) duplicate diet method, and (4) food checklist

## 4.4 TOOLS

According to the ASN (the American Society for Nutrition), there are several tools that are necessary for the improvement of nutrition research including (1) omics, (2) bioinformatics, (3) databases, (4) biomarkers, and (5) cost-effectiveness analysis.

- **Omics**—By using nutrigenetics and nutrigenomics research several nutrients interact by using genetic factor, proteins, and metabolites that will aid in defining an individual's health (Combs, 2013). Using genomics, individualized nutrient requirements, as well as how nutrients interact in the body are being studied. It will aid in the development of new nutritional and illness biomarkers, as well as determining and reflecting an individual's nutritional condition (Ohlhorst, 2013).
- **Bioinformatics**—It is a discipline of computer science that studies the information technology to create and improve tools for acquiring, storing, organizing, retrieving, and analyzing biological data. For managing and analyzing nutritional data nutrition researchers will use this. There is a correlation between nutrition and health with these techniques. Databases are necessary to gain the full benefits of bioinformatics since they make nutritional data readily accessible in a machine-readable way (Danielle, 2007).
- **Databases**—In this tool food and additional composition as well as ingestion data are linked. In this database, foods and their bioactive constituents, containing nonessential nutrients, are included. Improvement in data collection enhanced with accompaniments such as food intake photographic records, etc. (Praveen, 2019).
- **Biomarkers**—Individuals and subpopulations healthiness and dietary position determined and supervised via intake, outcome, also exposure biomarkers. For assessing development and management

of diseases these biomarkers used that response to diet and nutrition. Also, initial detection and prevention is possible (Dragsted, 2017). Development and validation to correctly track food and nutrient intake is possible if food supply is continually altered (Pico, 2019).

- **Cost-effectiveness analysis**—It is a method of computing and associating the relative costs plus profits of different nutrition research treatments. Cost-effectiveness analysis aids in determining the best cost-effective choice with the maximum public health advantage (Jaspreet, 2013).

## 4.5 CONCLUSION

Different nutritional assessment methods are used, such as domestic food production (Level I), total food availability (Level II), and household measurements (Levels III and IV). Measurement of food consumption in individuals (Level V) is used. Omics, bioinformatics, records, biomarkers, and cost-effectiveness analysis are also key instruments for the improvement of nutrition research, according to the American Society for Nutrition (ASN). Microarray and nanotechnology are two developing technologies that have the potential to promote nutrition research by supporting in the discovery, progress, and delivery of a variety of intervention techniques to improve health and minimize the risk and complications of a variety of diseases. The study's key finding is that human health requires enough nourishment daily. Previous studies and review papers have found that both forms of malnutrition such as under nutrition and over nutrition are harmful to one's health.

## KEYWORDS

- **food**
- **nutrition**
- **noncommunicable disease**
- **health**
- **aging**
- **malnutrition**
- **assessment**

## REFERENCES


Ahire, E. D.; Talele, S. G.; Shah, H. S. Nanoparticles as a Promising Technology in Microbial Pharmaceutics. In *Applied Pharmaceutical Science and Microbiology* Apple Academic Press, 2020; pp 133–158.

American Chemical Society National Historic Chemical Landmarks. The Discovery of Vitamin C by Albert Szent-Györgyi, 2002 [Online]. https://www.acs.org/content/acs/en/education/ whatischemistry/ landmarks/szentgyorgyi.html (assessed Apr 12, 2022)

Backstrand, J. R. The History and Future of Food Fortification in the United States: A Public Health Perspective. *Nutr. Rev.* **2002,** *60,* 15–26. DOI: 10.1301/002966402760240390

Bingham, S. A. The Dietary Assessment of Individuals: Methods, Accuracy, New Techniques and Recommendations. *Nutr. Abstr. Rev.* **1987,** *57*, 705–742.

Bingham, S. A. Limitations of Various Methods for Collecting Dietary Intake Data. *Ann. Nutr. Metab.* **1991,** *35*, 117–127.

Biró, G.; Hulshof, K. F. A. M.; Ovesen, L.; et al. Selection of Methodology to Assess Food Intake. *Euro. J. Clin. Nutr.* **2002,** *56* (Suppl 2), S25–S32.

Bishai, D.; Nalubola, R. The History of Food Fortification in the United States: Its Relevance for Current Fortification Efforts in Developing Countries. *Econ. Dev. Cult. Change* **2002,** *51*, 37–5310. 1086/345361

Burke, B. S. The Dietary History as a Tool in Research. *J. Am. Diet. Assoc.* **1947,** *23*, 1041–1046.

Cameron, M. E.; Staveren, W. A. Eds.; Manual on Methodology for Food Consumption Studies; Oxford: Oxford University Press,1988.

Carpenter, K. J. A Short History of Nutritional Science: Part 3 (1912-1944). *J. Nutr.* **2003,** *133*, 3023–3032. DOI: 10.1093/jn/133.10.3023

Combs, G.; Trumbo, P. R.; Mckinley, M. C.; Milner, J. Biomarkers in Nutrition: New Frontiers in Research and Application. *Ann. N Y Acad. Sci.* **2013,** *1278* (1),1–10.

Lemay, D. G.; Zivkovic, A. M.; German, J. B. Building the Bridges to Bioinformatics in Nutrition Research. *Am. J. Clin. Nutr.* **2007,** *86* (5), 1261–1269.

Dragsted, L. O.; Gao, Q.; Pratio, G.; et al. Dietary and Health Biomarkers—Time for an Update. *Genes Nutr.* **2017,** *12*, 24.

Ehrlich, P. R.; Holdren, J. P. Impact of Population Growth. *Science* **1971,** *171*, 1212–1217. DOI: 10.1126/science.171.3977.1212

Elia, M. Ed.; In *Guidelines for Detection and Management of Malnutrition.* Malnutrition Advisory Group, Standing Committee of BAPEN; BAPEN: Maidenhead, 2000.

Elia, M.; Stratton, R. J. How much Undernutrition is there in Hospitals?. *Br. J. Nutr.* **2000,** *84*, 257–259. 10.1017/S0007114500001525

Ensminger, M. E.; Ensminger, A. H. In *Foods & Nutrition, Encyclopedia*. CRC Press, US, 1993.

Food and Agricultural Organization (FAO). FAOSTAT Database, Food Balance Sheets. Version Current 2013 [cited Apr 17, 2022]. http://faostat.fao.org/site/354/default.aspx.

Henderson, K.; Loreau, M. How Ecological Feedbacks between Human Population and Land Cover Influence Sustainability. *PLoS Comput. Biol.* **2018,** *14*, e1006389. DOI: 10.1371/journal.pcbi.1006389.

https://www.herc.ox.ac.uk/downloads/health_datasets/browse-data-sets/national-food-survey-nfs. (cited Apr 17, 2022).

Jaspreet, K. C.; Ahuja, A. J.; Moshfegh, J. M.; Holden, H. USDA Food and Nutrient Databases Provide the Infrastructure for Food and Nutrition Research, Policy, and Practice1. *J. Nutr.* **2013,** *143*, 241S–249S.

Jorgensen, S.; Pedersen, L. J. T. The Circular Rather than the Linear Economy. In *Restart Sustainable Business Model Innovation*, 2018; pp 103–120. DOI: 10.1007/978-3-319-91971-3_8

Keservani, R. K.; Kesharwani, R. K.; Sharma, A. K.; Vyas, N.; Chadoker, A. Nutritional Supplements: An Overview. *Int. J. Curr. Pharm. Rev. Res.* **2010b,** *1* (1), 59–75.

Keservani, R. K.; Kesharwani, R. K.; Vyas, N.; Jain, S.; Raghuvanshi, R.; Sharma, A. K. Nutraceutical and Functional Food as Future Food: A Review. *Der Pharmacia Lettre* **2010a,** *2* (1), 106–116.

Keservani, R. K.; Sharma, A. K. Flavonoids: Emerging Trends and Potential Health Benefits. *J. Chinese Pharm. Sci.* **2014,** *23* (12), 815.

Keservani, R. K.; Sharma, A. K.; Jarouliya, U. Protein and Peptide in Drug Targeting and its Therapeutic Approach. *Ars Pharm.* **2015,** *56* (3), 165–177.

Keservani, R. K.; Sharma, A. K.; Kesharwani, R. K. Eds.; In *Nutraceuticals and Dietary Supplements: Applications in Health Improvement and Disease Management*; CRC Press, 2020; ISBN: 9781771888738.

Lawrence, T. L. J.; Pearce, J. Some Effects of Wintering Yearling Beef Cattle on Different Planes of Nutrition: I. Live-Weight Gain, Food Consumption and Body Measurement Changes During the Winter Period and the Subsequent Grazing Period. *J. Agric. Sci.* **1964,** *63* (1), 5–21.

Merli, R.; Preziosi, M.; Acampora, A. How do Scholars Approach the Circular Economy? A Systematic Literature Review. *J. Cleaner Prod.* **2018,** *178*, 703–722. DOI: 10.1016/j.jclepro.2017.12.112

Mozaffarian, D. Dietary and Policy Priorities for Cardiovascular Disease, Diabetes, and Obesity: A Comprehensive Review. *Circulation* **2016,** *133*, 187–225. DOI: 10.1161/Circulationaha 115. 018585

Mozaffarian, D.; Wu, J. H. Omega-3 Fatty Acids and Cardiovascular Disease: Effects on Risk Factors, Molecular Pathways, and Clinical Events. *J. Am. Coll. Cardiol.* **2011,** *58*, 2047–2067. DOI: 10.1016/j. jacc.2011.06.063

Muller, M.; Kersten, S. Nutrigenomics: Goals and Strategies. *Nat. Rev. Genet.* **2003,** *4*, 315–322.

Nagar, A.; Bendale, A.; Bhattacharya, S. In *Fundamentals of Social Pharmacy for D.Pharm Students*; Everest Publishing House: Pune, 2022; pp 46–48.

Nelson, M.; Bingham, S. A. Assessment of Food Consumption and Nutrient Intake. In *Design Concepts in Nutritional Epidemiology*, 2nd Ed.; Oxford: Oxford University Press, 1997; pp 123–169.

Ohlhorst, S. D.; Russell, R.; Bier, D.; Klurfeld, D. Nutrition Research to Affect Food and a Healthy Lifespan. *Adv. Nutr.* **2013,** *4* (5), 579–578.

Page, G. P.; Edwards, J. W.; Barnes, S.; Weindruch, R.; Allison, D. B. A Design and Statistical Perspective on Microarray Gene Expression Studies in Nutrition: The Need for Playful Creativity and Scientific Hard-Mindedness. *Nutrition* **2003,** *19*, 997–1000.

Pico, C.; Serra, F.; Rodriguez, A. M.; Keijer, J. Biomarkers of Nutrition and Health: New Tools for New Approaches. *Nutrients* **2019,** *11* (5), 1092.

Praveen, S.; Kumar, R. R.; Upadhyay, S.; Kumar, S. Omics Tools and Techniques For Nutritional Evaluationand Enhancement, ICAR Funded National Agricultural and Higher Educational Report, 2019.

Singh, P.; Kesharwani, R. K.; Keservani, R. K. Antioxidants and Vitamins: Roles in Cellular Function and Metabolism. In *Sustained Energy for Enhanced Human Functions and Activity*; Academic Press, 2017a; pp. 385–407, ISBN: 9780128054130.

Singh, P.; Kesharwani, R. K.; Keservani, R. K. Protein, Carbohydrates, and Fats: Energy Metabolism. In *Sustained Energy for Enhanced Human Functions and Activity*;Academic Press, 2017b; pp 103–115, ISBN: 9780128054130.

Skinner, A. C.; Steiner, M. J.; Perrin, E. M. Self-reported Energy Intake by Age in Overweight and Healthy-Weight Children in NHANES, 2001–2008. *Pediatrics* **2012,** 130, e936 –e942.

Surana, K. R.; Ahire, E. D.; Sonawane, V. N.; Talele, S. G. Biomolecular and Molecular Docking: A Modern Tool in Drug Discovery and Virtual Screening of Natural Products. In *Applied Pharmaceutical Practice and Nutraceuticals*; Apple Academic Press, 2021; pp 209–223.

Surana, K. R.; Ahire, E. D.; Sonawane, V. N.; Talele, S. G.; Talele, G. S.; Molecular Modeling: Novel Techniques in Food and Nutrition Development. In *Natural Food Products and Waste Recovery: Healthy Foods, Nutrition Design, and Extraction of Valuable Compounds*, 2021; p 17.

Valoppi, F.; Agustin, M.; Abik, F.; de Carvalho, D. M.; Sithole, J.; Bhattarai, M.; Varis, J. J.; Arzami, A. N. A. B.; Pulkkinen, E.; Mikkonen, K. S. Insight on Current Advances in Food Science and Technology for Feeding the World Population. *Front. Sustain. Food Syst.* **2021,** *5*, 626227. DOI: 10.3389/fsufs.2021.626227

World Health Organization, Global Health Risks: Mortality and Burden of Disease Attributable to Selected Major Risks, WHO, Geneva, 2009 [Online]. https://www.who.int/healthinfo/global_burden_disease/GlobalHealthRisks_report_full.pdf. (assessed Apr 12, 2022).

World Health Organization, Global Status Report on Noncommunicable Diseases 2014, Geneva, 2014 [Online]. http://www.who.int/nmh/publications/ncd-status-report-2014/en/ (assessed Apr 12, 2022).

## CHAPTER 5

# NUTRIGENOMICS RESEARCH-BASED NEW APPROACHES TO FOOD FUNCTIONS

ATUL R. BENDALE,[1] EKNATH D. AHIRE,[2] and SHUBHAM J. KHAIRNAR[2]

[1]*Sandip Institute of Pharmaceutical Sciences, Nashik, Maharashtra, India*

[2]*METs, Institute of Pharmacy, BKC, Nashik, Maharashtra, India*

## ABSTRACT

Nutrient-induced gene expression alterations can have a range of implications, including changes in metabolism and disease susceptibility. Several nutrients and therapeutic dietary ingredients have been studied for their effects on the epigenetic processes of DNA methylation and other histone modifications. Because a person's epigenetic model is set from early gestation and is transformed and customized by environmental effects, epigenetic mechanisms play a crucial role in the development of obesity and diabetes mellitus. As a result, epigenetic regulation has been identified as a major factor in the ageing process, obesity, and diabetes mellitus. Furthermore, both ageing and cancer show large changes in genome-wide DNA methylation, suggesting that nutrition may influence the ageing process and cancer progression through epigenetic mechanisms. As a result, various dietary components are emerging as important biological age moderators, and epigenetic clock models are assisting in unraveling the complex interplay between nutrition and the most common diseases. Folate and related B-vitamins are intriguing epigenetic age regulators because they are

Nutrigenomics and Nutraceuticals. Raj K. Keservani, Eknath D. Ahire, Shubham J. Khairnar, Sanjay J. Kshirsagar, & Rajesh K. Kesharwani (Eds.)

essential cofactors in 1-carbon metabolism, the main metabolic pathway for generating methyl groups for DNA methylation. More long-term study and randomized nutritional interventions are needed to advance the area.

## 5.1 INTRODUCTION

The evaluation of the impact of food and dietary ingredients on gene expression, as well as how genetic differences affect the nutritional environment, is known as nutrigenomics. To ascertain how certain nutrients or dietary regimens may affect human health, it focuses on figuring out how nutrients and other dietary bioactive interaction at the molecular level with the genome.

Using molecular techniques like proteomics, metabolomics, and transcriptomics, nutrigenomics analyzes how the elements of a specific diet may change the gene's expression in both health and sickness (Sales et al., 2014). The term "personalized nutrition" refers to the provision of individuals with tailored nutritional guidance, products, or services with the aim of improving or maintaining optimum health. Typically, this advice considers a person's genetic, phenotypic, and medical data (Ordovas et al., 2018).

In recent years, consumer interest in nutrigenomics has fueled the rise of companies like DNAFit, Habitat, and EasyDNA. Based on the results of home nutrigenetic testing kits, these businesses provide individualized dietary recommendations. With the development of the food industry and the public's interest in nutrigenomics, "functional foods" are increasingly being produced. These foods include, for instance, yoghurts with probiotics and stanol esters and plant sterols added to margarines (British Nutrition Foundation, 2018; EUFIC, 2018; Surana et al., 2021).

As new gene–diet interactions are discovered every day through candidate gene research and genome-wide association studies, the topic of how customized nutrigenomic meals may be developed and implemented into the functional food market remains unanswered. This article will look at some of the issues that these foods have raised as well as where nutrigenomic foods are at this point in their development.

Consumers have made it clear that they want customized nutrigenomic meals introduced for optimal health. Fragmentation and personalization were named one of the ten main themes in Food, Nutrition, and Health for 2019, according to recent research by New Nutrition Business (New nutrition Business, 2018). Changing population demographics, growing consumer knowledge of the relationship between nutrition and genetics, and increased clinical data supporting the usefulness of functional foods appear to be some of the primary market drivers for customization of food items (German et al., 2011). Considering that most people can self-regulate every essential nutrient across a wide range of dietary consumption, the current public health method to ensuring that the population reaches the reference nutrient intake (RNI) of each critical to overload everyone on nutrients (German et al., 2011). However, a fundamental flaw in this "one size fits all" approach is that it ignores individual genetic variants that might alter nutritional requirements, as well as other factors including total food choices, drug–nutrient interactions, and medical conditions that can impair nutrient absorption (Surana et al., 2021).

The metabolism of folic acid among carriers of the methylene tetrahydrofolate reductase (MTHFR) gene variants—MTHFR C677T and MTHFR A1298C—provides the most persuasive data to date that supports interindividual variation in nutritional requirements. Carriers of MTHFR gene mutations require a higher dietary consumption of folate to obtain the optimal MTHFR levels for adequate homocysteine metabolism (Van der Put et al., 1998). The risk of phenotypic effects has been linked to inadequate folate intake such as heart disease, cancer, and an increased incidence of neural tube malformations among carriers of these mutations (Gohil et al., 2009; Galvan-Portillo et al., 2009; Yaliwal and Desai, 2012). Folate supplementation has been proven to lessen the incidence of these unfavorable health consequences in genetically vulnerable persons with MTHFR mutations. Similarly, studies on folic acid supplementation have spurred interest in the role of L-methylfolate in mood regulation and better antidepressant response mechanisms (Stahl, 2007). The only form of folate or folic acid that can cross the blood–brain barrier and aid in the synthesis of serotonin, norepinephrine, and dopamine is L-methylfolate (Shelton et al., 2013). The findings in relation to L-methylfolate and mental health have piqued the interest of nutraceutical businesses. To support this developing evidence foundation, nutraceuticals including MethylPro, Deplin, and XaQuil XR have been created. The microbiome depends on individual differences in gut microbiota, as is now well documented. However, the idea that bacteria or their direct products can be useful to the diet in ways other than giving certain vitamins has largely gone unexplored to date. Personalized nutrition, according to several experts, can be achieved by altering gut flora. Gut microbiota and its by-products are recognized to have the ability to change the host's metabolome, genome, transcriptome, and proteome (Lewis and Burton-Freeman, 2010). Despite scientific progress in this subject, it is still unclear how different dietary components influence the selective growth of specific microbial populations over others, resulting in beneficial metabolic outcomes (Lewis and Burton-Freeman, 2010).

As a result, the food industry is now concentrating on functional foods that come in the form of probiotics, living cultures of beneficial bacteria. On the other hand, prebiotics may benefit human health and should not be ignored. For instance, it has been discovered that human milk oligosaccharides specifically encourage the development of the good gut flora. Infant is Bifidobacterium longum subsp (Sela et al., 2008). In vitro studies revealed that the prebiotic panose inhibited the growth of Clostridium, a Gram-positive bacteria linked to many human illnesses (Makelainen et al., 2009). By learning about microbial–food interactions, food companies may be able to create unique metabolically targeted foods that aim to enhance human health.

To date, most food firms have concentrated their efforts on manufacturing probiotic drinks and smoothies. However, occurrences have a push in recent years to develop more individualized probiotic meals. Companies like uBiome, Rhythm Health, and Microba have created home microbiome testing kits, and many of them provide specific probiotic supplements based on the findings of these tests.

Food consumption, unlike pharmacogenetics, cannot be connected to "single exposure – single time point" effects, which are most common difficult aspects of investigating gene–diet relationships. Because a person's diet is very changeable and contains a range of nutrients and other bioactive chemicals that might interact with one another, determining which nutrient is impacting gene expression is difficult. Similarly, because epistatic interactions between genes can occur, the contribution of single nucleotide polymorphisms (SNPs) to total illness risk can be influenced by the existence of additional gene variants. Another significant problem related to the commercialization of nutrigenomic meals is ethical, legal, and social concerns (ELSIs). The present "biohype" surrounding nutrigenomic food has prompted international nutrigenomics professionals to highlight concerns about:

1. the health claims that could be exploited to advertise nutrigenomic products
2. commercial companies' management of private nutrigenomic information
3. the provision of nutritional genomics services
4. the development of new nutrigenomic products
5. Given the high cost of nutrigenomic products, they have the potential to exacerbate social inequities in health (Oliver et al., 2005).

Additionally, the lines between food and medications may become fuzzier as functional foods are introduced to the market. Similar to pharmaceuticals, nutrigenomic meals need to be examined for both safety and efficacy. As a result of this ambiguity, more open law and control of these foods is required. The only source of guidance in this field inside the European Union at the moment is the European Nutrigenomics Organization's (NuGO) ethical guidelines for nutrigenomic researchers (European Commission, 2007). The success of customized nutrigenomic goods depends on the future regulation of these commodities' implementation. It is worth mentioning, however, that while scientific regulation increases customer confidence, it also adds to the challenges of food production. It is still unknown what the future nutrigenomic foods will resemble. According to some theories, nutrigenomic food items would be created that would precisely fit a person's nutritional needs (Keservani et al., 2020). This seems a little premature, though, as diet–gene links are yet unclear and nutrigenomics is a very young field of study. A more accurate forecast of the food effects of nutrigenomic research is food products that identify food components that are advantageous to people who are genetically susceptible to medical problems. Food companies could need to expand their offerings to include nutrigenetic testing and specialized product recommendations based on the findings of these tests to discover these genetic predispositions (Oliver et al., 2005).

## 5.2 NUTRIGENOMICS AND FOOD

Unquestionably, the development of genetics is altering how we view people and influencing how diseases will be treated in the future. However, as an interdisciplinary field of study, food science will have a lot more opportunities thanks to genomics. The same technologies have the ability to look at the genetic and metabolic variety of people and how it affects their health-related responses to food, as well as the diversity of individual sensory preferences for food and sensitivities and intolerances to food ingredients. With the use of nutrigenomics, researchers will have a better grasp of human nutritional demands and responses as well as how genetics underlies their variation in a variety of features (Nestle Nutrition Institute, 2018). The biology, chemistry, and functioning of all the biomaterials that make up food will be clarified by agricultural genomics of food commodities.

The genomes of production animals will direct research into everything from absolute production yield efficiency to reshaping the composition of consumable tissues to disease and toxin defense (Lemay et al., 2009). Metagenomics, or the genomics of microbial ecosystems, is a last subset of genomics that is becoming increasingly important to food scientists.

The science of food microbiology and its applications to professional microorganisms are beginning to be guided by the microorganism genomes, which are already redefining all aspects of microbial food safety. These applications range from improved food bioprocessing to the explicit inclusion of live microorganisms for their food health value (Sela et al., 2008). Systems biology is the science of comprehending living species and entire living ecosystems, and genomics lies at the heart of that research.

Food is the most significant industrial system in civilization, and many of the ideas being developed in other systems biology fields will apply to it (Keservani et al., 2010a; Keservani et al., 2010b). One of the most obvious examples of a systems approach to eating is lactation and milk. In response to the Darwinian urge to produce a complete system of nutrition for offspring, mammals developed the mammary gland, an unrivalled bioreactor (Lemay et al., 2009). The structures, purposes, and advantages of milk components that have evolved via evolution have guided nutrition research for almost a century.

The arrival of mammalian genomes, as well as the insights into molecular evolution that they bring, will usher in a new era of research in which the biology of the mammary gland and breastfeeding will inspire a far broader vision of food research, ranging from ingredient functioning to bioprocessing.

## 5.3 NUTRIGENOMICS: ESSENTIAL AND NONESSENTIAL NUTRIENTS

Science has finished the process of determining the nutrients that are absolutely necessary for human health. Making sure that everyone in the population receives a diet full of necessary nutrients is the current focus of public health. Overdosing everyone is the food strategy for ensuring adequacy of all essential nutrients in a diverse population. This strategy is founded on a significant biological advantage. Because individuals generally regulate each important nutrient throughout a wide range of intakes, the problem of critical nutrients can be solved by ostensibly slightly overdosing the majority of the population.

Despite this, many sections of the population still consume insufficient amounts of these essential nutrients. The recurring issue of nutritional scarcity is exacerbated by overt poverty, food preferences, genetic polymorphisms that change or increase demand, disorders of malabsorption such as celiac disease, illnesses, or drugs that interfere with food uptake or metabolism. Isolated shortfalls can also be brought on by odd diets and lifestyle decisions that prevent alternate sources (like sun exposure), such as a lack of vitamin D. Additionally, population-based fortification of a single nutrient with a one-size-fits-all approach can have unfavorable effects.

Folic acid fortification in the food supply, for example, has not only masked vitamin B12 deficiency–anemia, but evidence is accumulating that it may also result in reduced cognitive performance in some subgroups of the population (Refsum and Smith, 2008; Winkels et al., 2008). The biological fact that some persons do not achieve effective nutrient sufficiency while adhering to conventional dietary recommendations will be addressed in what way by food recommendations? Nutrigenomics is actively finding many of the genetic causes due to the effects of dietary fat composition on plasma lipids (Ferguson et al., 2010; Joffe et al., 2010) and obesity risk to the effect of epigenetics on the widespread increased rates of food allergy (Allen and Martin, 2010).

Almost often, the use of this science will necessitate a more tailored strategy to giving necessary nutrients. Because necessary nutrients may be supplied as supplements, solving the challenge of tailoring vital nutrient intakes will be simple. It will be more difficult to personalize the total diet. Nutritional wellness is determined by consuming more than just needed nutrients. In truth, the global epidemic of noncommunicable illnesses is mostly driven by diets that provide adequate fundamental nutrition, but chronic dietary imbalances, along with a variety of lifestyles and genetics, cause metabolic diseases. The genome and postgenomic products are affected by nonessential nutrients, other environmental factors, and other factors in addition to necessary nutrients. The genome and postgenomic products are affected by nonessential nutrients, other environmental factors, and other factors in addition to necessary nutrient, antioxidant capacity, fibers, (Dom´ınguez et al., 2010; O'Sullivan et al., 2010; Papathanasopoulos and Camilleri, 2009; Puchau et al., 2009; Vijay-Kumar et al., 2010), furthermore to gut microbiota makeup, prebiotics (Cani et al., 2009), metabolic phenotype, and environmental and metabolic regulation (Peppa et al., 2010). The goal of tailoring nutrition based on an individual's genotypic and metabolic differences will necessitate first separating dieters from non-dieters. To meet the rising needs of the consumer's health, metabolism, performance, and cognitive abilities, food commodities will be produced both separately and collectively.

## 5.4 NUTRIENT SENSORS AND DIETARY SIGNALS

Nutrigenomics can comprehend the variability of the human genome in terms of a wide range of nutrient concentrations and a variety of food nutrition or perceiving receptors by identifying specific dietary signals, signal sensing, and other ways. Ruden et al. (2005) conducted nutrigenomics research including an experiment with Drosophila, a model organism, and found that each nutrient has many target sites with different affinities and specificities. Drosophila has a lipid transport system and tissues that resemble adipose tissue, and it has more in common with humans in terms of obesity and related disorders than any other model organism. Müller and Kersten (2003) also identified unique cellular-sensing systems, as well as nutrients and dietary metabolites as signaling factors.

The molecular structure of nutrients is naturally intended to carry information about how to activate specific signaling pathways to reach the target site. Minor structural differences (such as saturated vs unsaturated fatty acids or cholesterol vs plant sterols) can have a significant impact on which sensor pathways are engaged. Identifying the molecular pathways and upstream/downstream regulation of each nutrient is a difficult task for scientists. Nutrigenomics research can lead to the discovery of molecular pathways through genome-wide characterisation of dietary target genes. This type of data can aid researchers in better understanding the mechanism of action of particular nutrients and how they are linked to food, which plays a significant role in overall health.

For example, Patsouris et al. (2004) found that while the involvement of PPAR in obesity is unknown, there is some evidence that it plays a role in the obesity-linked pathophysiology of type 2 diabetes. It was recently shown that PPAR modulates genes directly expressed in hepatic gluconeogenesis and glycerol metabolism (Mandard et al., 2004; Moller and Kaufman, 2005). The liver may interpret elevated plasma free fatty acid levels as signals of "hunger" or "in need of glucose" when they connect with the PPAR. This may boost gluconeogenesis in a PPAR-dependent way, especially in cases of hepatic insulin resistance. Fasted PPAR null mice mutant (lack of functioning PPAR) suffers from a number of metabolic abnormalities, including hypothermia, hypoglycemia, hypoketonemia, and high plasma-free fatty acid levels, according to Mandard et al. (2004) and Kersten et al. (1999).

## 5.5 NUTRIGENOMIC DISEASES AND BIOMARKERS

Nutrigenomics research is focused on the principle of an individual's nutrition–gene–disease relationship and how to safeguard humanity from noncommunicable diseases (NCDs) such as CVS and metabolic diseases, obesity, diabetes, respiratory illnesses, metabolic syndrome, and malignancies around the world. These are essentially busy junk food eating lifestyle diseases of cities that are mediated by prolonged exposure to specific dietary components. Biomarkers can diagnose nutritional disorders like these. It could be aberrant insulin sensitivity, elevated blood pressure, or disturbed lipid profiles to examine cholesterol and/or triglyceride levels as indicators of NCDs such as cardiovascular disease or metabolic syndrome.

These biomarkers are usually single proteins (Jarouliya et al., 2019b), metabolites, or body functions that act as a detector for proteomics and metabolic changes in an individual's body, which can be a cause of a number of chronic diseases depending on the genotype of the individual. The molecular elements of an individual's genetic material (DNA) damage can be diagnosed in a variety of methods, including:

i. single-base damage (e.g., DNA adducts caused by oxidative stress, such as the addition of a hydroxyl radical to guanine);
ii. abasic site in the DNA sequence (measurable with the aldehyde-reactive probe);
iii. DNA strand breaks (commonly measured with the Comet assay);
iv. telomere shortening (measured by terminal restriction fragment length analysis, quantitative PCR, or flow cytometry).

The use of damaged DNAs as biomarkers has recently been validated at multiple levels based on nutrient-related evidence (cross-sectional epidemiology and intervention studies) and disease-related evidence (cross-sectional epidemiology and prospective cohort studies). The micronucleus assay in cytokinesis-blocked lymphocytes is now the best validated indicator for nutritional genomic studies of DNA damage. Transcriptomics, which uses a microarray assay to evaluate the mRNA copies for all actively transcribed genes, is another well-validated nutrigenomics technique. This method has the benefit of allowing for the simultaneous measurement of the expression levels of thousands of genes in a single test. Peripheral blood cell gene expression patterns have been shown through studies to be very specific for various disease states.

But, according to Martin et al. (2001), Valk et al. (2004), disease-specific gene-expression patterns in blood cells have been established for breast cancers and leukemia, and such patterns are being used as biomarkers for disease diagnosis.

### 5.5.1 GENE–DIET–DISEASE INTERACTION

SNPs (single-nucleotide polymorphisms) and SNVs (single-nucleotide variations) are the most frequently accepted markers for genetic variation these days. SNPs and SNVs can detect genotypic changes, enabling us to provide a suitable food strategy to prevent noncommunicable diseases (NCDs) as cancer, diabetes, obesity, and cardiovascular disease. In this regard, Ramos-Lopez et al. (2016) identified the receptor for sweetness (TAS1R2), which is involved in taste perception, and Ramos-Lopez et al. (2015) found the differentiation cluster (CD36), which is linked to dyslipidemia in Mexicans who consume a lot of carbohydrates and fats.

Furthermore, common variations of genes that govern homocysteine metabolism, such as methylene tetrahydrofolate reductase (MTHFR) and methionine synthase (MTR), have been related to a higher risk of breast cancer in people who consume less folate, vitamin B6, and vitamin B12 (Singh et al., 2017a; Jiang-Hua et al., 2014). Vitamin D levels are polymorphic across the population and have the ability to affect many metabolic processes in the body. SNPs in the vitamin D receptor (VDR) gene, for example, influence vitamin D availability and lead to osteoporosis in postmenopausal women with low calcium intakes (Desmarchelier et al., 2016; Stathopoulou et al., 2011).

In addition, SNPs in genes producing lipid proteins (Keservani et al., 2015; Singh et al., 2017b) including apolipoprotein C3 (APOC3) and apolipoprotein A1 (APOA1) were linked to a higher risk of metabolic syndrome in people who ate a Western diet (Hosseini-Esfahani et al., 2014; Hosseini-Esfahani et al., 2015). Similarly, moderate and heavy coffee drinkers had an elevated risk of hypertension and CVD, which was linked to a genetic polymorphism in the cytochrome P450 family 1 subfamily A member 2 (CYP1A2) gene (Palatini et al., 2009; Cornelis et al., 2006). Furthermore, research utilizing genetic risk scores (GRS) has looked at the cumulative influence of SNPs on food interactions and disease vulnerability. Macronutrients have the ability to change the obesity (Jarouliya et al., 2019a) GRS with higher adiposity levels (Goni et al., 2015). In several cohort studies, obesity GRS correlated with sugar-sweetened beverage consumption (Qi et al., 2012) and fried food consumption (Qi et al., 2014) in relation to BMI and obesity.

## 5.6 NUTRIGENOMICS' BENEFITS

When incorporated into an integrative dietary plan, nutrigenomics can highlight issues that a simple health history cannot. This might be anything from a genetic predisposition to heart disease to the fact that you have tried everything and still cannot lose weight. Genomic testing is extremely beneficial for anyone looking to take control of their health, according to a dietitian who is also a trained genomic medical clinician. Genomic analysis can shed light on the causes of a patient's condition and suggest the best eating regimen for them. A practitioner may be able to advise you on which dietary habits will or will not work for you based on your genetic composition. Due to genomic preferences for fat metabolism, gene polymorphisms may indicate that your body would not profit from a vegan diet or would not adapt well to a keto diet.

A nutrigenomic test can also reveal your personal best macro- and micronutrient sources. Perhaps your body cannot use omega-3 fatty acids from plants to their full potential, or you cannot turn sunlight into vitamin D. A qualified practitioner can use this information to advise you on the foods to consume or supplements to take to satisfy your needs. Similarly, a nutrigenomics test may reveal predispositions to particular diseases.

Genes involved in oestrogen metabolism may allow us to detect gene alterations that increase a person's chance of developing breast cancer. Coronary artery disease, Diabetes Information from a Reliable Source Obesity, mental health, and genetic manifestations have all been connected, and each has dietary preventative techniques. With this knowledge, one can make informed decisions about how to reduce risk through diet.

## 5.7 METABOLOMICS—THE NEWEST TOOL FOR NUTRIGENOMICS?

The most recent "omics" technology being used in nutrigenomics is metabolomics. It makes use of analytical chemistry techniques like nuclear magnetic resonance (NMR) spectroscopy and mass spectrometry (MS) to collect comprehensive data on the metabolome, or the collection of low molecular weight metabolites, nutrients, and other chemicals in various human biofluids. In a recent review, the advantages and disadvantages of using metabolomics in nutrition research were discussed. The metabolome is thought to be the functional result of protein and gene expression that is easily accessible in body fluids. To measure nutrient status, metabolic reactions, illness risk, etc., and so characterize the nutritional phenotype, it is suggested that metabolomic technology may give the comprehensive biomarker of various metabolites. In a recent study, the effects of n-6 PUFA and n-3 PUFA therapies were investigated using an integrated metabolism approach. To establish the impact of meals supplemented with an arachidonic acid (AA)-enriched oil or an eicosapentaenoic acid (EPA)/docosahexaenoic acid (DHA)-enriched fish oil, a thorough analysis of the hepatic transcriptome and the lipid metabolome was carried out. The goal of the study was to ascertain if variations in the gene expression of pertinent biosynthetic gene products might explain the metabolomic responses to a dietary fatty acid intervention.

It proved that the single most significant gene expression event to explain the alterations in hepatic metabolome after the AA-rich meal was the down-regulation of hepatic SCD1, but not SCD2 or SCD3. Given these information, one can infer that hepatic SCD1 is a crucial nutrient-sensitive gene that could be affected by fatty acid interventions or their functional analogues. As a result, hepatic SCD1 may represent a promising therapeutic target to modify metabolism and thwart the onset of CVD and diet-induced obesity. Very few research has used the metabolomics technique in the context of nutrition, according to the literature to date. To create a more thorough biological understanding of lipid metabolism—hepatic gene expression profiles reflecting the metabolic profile—the aforementioned study established analytic synergy by merging metabolomic data with a transcriptomics approach.

## 5.8 PERSPECTIVE FOR THE FUTURE

The goal of this review was to highlight the emerging technologies that can be used to find cellular homeostasis and metabolism biomarkers that are relevant to food, nutrition, and health. In the end, we require early indicators of dietary perturbations that mark critical turning points in the transition from health to disease. Strong nutritional genomics research programmes must be developed and put into practise by nutrigenomics that are sensitive enough to account for both the diversity of dietary exposure and human genetic variation. To uncover the nutrient-sensitive gatekeepers of various disease processes, nutrigenomics can benefit from a systems biology approach that makes use of a number of high-throughput "omic" technology platforms. To fully understand the mechanisms underpinning diet and health, each analytic platform (genomics, proteomics, transcriptomics, or metabolomics) offers complementary information that is necessary and implicit.

## 5.9 CONCLUSION

The field of nutrigenomics is still developing despite extensive study on the connection between genetics, nutrition, and health. As a relatively new area of study, nutrigenomics still has a lot to teach us.

This is not to say that genetics should not be taken into account when giving nutritional advice. Just keep in mind that nutrigenomics cannot fix every dietary issue and those genes are simply one of many variables impacting health and the best foods to eat.

In our work, we must take into account lifestyle, health history, health condition, personal preferences, cultural identification, patient readiness to change, and their own health goals. Genomic testing should not be used as the primary foundation for recommendations. Direct-to-consumer genetic testing for diet-related objectives may seem interesting and handy, but it also has the potential to be detrimental.

The most significant disadvantage of these tests is that they are not analyzed by a trained doctor. Skilled practitioners employ a polygenic approach, which considers how all genes are connected to larger systems in the body. They interpret how these systems interact in the context of one's overall health.

## KEYWORDS

- **DNA**
- **nutrition**
- **cancer**
- **diabetes mellitus**
- **nutrients**
- **genetics**
- **health**
- **human genome project**
- **malnutrition**
- **assessment**
- **food function**
- **nutrigenomics**
- **RNA**
- **metabolites**

## REFERENCES


Ahn, J.; Yu, K.; Stolzenberg-Solomon, R.; Simon, K. C.; McCullough, M. L.; Gallicchio, L.; Jacobs, E. J.; Ascherio, A.; Helzlsouer, K.; Jacobs, K. B.; Li, Q.; Weinstein, S. J.; Purdue, M.; Virtamo, J.; Horst, R.; Wheeler, W.; Chanock, S.; Hunter, D. J.; Hayes, R. B.; Kraft, P.; Albanes, D. Genome-Wide Association Study of Circulating Vitamin D Levels. *Hum. Mol. Genet.* **2010,** *19*, 2739–2745. DOI: 10.1093/hmg/ddq155.

Allen, K. J.; Martin, P. E. Clinical Aspects of Pediatric Food Allergy and Failed Oral Immune Tolerance. *J. Clin. Gastroenterol.* **2010,** *44*, 391–401.

Barry, E. L.; Rees, J. R.; Peacock, J. L.; Mott, L. A.; Amos, C. I.; Bostick, R. M.; Figueiredo, J. C.; Ahnen, D. J.; Bresalier, R. S.; Burke, C. A.; Baron, J. A. Genetic Variants in CYP2R1, CYP24A1, and VDR Modify the Efficacy of Vitamin D3 Supplementation for Increasing Serum 25-hydroxyvitamin D Levels in a Randomizedcontrolled Trial. *J. Clin. Endocrinol. Metab.* **2014,** *99*, E2133–E2137. DOI: 10.1210/jc.2014-1389.

British Nutrition Foundation. Functional Foods, 2018 [Online]. https://www.nutrition.org.uk/nutritionscience/foodfacts/functional-foods.html (accessed Feb. 04, 2021].

Cani, P. D.; Lecourt, E.; Dewulf, E. M.; Sohet, F. M.; Pachikian, B. D.; et al. Gut Microbiota Fermentation of Prebiotics Increases Satietogenic and Incretin Gut Peptide Production with Consequences for Appetite Sensation and Glucose Response after a Meal. *Am. J. Clin. Nutr.* **2009,** *90*, 1236–1243.

Cornelis, M. C.; El-Sohemy, A.; Kabagambe, E. K.; Campos, H. Coffee, CYP1A2 Genotype, and Risk of Myocardial Infarction. *JAMA* **2006,** *295*, 1135–1141. DOI: 10.1001/jama.295.10.1135.

Desmarchelier, C.; Borel, P.; Goncalves, A.; Kopec, R.; Nowicki, M.; Morange, S.; Lesavre, N.; Portugal, H.; Reboul, E. A Combination of Single-Nucleotide Polymorphisms is Associated with Interindividual Variability in Cholecalciferol Bioavailability in Healthy Men. *J. Nutr.* **2016,** *146*, 2421–2428. DOI: 10.3945/jn.116.237115.

Dom´ınguez, C.; Cabrera, L.; Rodr´ıguez, P.; Borges, ´A.; Carrillo, F.; et al. Association between Glycemic Index, Glycemic Load, and Fructose with Insulin Resistance: The CDC of the Canary Islands Study. *Eur. J. Nutr.* **2010,** 1–8.

EUFIC. Functional Foods, 2018 [Online]. https://www.eufic.org/en/food-production/article/functional-foods (accessed Apr 04, 2022).

European Commission. EU Project Publishes Bioethics Guidelines for Nutrigenomics Research, 2007 [Online]. https://cordis.europa.eu/news/rcn/28093_en.html (accessed Apr 04, 2022).

Ferguson, J. F.; Phillips, C. M.; McMonagle, J.; P´erez-Mart´ınez, P.; Shaw, D. I.; et al. NOS3 Gene Polymorphisms are Associated with Risk Markers of Cardiovascular Disease, and Interact with Omega-3 Polyunsaturated Fatty Acids. *Atherosclerosis* **2010,** *211*, 539–544.

Galvan-Portillo, M.; et al. Dietary Folate and Vitamin B12 Intake before Diagnosis Decreases Gastric Cancer Mortality Risk among Susceptible MTHFR 677TT Carriers. *J. Nutr.* **2009,** *26*, 201–208.

German, B. J.; et al. Nutrigenomics and Personalised Diets: What will they Mean for Food? *Ann. Rev. Food Sci. Technol.* **2011,** *2*, 97–123.

Gohil, R.; Peck, G.; Sharma, P. The Genetics of Venous Thromboembolism. A Meta-analysis Involving ~ 120,000 Cases and 180,000 Controls. *J. Thromb. Haemost.* **2009,** *102*, 360–370.

Goni, L.; Cuervo, M.; Milagro, F. I.; Martínez, J. A. A Genetic Risk Tool for Obesity Predisposition Assessment and Personalized Nutrition Implementation based on Macronutrient Intake. *Genes Nutr.* **2015,** *10*, 445. DOI: 10.1007/s12263-014-0445-z.

Hosseini-Esfahani, F.; Mirmiran, P.; Daneshpour, M. S.; Mehrabi, Y.; Hedayati, M.; Soheilian-Khorzoghi, M.; Azizi, F. Dietary Patterns Interact with APOA1/APOC3 Polymorphisms to Alter the Risk of the Metabolic Syndrome: The Tehran Lipid and Glucose Study. *Br. J. Nutr.* **2015,** *113*, 644–653. DOI: 10.1017/S0007114514003687.

Hosseini-Esfahani, F.; Mirmiran, P.; Daneshpour, M. S.; Mehrabi, Y.; Hedayati, M.; Zarkesh, M.; Azizi, F. Western Dietary Pattern Interaction with APOC3 Polymorphism in the Risk of Metabolic Syndrome. Tehran Lipid and Glucose Study. *J. Nutrigenet. Nutrigenomics*. **2014,** *7*, 105–117. DOI: 10.1159/000365445.

Jarouliya, U.; Keservani, Raj K. Pathways Leading to Child Obesity, Chapter 12. In *Global Perspectives on Childhood Obesity: Current Status, Consequences and Prevention*; Second Ed.; Bagchi, D., Ed,; Academic Press, Elsevier, 2019a; pp 137–146, ISBN: 9780128128404.

Jarouliya, U.; Keservani, Raj K. Protein Function as Cell Surface and Nuclear Receptor in Human Diseases, Chapter 1. In *Medicinal Chemistry with Pharmaceutical Product Development*; Bharti, S.; Mahapatra, D. kar, Apple Academic Press, CRC Press,Taylor & Francis Group, 2019b; pp 1–32, ISBN: 9781771887106.

Jiang-Hua, Q.;, De-Chuang, J.; Zhen-Duo, L.; Shu-de, C.; Zhenzhen, L. Association of Methylenetetrahydrofolate Reductase and Methionine Synthase Polymorphisms with Breast Cancer Risk and Interaction with Folate, Vitamin B6, and Vitamin B12 Intakes. *Tumour Biol.* **2014,** *35*, 11895–11901. DOI: 10.1007/s13277-014-2456-1

Joffe, Y.; van der Merwe, L.; Carstens, M.; Collins, M.; Jennings, C.; et al Tumor Necrosis Factor-alpha Gene308 G/A Polymorphism Modulates the Relationship between Dietary Fat Intake, Serum Lipids, and Obesity Risk in black South African Women. *J. Nutr*. **2010,** *140*, 901–907.

Kersten, S.; Seydoux, J.; Peters, J. M.; Gonzalez, F. J.; Desvergne, B.; Wahli, W. Peroxisome Proliferator-Activated Receptor Alpha Mediates the Adaptive Response to Fasting. *J. Clin. Invest.* **1999,** *103*, 1489–1498. DOI: 10.1172/JCI6223

Keservani, R. K.; Kesharwani, R. K.; Sharma, A. K.; Vyas, N.; Chadoker, A. Nutritional Supplements: An Overview. *Int. J. Curr. Pharm. Rev. Res.* **2010b,** *1* (1), 59–75.

Keservani, R. K.; Kesharwani, R. K.; Vyas, N.; Jain, S.; Raghuvanshi, R.; Sharma, A. K. Nutraceutical and Functional Food as Future Food: A Review. *Der Pharmacia Lettre* **2010a,** *2* (1), 106–116.

Keservani, R. K.; Sharma, A. K.; Jarouliya, U. Protein and Peptide in Drug Targeting and its Therapeutic Approach. *Ars Pharm.* 2015, *56* (3), 165–177.

Keservani, R. K.; Sharma, A. K.; Kesharwani, R. K. Eds.; In *Nutraceuticals and Dietary Supplements: Applications in Health Improvement and Disease Management*; CRC Press, 2020; ISBN: 9781771888738.

Lemay, D.; Lynn, D.; Martin, W.; Neville, M.; Casey, T.; et al. The Bovine Lactation Genome: Insights into the Evolution of Mammalian Milk. *Genome Biol.* **2009,** *10*, R43.

Lewis, K. D.; Burton-Freeman, B. M. The Role of Innovation and Technology in Meeting Individual Nutritional Needs. *J. Nutr.* **2010,** *140*, 26–36.

Makelainen, H.; et al. Panose, A New Prebiotic Candidate. *Lett. Appl. Microbiol.* **2009,** *49*, 666–672.

Mandard, S.; Müller, M.; Kersten, S. Peroxisome Proliferator-Activated Receptor Alpha Target Genes. *Cell Mol. Life Sci.* **2004,** *61*, 393–416. DOI: 10.1007/s00018-003-3216-3

Martin, K. J.; Graner, E.; Li, Y.; Price, L. M.; Kritzman, B. M.; Fournier, M. V.; Rhei, E.; Pardee, A. B. High-sensitivity Array Analysis of Gene Expression for the Early Detection of Disseminated Breast Tumor Cells in Peripheral Blood. *Proc. Natl. Acad. Sci. USA* **2001,** *98*, 2646–2651. DOI: 10.1073/pnas.041622398

Moller, D. E.; Kaufman, K. D. Metabolic Syndrome: A Clinical and Molecular Perspective. *Annu. Rev. Med.* **2005,** *56*, 45–62. DOI: 10.1146/annurev.med.56.082103.104751

Müller, M.; Kersten, S. Nutrigenomics: Goals and Strategies. *Nat. Rev. Genet.* **2003,** *4*, 315–322. DOI: 10.1038/nrg1047

Nestle Nutrition Institute. Nutrigenetics [Online] 2018. https://www.nestlenutrition-institute.org/resources/publication-series/publications/article/nutrigenetics (accessed Apr 04, 2022).

Newnutrition Business. 10 Key Trends in Food, Nutrition & Health 2019, 2018 [Online]. https://www.new-nutrition.com/keytrend?id=90 (accessed Apr 04, 2022).

O'Sullivan, T. A.; Bremner, A. P.; O'Neill, S.; Lyons-Wall, P. Comparison of Multiple and Novel Measures of Dietary Glycemic Carbohydrate with Insulin Resistant Status in Older Women. *Nutr. Metab.* **2010,** *7*, 25.

Oliver, D. Ed.; In *The Future of Nutrigenomics. From the Lab to the Dining Room*; Institute for the Future: California, 2005.

Ordovas, J.; et al. Personalised Nutrition and Health. *Br. Med. J.* **2018**. DOI: 361:bmj.k2173

Palatini, P.; Ceolotto, G.; Ragazzo, F.; Dorigatti, F.; Saladini, F.; Papparella, I.; Mos, L.; Zanata, G.; Santonastaso, M. CYP1A2 Genotype Modifies the Association between Coffee Intake and Therisk of Hypertension. *J. Hypertens.* **2009,** *27*, 1594–1601. DOI: 10.1097/HJH.0b013e32832ba850

Papathanasopoulos, A.; Camilleri, M. Dietary Fiber Supplements: Effects in Obesity and Metabolic Syndrome and Relationship to Gastrointestinal Functions. *Gastroenterology* **2009,** *138*, 65–72.

Patsouris, D.; Müller, M.; Kersten, S. Peroxisome Proliferator Activated Receptor Ligands for the Treatment of Insulin Resistance. *Curr. Opin. Investig. Drugs* **2004,** *5*, 1045–1050. PMID: 15535425

Peppa, M.; Koliaki, C.; Nikolopoulos, P.; Raptis, S. A. Skeletal Muscle Insulin Resistance in Endocrine Disease. *J. Biomed. Biotechnol.* **2010,** *2010*, 527850.

Puchau, B.; Zulet, M. A.; de Ech´avarri, A. G.; Hermsdorff, H. H.; Martinez, J. A. Dietary Total Antioxidant Capacity: A Novel Indicator of Diet Quality in Healthy Young Adults. *J. Am. Coll. Nutr.* **2009,** *28*, 648–656.

Qi, Q.;, Chu, A. Y.; Kang, J. H.; Huang, J.; Rose, L. M.; Jensen, M. K.; Liang, L.; Curhan, G. C.; Pasquale, L. R.; Wiggs, J. L.; De Vivo, I.; Chan, A. T.; Choi, H. K.; Tamimi, R. M.; Ridker, P. M.; Hunter, D. J.; Willett, W. C.; Rimm, E. B.; Chasman, D. I.; Hu, F. B.; Qi, L. Fried Food Consumption, Genetic Risk, and Body Mass Index: Gene-diet Interaction Analysis in Three US Cohort Studies. *BMJ* **2014,** *348*, g1610. DOI: 10.1136/bmj.g1610.

Qi, Q.; Chu, A. Y.; Kang, J. H.; Jensen, M. K.; Curhan, G. C.; Pasquale, L. R.; Ridker, P. M.; Hunter, D. J.; Willett, W. C.; Rimm, E. B.; Chasman, D. I.; Hu, F. B.; Qi, L. Sugar-Sweetened Beverages and Genetic Risk of Obesity. *N. Engl. J. Med.* **2012,** *367*, 1387–1396. DOI: 10.1056/NEJMoa1203039

Ramos-Lopez, O.; Panduro, A.; Martinez-Lopez, E.; Fierro, N. A.; Ojeda-Granados, C.; Sepulveda-Villegas, M.; Roman, S. Genetic Variant in the CD36 gene (rs1761667) is Associated with Higher Fat Intake and High Serum Cholesterol among the Population of West Mexico. *J. Nutr. Food Sci.* **2015,** *5*, 353. DOI: 10.4172/2155-9600.1000353

Ramos-Lopez, O.; Panduro, A.; Martinez-Lopez, E.; Roman, S. Sweet Taste Receptor TAS1R2 Polymorphism (Val191Val) is Associated with a Higher Carbohydrate Intake and Hyper Triglyceridemia among the Population of West Mexico. *Nutrients* **2016,** *8*, 101. DOI: 10.3390/nu8020101

Refsum, H.; Smith, A. D. Are we Ready for Mandatory Fortification with Vitamin B-12? *Am. J. Clin. Nutr.* **2008,** *88*, 253–254.

Ruden, D. M.; De Luca, M.; Garfinkel, M. D.; Bynum, K. L.; Lu, X. Drosophila Nutrigenomics can Provide Clues to Human Gene-nutrient Interactions. *Annu. Rev. Nutr.* **2005,** *25*, 499–522. DOI: 10.1146/annurev.nutr.25.050304.092708

Sales, N. M. R.; Pelegrini, P. B.; Goersch, M. C. Nutrigenomics: Definitions and Advances of This New Science. *J. Nutr. Metab.* **2014**. DOI: 10.1155/2014/202759.

Sela, D. A.; Chapman, J.; Adeuya, A.; Kim, J. H.; Chen, F.; et al. The Genome Sequence of Bifidobacterium Longum subsp. Infantis Reveals Adaptations for Milk Utilization within the Infant Microbiome. *Proc. Natl. Acad. Sci. USA* **2008,** *105*, 18964–18969.

Shelton, R.; et al. Assessing Effects of l-Methylfolate in Depression Management: Results of a Real-World Patient Experience Trial. *Prim. Care Companion CNS Disorders* **2013,** *15* (4).

Singh, P.; Kesharwani, R. K.; Keservani, R. K. Antioxidants and Vitamins: Roles in Cellular Function and Metabolism. In *Sustained Energy for Enhanced Human Functions and Activity*; Academic Press, 2017a; pp. 385–407, ISBN: 9780128054130.

Singh, P.; Kesharwani, R. K.; Keservani, R. K. Protein, Carbohydrates, and Fats: Energy metabolism. In *Sustained Energy for Enhanced Human Functions and Activity*; Academic Press, 2017b; pp 103–115, ISBN:9780128054130.

Stahl, S. M. Novel Therapeutics for Depression: l-methylfolate as a Trimonoamine Modulator and Antidepressant-Augmenting Agent. *CNS Spectr.* **2007,** *12* (10), 739–744.

Stathopoulou, M. G.; Dedoussis, G. V.; Trovas, G.; Theodoraki, E. V.; Katsalira, A.; Dontas, I. A.; Hammond, N.; Deloukas, P.; Lyritis, G. P. The Role of Vitamin D Receptor Gene Polymorphisms in the Bone Mineral Density of Greek Postmenopausal Women with Low Calcium Intake. *J. Nutr. Biochem.* **2011,** *22*, 752–757. DOI: 10.1016/j.jnutbio.2010.06.007

Surana, K. R.; Ahire, E. D.; Sonawane, V. N.; Talele, S. G. Biomolecular and Molecular Docking: A Modern Tool in Drug Discovery and Virtual Screening of Natural Products. In *Applied Pharmaceutical Practice and Nutraceuticals*; Apple Academic Press, 2021; pp 209–223.

Surana, K. R.; Ahire, E. D.; Sonawane, V. N.; Talele, S. G.; Talele, G. S. Molecular Modeling: Novel Techniques in Food and Nutrition Development. In *Natural Food Products and Waste Recovery: Healthy Foods, Nutrition Design, and Extraction of Valuable Compounds*, 2021; p 17.

Valk, P. J.; Verhaak, R. G.; Beijen, M. A.; Erpelinck, C. A.; Barjesteh van Waalwijk van Doorn Khosrovani, S.; Boer, J. M.; Beverloo, H. B.; Moorhouse, M. J.; van der Spek, P.J.; Lowenberg, B.; Delwel, R. Prognostically Useful Gene-expression Profiles in Acute Myeloid Leukemia. *N. Engl. J. Med.* **2004,** *350*, 1617–1628. DOI: 10.1056/NEJMoa040465

Van der Put N. M.; et al. A Second Common Mutation in the Methylenetetrahydrofolate Reductase Gene: An Additional Risk Factor for Neural Tube Defects? *Am. J. Human Genet.* **1998,** *62*, 1044–1051.

Vijay-Kumar, M.; Aitken, J. D.; Carvalho, F. A.; Cullender, T. C.; Mwangi, S.; et al. Metabolic Syndrome and Altered Gut Microbiota in Mice Lacking Toll-like Receptor 5. *Science* **2010,** *328*, 228–231.

Winkels, R. M.; Brouwer, I. A.; Clarke, R.; Katan, M. B.; Verhoef, P. Bread Cofortified with Folic Acid and Vitamin B-12 Improves the Folate and Vitamin B-12 Status of Healthy Older People: A Randomized Controlled Trial. *Am. J. Clin. Nutr.* **2008,** *88*, 348–55.

Yaliwal, L. V.; Desai, R. M. Methylenetetrahydrofolate Reductase Mutations, A Genetic Cause for Familial Recurrent Neural Tube Defects. *Indian J. Human Genet.* **2012,** *18* (1), 122–124.

## CHAPTER 6

# DIETARY, GENOMICS, AND CANCER-RELATED FACTS AND DEBATES OR CONTROVERSIES

NEELAM L. DASHPUTRE, RAHUL R. SABLE, SHUBHAM J. KHAIRNAR, JAYESH D. KADAM, and SURABHI B. PATIL

*Department of Pharmacology, MET's Institute of Pharmacy, Adgaon, Nashik, India*

### ABSTRACT

The study of connections between dietary and genetic factors is a modern and important research area. Dietary habits, nutrients, and other constituents of food are crucial parts of environmental effects that contribute to cancer risk. The diet-cancer relationship is widely documented, and recent genomic technologies have enabled researchers to investigate how nutrients, micronutrients, and phytochemicals influence carcinogenesis. Nutritional effects on DNA damage and its repair, DNA methylation, which affect gene transcription and cellular phenotypes, antioxidant reconfiguration and reactive oxygen species (ROS), signaling pathways and target receptors, cell cycle controls and checkpoints, and apoptosis processes, are being evaluated in an ongoing study of nutrient-modulated carcinogenesis. Diet-gene interactions are likely to play a significant role in the known interindividual variability of the risk of cancer as an aftereffect of dietary factors that either develop or prevent tumors. Dietary modularity, therefore, has to be adapted to distinguish the attributes of tumors and their treatments. Only 5–10% of all cancer cases are due to genetic abnormalities, while 90–95% due to environmental and lifestyle factors. Cigarette addiction, eating habits (fried foods, meat, etc.),

Nutrigenomics and Nutraceuticals. Raj K. Keservani, Eknath D. Ahire, Shubham J. Khairnar, Sanjay J. Kshirsagar, & Rajesh K. Kesharwani (Eds.)

liquor, sun damage, environmental toxins, infections, stress, overweight, and low physical activity are a few of the lifestyle factors. According to statistics, smoking is responsible for 25–30% of all cancer-related mortality, whereas diet for 30–35%, infections for 15–20%, and other factors are responsible for the remaining proportion. By the end of the century, cancer is expected to be the top cause of mortality in every country on the planet. However, cancer is a preventable disease that can be avoided with significant lifestyle modifications.

## 6.1 INTRODUCTION

The DNA structure was discovered by Watson and Crick half a century ago. Science profited from well-developed recombinant DNA technology within 20 years of its discovery. HGP and Celera Genomics delivered the first complete manuscript of human genetic material in 2001. The start of the post-genomic era was signaled by this sequencing breakthrough (Go et al., 2003). Nutrigenomics is the study of the modifying effects of chemical substances in foods on DNA synthesis, gene expression, and stability. Nutrients can alter both genes and how they are expressed via nucleotide synthesis, DNA damage prevention and repair, and epigenetic mechanisms such as methylation of histone proteins (Keservani et al., 2015) for the structure of chromatin, which has a major effect on gene expression regulation. Nutrigenomics is an interdisciplinary field of study that uses genetic methodologies in addition to biochemical and epidemiological components to better understand the etiology of chronic conditions such as heart diseases (Keservani et al., 2016a), diabetes, obesity (Jarouliya et al., 2019a), and cancer (Nepomuceno, 2013). Although the molecular grounds for these correlations are typically poorly understood, it's indeed clear that food choices can increase the probability of acquiring certain cancers (Tajan and Vousden, 2020).

Cancer is, at its core, a genetic disease. A single cell's DNA gets altered at the start of the process. Several factors, including chemical, physical, and biological phenomena, can trigger this alteration in DNA. The second stage of carcinogenesis is the promotion stage. Carcinogens classified as oncopromoters affect genetically altered cells, aka "initiated cells." The initiated cell undergoes a slow and steady transformation into a malignant cell. A very long and persistent exposure to carcinogen promoters is required for this change to occur. The third and last stage is the progression stage, which is marked by uncontrolled cell proliferation and irreversible changes. Cancer has already gained a foothold at this point, leading to the development of the disease's initial clinical manifestations. In this way, the diet plays a major role in different phases of cancer development. Nutritional variables may affect the progression of carcinogenesis through processes that either facilitate or hinder it. Some foods may include not only carcinogens but also chemicals that protect the genetic material of cells against the effects of environmental mutagens. The occurrence of a tumor in a person does not indicate the causative agent(s). However, population-based epidemiological evidence implies that a considerable percentage of human tumors are associated with lifestyle or diet. Such research may also assist in the detection of causative agents, although the likelihood of identifying the agent is low unless there is strong dose-response data from people and/or animal models. Cancer can develop due to endogenous reactions like oxidation or exogenous factors like cigarette smoke (lung cancer), sun exposure (melanoma), aflatoxin (hepatocellular carcinoma), and relative radiation exposure (many types of cancers) (Nepomuceno, 2013).

## 6.2 DIET OR FOODS THAT PROMOTE CANCER

For decades, scientists have concluded that diet has a significant impact on the probability of developing cancer. Cancer rates differ significantly among populations, according to epidemiological research dating back to the 1960s, and cancer rates in migrants transferring from low- to high-risk countries may match or even exceed the rates in the local community. These results suggest that there may be significant environmental causes of cancer, and further research has found considerable connections between various cancers (Key et al., 2020). Dietary elements are responsible for over 30% of total cancers in developed countries, with diet ranking second only to smoking as a potentially avoidable cause of cancer. In developing countries, this figure is said to be approximately 20% but might increase if diet and lifestyle change. The few dietary elements (Keservani et al., 2017) influencing the risk of digestive system cancer and other common kinds of cancer are discussed below (Key et al., 2020).

### *6.2.1 ORAL CAVITY AND PHARYNGEAL CANCERS*

Approximately 60% of oral cavity, pharyngeal, and throat cancers are estimated to be caused by vitamin deficiencies generated by a restricted diet deficient in veggies, fruits, and animal products in developing nations. The relative influence of several micronutrients is still unclear. Consuming drinks and foods at extremely high temperatures has also been related with a significant probability of these cancers. Nasopharyngeal cancer is prominent in a few populations around the world, such as the Cantonese in southern parts of China and certain indigenous people in Asia's southeast, the Arctic, North Africa, and the Middle East. This cancer has been related to the use of salt-preserved foods, with the mechanism being either nitrosamine generation or the recurrence of the Epstein-Barr virus. The World Health Organization's IARC has categorized Chinese-style salted fish as a carcinogen based on case-control studies (Key et al., 2020).

### 6.2.2 STOMACH CANCER

The most prominent disease in the world, approximately 20 years ago, was stomach cancer, but mortality rates have been declining in all developed countries, and Asia today has a considerably higher incidence of stomach cancer than North America or Europe. Consumption of huge quantities of foods that have been salted, such as fish that are salt-preserved, is linked to a greater risk of cancer; this can be because of the salt itself or carcinogens produced from nitrites present in preserved foods. Salty foods may enhance the chances of *Helicobacter pylori* infection (a known cause of stomach cancer) and work together to accelerate the disease's progression. According to some data, eating lots of pickled vegetables enhances the risk of stomach cancer because N-nitroso molecules are produced by mold or fungi, which are sometimes found in these foods (Key et al., 2020).

### 6.2.3 COLORECTAL CANCER

The world's third most common cancer is colorectal cancer. Being overweight or obese, as well as drinking and smoking, may enhance the risk (Key et al., 2020). Ecological studies establish a greater link between the consumption of meat and colorectal cancer. Processed meat was designated as toxic to humans by the IARC in 2015, whereas unprocessed red meat was classified as probably carcinogenic, based in part on a meta-analysis that discovered a 17% increase in risk for each 50 g daily increase in intake of meat and an 18% increase in risk for each 100 g increase in consumption of red meat. Unprocessed red meat has been related with a lesser increase in risk in recent systematic evaluations. Nitrates and nitrites are preservatives used in processed meat, which may expose the gut to carcinogenic N-nitroso compounds, greatly increasing cancer risk. Heme iron is found in red meat, both cooked and raw, and it has a toxic effect on the stomach and enhances the production of N-nitroso compounds. High-temperature cooking of meat can produce oncogenic heterocyclic hydrocarbons and aromatic amines. It's unknown whether any of these theories can explain the connection between colorectal cancer and red or processed meat intake (Key et al., 2020).

### 6.2.4 LIVER CANCER

The crucial diet-related danger for liver cancer is alcohol, which causes cirrhosis and alcoholic hepatitis (Key et al., 2020). Intake of foods contaminated with the mycotoxin aflatoxin, along with active hepatitis virus infection, is a risk for people. Obesity and being overweight enhance the risk.

IARC has categorized a mutagenic chemical formed by the fungus *Aspergillus* in foods, such as dry fruits, nuts, and cereals, when stored in warm and damp conditions, as a carcinogen called aflatoxin (Key et al., 2020).

### 6.2.5 PANCREATIC CANCER

Obesity elevates pancreatic cancer risk by approximately 20%. Diabetes is linked to a greater threat of death, and randomized research based on Mendelian principles shows that this is attributable to elevated insulin levels instead of just the disease itself (Key et al., 2020). High meat consumption can lead to an elevated risk, according to several studies.

### 6.2.6 BREAST CANCER

It is the most prevalent cancer among women and also the world's second most common cancer. Obesity gives rise to an increase in the risk of breast cancer in postmenopausal women, most likely due to a rise in the level of circulating estrogens generated by aromatase in adipose tissue. Obesity in premenopausal females is linked to lower risk, according to most research, possibly due to lower hormone levels linked to greater anovulation frequency.

If alcohol is taken daily, the risk increases by around 10%; the mechanism may entail higher estrogens (Key et al., 2020).

### *6.2.7 PROSTATE CANCER*

The world's fourth most frequent cancer is prostate cancer. Age, family history, black ethnicity, and genetic factors are the only recognized risk factors. Obesity is considered to raise the severity of prostate cancer (Key et al., 2020). High hormone levels of IGF-1, which promote cell growth, have been related to a higher threat of prostate cancer. (Key et al., 2020).

### *6.2.8 KIDNEY CANCER*

Obesity and overweight are well-known factors of risk for kidney cancer, accounting for up to 30% of kidney malignancies in both males and females.

**TABLE 6.1** Dietary Components: Carcinogenic Components, Polymorphic Genes, and Cancer Sites.

| Dietary component | Polymorphic gene/phenotype1 | Cancer site | Probably decrease risk |
|---|---|---|---|
| | | **Carcinogens** | |
| Heterocyclic amines | NAT2, NAT1, and CYP1A2 | Colorectal and breast | Decreasing cooking temperature, time, or manner |
| Polycyclic hydrocarbons | CYP1A1<br>CYP1A1 and GSTM1 | Other sites<br>Gastrointestinal tract, nasopharyngeal, and stomach | Decreasing consumption of smoked and charbroiled foods |
| Nitrosamines | CYP2E1 | Other sites | Eat a diet high in antioxidants |
| Aflatoxins | GSTM1 and EPHX | Liver | Discarding nuts and nut butter that look moldy, discolored, or shriveled |
| Alcohol | ADH (ALDH and CYP2E1) | Colorectal and oral | Stopping alcohol consumption |

*Source*: Modified from Nepomuceno (2013).

## 6.3 NUTRITION AND EPIGENETICS

The procedure by which humans utilize food for life, growth, and the regular functioning of all organs and tissues is known as nutrition. In this century, the link between required nutrition and health or disease risk has been well established, but other dietary (Keservani et al., 2010; Keservani et al., 2010b; Keservani et al., 2020) components are also being recognized as having significantly huge effects on health and the risk of disease (Rock et al., 2000). Nutritional or dietary components may affect the threat of cancer progression, as per evidence from epidemiological, human, and lab investigations (Rock et al., 2000). Of all the cancers in the US, as per current estimates, 20–60% are linked to nutrition and dietary variables. In various stages of cancer development, food plays a very important role. Nutritional elements may impact the formation of carcinogenesis through processes that enhance or inhibit it. Some meals may include not only carcinogens but also elements that protect the genetic material of cells against the impacts of environmental mutagens (Nepomuceno, 2013).

The genomes of all cells in the human body are identical. Each cell, on either side, has one of several "epigenomes," or individual groups of epigenetic guidelines for developing and sustaining lineage-specific expression profiles. The genome is organized to express a specific group of genes in specified tissues at specific times during a human's life (Nepomuceno, 2013). Epigenetic events support genomic activities such as cell identity maintenance throughout differentiation, the replication of important chromosomal structural traits, and dosage compensation by creating a memory of cell identity. Epigenetic techniques can control gene expression by modifying the structure of the chromosomes. DNA methylation is a good epigenetic mechanism example (Nepomuceno, 2013).

For proliferation, differentiation, and embryonic development, DNA methylation is crucial. Similarly, methylation influences gene expression regulation in some circumstances. Except for CpG islands and certain stretches (about 1 kb in length) where large CpG contents are observed, methylation is found sparingly but worldwide in mammals, arranged in CpG sequences that are defined throughout the genome (Nepomuceno, 2013). Methylation of these areas can result in undesirable gene silence, such as TSG silencing in cancer cells. Cancer and tumor suppressor genes have received a lot of attention in the field of DNA methylation and illness. Hypermethylation usually silences TSG in cancer cells. Unlike normal cells, cancer cells' genomes are hypomethylated in general, except for the hypermethylation process at genes participating in the cell cycle, invasion of tumor cells, DNA repair, and

**TABLE 6.2** Epigenetic Nutrition Influencing the Physiological and Pathological Processes.

| | Nutrients | Epigenetic process |
|---|---|---|
| **Embryo development** | Folate | DNA methylation and Imprinting |
| | Choline | DNA methylation |
| | Protein restriction | DNA methylation and histone modifications |
| | Alcohol | DNA methylation |
| **Stem cell** | Butyrate | Histone acetylation and DNA methylation |
| | Retinoic acid | PRC |
| **Aging** | Folate | DNA methylation |
| | Calorie restriction | Histone acetylation |
| **Immune function** | Folate | DNA methylation |
| **Cancer** | Methyl-deficient diet | Histone modification and microRNA |
| | Genistein | DNA methylation and microRNA |
| | (-)-Epigallocatechin-3-gallate | DNA methylation and PRC |
| | Curcumin | microRNA |
| **Obesity and insulin resistance** | High-fat diet | DNA methylation and microRNA |
| | Methyl-deficient diet | DNA methylation |
| | Curcumin | Histone acetylation |
| **Inflammation** | Resveratrol | Histone acetylation |
| | AdoMet | Histone acetylation |
| | Methyl-deficient diet | microRNA |
| **Neurocognition** | Choline | DNA methylation and histone methylation |

*Source*: Data adapted from Nepomuceno (2013).

other events that encourage metastasis. Hypermethylation can be detected early in some cancers, like colon cancer, and utilized as a disease biomarker (Nepomuceno, 2013). Epigenetics is particularly significant in the nutritional sector, as functional foods (bioactive foods) and nutrients may influence epigenetic events and gene expression at the transcriptional level. Nutrients can modify the expression of essential gene variants in physiological and pathological processes such as embryonic development, aging, and cancer by reversing or modifying epigenetic events like DNA methylation alterations (Nepomuceno, 2013).

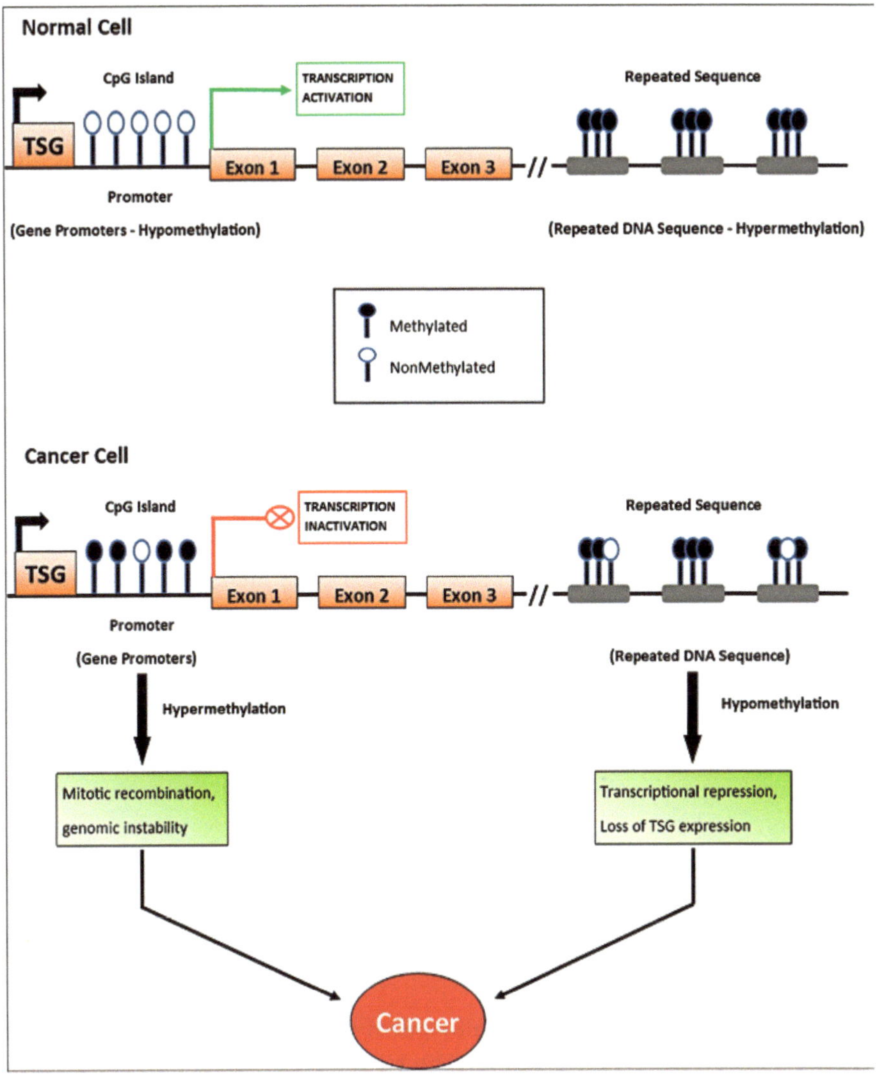


**FIGURE 6.1** DNA methylation and cancer.

## 6.4 THE ROLE OF DIET IN CANCER PREVENTION

Current cancer models are those that are transmitted through generations and account for just approximately 5% of all human cancer incidences. Mutational events are the reason for the occurrence of these cancers. The other 5% are occasional episodes that develop as a result of environmental exposure, which includes exposure to both environmental pollutants and dietary factors (Nepomuceno, 2013).

A cancer-treatment strategy for dietary intervention that selectively removes a specified diet based on mechanistic knowledge of the tumor's metabolic requirements is growing rapidly (Tajan and Vousden, 2020). Nutrigenomics is a method that could be utilized to analyze and prevent several diseases, including cancer. In the absence of DNA sequence changes, histone modifications and DNA methylation, like epigenetic processes, mediate gene expression alterations and chromatin structure (Nepomuceno, 2013). The increased vulnerability to cancer with age may be because of the buildup of epigenetic alterations, and this could be a target for bioactive chemical therapy. Nuclear receptors and transcriptional factors, which act as sensors for dietary components and determine changes in transcript patterns, are two factors that mediate the response to dietary factors (Nepomuceno, 2013).

Human genetic differences, epigenetic alterations that specifically and irreversibly modify gene expression, complicated interactions/associations among food components, and cell heterogeneity are all factors to consider within a tumor that influence the possibility of focusing nutrients and genetic interactions to impact the cancer process, according to Milner and Romagnolo. To fully comprehend how nutrigenomics might be employed in cancer prevention, it is necessary to integrate information about gene polymorphisms, identify gene targets that govern cell and tissue-specific pathways, and create diagnostic techniques to account for clinical heterogeneity (Nepomuceno, 2013).

## 6.5 BALANCED CARB AND FATTY DIET

For over 50 years, scientists have disputed the best fat, protein, and carbohydrate balance for human health. The NCI, CRUK, and ESPEN are a few of the national cancer organizations that provide dietary guidelines for cancer patients, emphasizing a healthy and balanced diet and appropriate nutrient consumption. Cancer patients are suggested to maintain their calorie intake, and in particular, to eat a high-protein (Jarouliya et al., 2019b) diet. These suggestions, on the other hand, are aimed at enhancing overall health and reducing the consequences of the tumor or treatments (Tajan and Vousden, 2020).

**TABLE 6.3** Matrix Summary of the WCRF/AICR Judgments on the Role of Various Foods in the Risk of Cancer.

| | Vegetables | Fruits | Grains, fibers | Tea | Sugar | Alcohol | Salt and salting | Meat | Eggs | Milk and dairy |
|---|---|---|---|---|---|---|---|---|---|---|
| **Mouth, pharynx** | −−− | −−− | | | +++ | | | | | |
| **Nasopharynx** | | | | | | | +++ | | | |
| **Esophagus** | −−− | −−− | + | | | +++ | | | | |
| **Stomach** | −−− | −−− | − | − | | | ++ | | | |
| **Pancreas** | −− | −− | − | | | | | | + | |
| **Gallbladder** | | | | | | | | | | |
| **Liver** | − | | | | | +++ | | | | |
| **Colon, rectum** | −−− | | − | | | | | ++ | + | |
| **Larynx** | −− | −− | | | | +++ | | | | |
| **Lungs** | −−− | −−− | | | | + | | | | |
| **Breast** | −− | −− | − | | | ++ | | + | | |
| **Cervix** | − | − | | | | | | | | |
| **Endometrium** | − | − | | | | | | | | |
| **Ovary** | − | − | | | | | | | | |
| **Prostate** | − | | | | | | | + | | + |
| **Bladder** | −− | −− | | | | | | | | |
| **Kidney** | − | | | | | | | + | | + |
| **Thyroid** | − | | | | | | | | | |

*Increased risk*: +++, convincing; ++, probable; +, possible; decreased risk: −−−, convincing; −−, probable; −, possible.

*Source*: Data adapted from Nepomuceno (2013).

Glucose absorption and glycolysis are enhanced in many malignancies, which are considered to assist biomass formation, energy production, and antioxidant defense. Increased systemic fructose stimulates glycolysis and exacerbates the onset of cancer. Various dietary treatments, such as the keto diet, fasting, or supplementing with mannose, can inhibit cancer cell development by restricting glucose supply and limiting glycolysis. Fasting or a low-carb diet can decrease the amount of insulin and IGF-1 available to the tumor by lowering circulating glucose levels. This reaction reduces PI3K pathway activation, which is dominant in many malignancies, and hence improves the medicinal response to PI3K inhibitors (Tajan and Vousden, 2020).

Several dietary strategies for cancer therapy are now being evaluated in clinical trials, the most basic of which is eliminating eating patterns through FMD, or fasting; chronic calorie restriction, which may lead to weight loss; and short-term intermittent fasting, which enables one to resume the daily normal diet, thereby assisting in body weight maintenance and compliance. Fasting's efficacy is assumed to be based on a combination of interactions, including physiological responses like reduced circulating glucose, insulin, and IGF-1 with increased IGF-binding proteins (IGFBPs), as well as normal cells being induced to provide defensive responses like proliferative arrest, AMPK induction, antioxidants, and autophagy (Tajan and Vousden, 2020).

The keto diet is a low-carbohydrate, high-fat, and low- or no-protein diet. Ketones become the energy source for cells in the body when fat and protein are used as sources of energy rather than glucose. Ketogenic diets have also been suggested as a way to take advantage of cancer patients' altered glucose metabolism. These high-fat, low-carbohydrate diets differ from the high-fat, high-carbohydrate diets that contribute to obesity and cancer growth, and keto diets have been related to cancer progression delay. The ketogenic diet has a low sugar content, which is similar to FMD (Tajan and Vousden, 2020). The ketogenic diet stimulates the liver to generate ketone bodies such as acetoacetate, beta-hydroxybutyrate, and acetone by beta-oxidation of fatty acids. Ketones go through the blood to tissues, where they are transformed into acetyl-CoA, which is used to activate the tricarboxylic acid cycle. A drop in circulating glucose and lower insulin and IGF-1 levels could be one response to the ketogenic diet, which may help to avoid the aberrant activation of oncogenic signaling pathways in the cascade of the insulin receptor in the tumor (Tajan and Vousden, 2020).

## 6.6 AMINO ACIDS

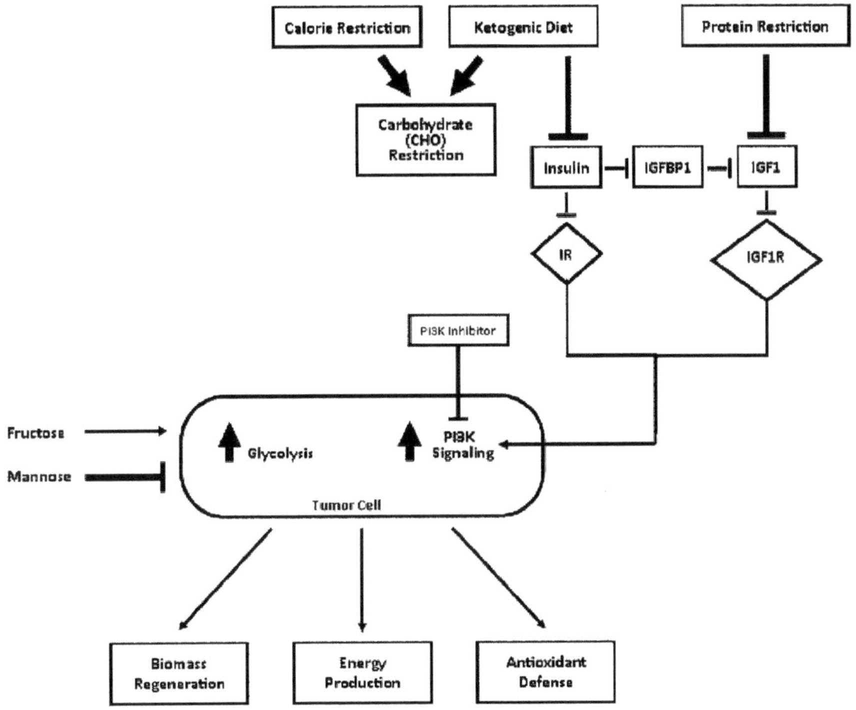


**FIGURE 6.2** Tumor metabolism by dietary carbs and fatty diet.

During the past couple of years, there's been a huge concern about determining how cancer cells fulfill their amino acid requirements. Proteinogenic amino acids contribute significantly to all proteins as well as being important players in macromolecular synthesis essential for growth, redox regulation, and proteins and DNA modification. Although EAAs cannot be produced by mammalian cells, both healthy and cancerous tissues rely on an outside supply. NEAAs, on the other hand, may be generated de novo, and most normal cells are not reliant on the diet of this group of amino acids. The dietary modification of NEAAs has gained a lot of attention since cancer cells typically indicate an elevated demand for them, which can be fulfilled in one of two ways (Tajan and Vousden, 2020). Some cancers can enhance their ability to manufacture such amino acids to the point where they become reliant on de novo NEAA production for growth and survival.

Small-molecule inhibitors of these pathways are a good target for this type of cancer. On the other hand, many cancers exhibit a high fraction of absorption and reliance on an external supply of NEAAs. In other words, these NEAAs become crucial to the cancer cell, opening up a magical therapeutic window for treatments that limit the access of these amino acids to cancer cells (Tajan and Vousden, 2020).

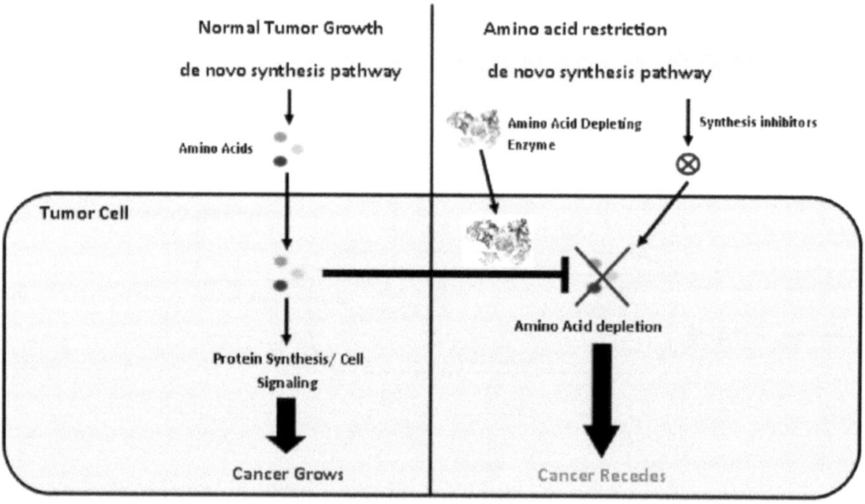


**FIGURE 6.3** Availability of NEAAs for cancer.

The variation in nutrient levels between the circulation and the tumor environment, as well as the importance of tumor form and site in adversely affecting nutrient availability, were highlighted in a recent study. Dietary-induced changes in nutrient concentration in the bloodstream and tumor interstitial fluids followed a similar pattern, indicating that serum metabolites might be utilized as a proxy for tumor interstitial fluid response. Utilizing catalysts that focus on amino acids for destruction or eliminating the amino acid from a typical diet, it is possible to restrict the systemic availability of particular amino acids selectively. These diets differ from low-protein diets in that they aim to alter just the structure of the reserve of amino acids while preserving total amino acid consumption. This method is effective in treating children with phenylketonuria, a metabolic disorder characterized by a deficiency in phenylalanine metabolism (Tajan and Vousden, 2020).

## 6.7 TRACE ELEMENTS

### 6.7.1 VITAMINS

Vitamin (Singh et al., 2017a) deficits are linked to the growth of some cancers, but judicious vitamin limitation could also be used to inhibit pathways involved in metabolism that tumors rely on. Vitamin B12 (cobalamin) and B9 (folate) deficiency, for example, limits 1C (one-carbon) metabolism and the potential to synthesize purines and thymidylate. While the resultant DNA damage has been related to the process of cancer, it's also been established that restricting these vitamins in mice inhibits cancer formation and promotes treatment with conventional chemotherapy. Cobalamin depletion may also lead to inaccurate methionine recycling, making cancer cells more prone to methionine deficiency. E and C vitamins have long been recommended as dietary supplements that could help prevent cancer, according to their antioxidant properties (Tajan and Vousden, 2020). The anticancer impact of C-vitamin in large doses is believed to be because of the intake of C-vitamin in an oxidized state, which leads to a rise in oxygen radicals (ROS) in cancer cells. Increased activities of epigenetic regulators like TET proteins and suppression of HIF1 activity are two alternative ways to increase vitamin C activity (Tajan and Vousden, 2020).

### 6.7.2 POLYPHENOLS

Polyphenols (Keservani and Sharma, 2014) are prominent ingredients in plant-based foods and important antioxidants in our diet. Fruits and beverages are the primary sources of polyphenols in the diet. Polyphenols could range up to 200–300 mg/100 g weight in fruits like apples, grapes, pears, cherries, and many other berries. A glass of red wine or a cup of tea or coffee generally contains approximately 100 mg of polyphenols. Polyphenols are also found in cereals, cocoa, and dry legumes. Polyphenols in red wine, which are made up of a variety of potent antioxidants such as flavones and stilbenes, are linked with cancer prevention and enhanced human health with no identified adverse effects. Red wine polyphenols such as resveratrol, quercetin, (+)-catechin, and gallic acid have been studied as potential cancer chemopreventive agents. However, because red wine carries a large percentage of polyphenols and protective benefits have yet to be identified for a specific fraction or molecule, it is unknown which chemicals possess protective activity in red wine. On the spread and progression of androgen-sensitive LNCaP-human prostate cancer, researchers examined the combination of the impacts of SPC and green and black tea components. The researchers have found that both black and green tea reduced the tumorigenicity of LNCaP tumors. In vivo, the combination of soy phytochemicals and tea reduced tumorigenicity, ultimate tumor weight, and lymph node metastasis. This research supports future research into the use of soy and tea mixtures as effective nutritional strategies for prostate cancer prevention. According to the authors, examinations of tea polyphenols indicate EGCG is the most active element in green tea, while black tea has less. Black tea contains other polyphenols, including theaflavins and thearubigins. It is also claimed that considerable in vitro as well as in vivo research has been done on the chemopreventive effects of the soy isoflavone, genistein (Tajan and Vousden, 2020).

### 6.7.3 DIETARY SUPPLEMENTS

In rare cases, increasing systemic amounts of other nutritional components may be useful. Dietary supplementation with histidine, for example, promotes the histidine breakdown process, limiting the availability of folate intermediates. This response made tumors more susceptible to methotrexate, a drug that reduces the formation of tetrahydrofolate, in animal models. Increased systemic levels of other nutritional factors may be helpful in certain situations. For example, dietary supplementation containing histidine accelerates histidine breakdown, decreasing the amount of folate pool intermediates. In animal models, these tumors are more sensitive to methotrexate, a chemical that prevents the formation of tetrahydrofolate (Surana et al., 2021a).

## 6.8 HEREDITARY FACTORS IN GENE EXPRESSION

There are a variety of hereditary factors that increase cancer risk. The most well-known mutations are the highly penetrant, predominant variants that have been related to extraordinarily high, relatively increased risks in epidemiological studies. BRCA1 is an example of this mutation type. This form of genetic variation is quite penetrating, although it is also very uncommon. BRCA1 mutations are probably responsible for 5% of total breast cancers in females aged below 40, 2%, and 1% in females of age 40–49 years and 50–70 years, respectively. Although the occurrence of this gene mutation helps boost an individual's breast cancer risk, the public health consequences are unclear because the majority of women who get breast cancer do not have this or other highly penetrant mutations that have been reported. Many women with BRCA1 (or BRCA2) do not get breast cancer, according to recent research in low-risk groups. Familial retinoblastoma, Wilms' tumor, familial polyposis coli, xeroderma pigmentosum A, and a subset of breast and ovarian malignancies (Li-Fraumeni syndrome) are other cancers related to these types of human germline gene alterations. Another component of the research on the diet-gene interaction is identifying the genomic alterations that indicate the mechanism through which foods and other dietary recommendations affect the development of carcinogenesis. Although epigenetic changes are not heritable risk factors, they can have an impact on cancer risk through altered transcriptional control. In epidemiological research, knowing about these DNA mutations is also crucial for establishing a cause-and-effect relationship between nutritional aspects related to cancer risk (Surana et al., 2021b).

## 6.8.1 EXAMPLES

### 6.8.1.1 XENOBIOTIC-METABOLIZING ENZYMES

Xenobiotic-metabolizing enzymes have a role in the activation and/or detoxification of foreign substances, such as carcinogens. Interindividual diversity in medication and toxin responses have been reported. Many dietetic factors (nonnutrient components and micronutrients of foods) are quite well involved in the enzyme substrates or stimulators systems metabolism. The addition of hydroxyl groups to a relatively nonpolar hydrocarbon, which converts the molecule into an electrophilic or more polar intermediate, is often the preliminary stage in the biotransformation of a foreign object or carcinogen. Phase I, or activating enzymes, the cytochromes P450, which are coded by CYP genes, carry out these oxidation processes. Several endogenous chemicals, such as steroid hormones and vitamin D metabolites, are oxidized by the cytochrome P450 enzymes. Phase II enzymes, also known as conjugating enzymes, catalyze conjugation events with substances like glutathione, making elimination easier. UDP-glucuronosyl-transferases, GSTs, NATs, mEH, and SULTs are a few of the phase II enzymes. Depending on the particular activity of the enzyme that is activated and the substrate involved, a variation of these enzymes enhances or decreases the risk of tumors.

#### 6.8.1.2 CYTOCHROME P450 AND CYP GENES

The cytochrome P450 (and related CYP genes) is an enzyme superfamily that is involved in a large variety of endogenous chemical oxidation processes as well as phase I initiation of a variety of substrates. Although a small group of P450s seems to be accountable for most xenobiotic oxidation processes in humans, approximately 30 of these enzymes have been fully described. Polymorphic CYP genes cause variations in the ability to oxidize substrates in the overall population. Carcinogenic chemicals, as well as certain meals and nutritional elements, have been proven to induce P450s in humans, as have micronutrient supplements in experimental animals.

#### 6.8.1.3 GLUTATHIONE S-TRANSFERASES

GSTs are cytosolic catalyzing enzymes for the conjugation of reduced glutathione to a diverse variety of electrophilic substances produced by cytochrome P450 enzymes. GSTs play a crucial role in shielding cells from the cytotoxic and carcinogenic effects of electrophiles and complexes, as well as possible DNA mutations. As a result, it's been postulated that GST induction reduces cancer susceptibility in general, whereas GST detoxification impairment increases disease susceptibility.

#### 6.8.1.4 OTHER METABOLIZING ENZYMES

The NAT gene's connection with food has been studied in various epidemiological studies, along with other enzymes that may be affected by dietetic factors. In humans, two NAT enzymes are found, namely, NAT1 and NAT2 that are polymorphic. Heterocyclic aromatic amine carcinogens are metabolically stimulated by these enzymes.

Nutritional interactions with several other metabolizing enzymes are also being examined. The occurrence of genetic modifications in the phase II enzyme sEH and cytochromes P4503A3 and P4503A4, for example, may change the possibility of developing liver cancer related to aflatoxin exposure, a fungal contaminant that is found in peanuts. Because aflatoxin B1 is triggered and linked via phases I and II enzymes, it has been postulated that hereditary factors play a key role in aflatoxin-exposed cancer risk. The cytochrome P450 enzymes transform aflatoxin B1 into various compounds, at least one of which, exo-8,9-epoxide, seems to be carcinogenic. As a mutagenic metabolite, the aflatoxin B1 epoxide metabolite interacts with DNA; however, it can also be metabolized (detoxified) by epoxide hydroxylase.

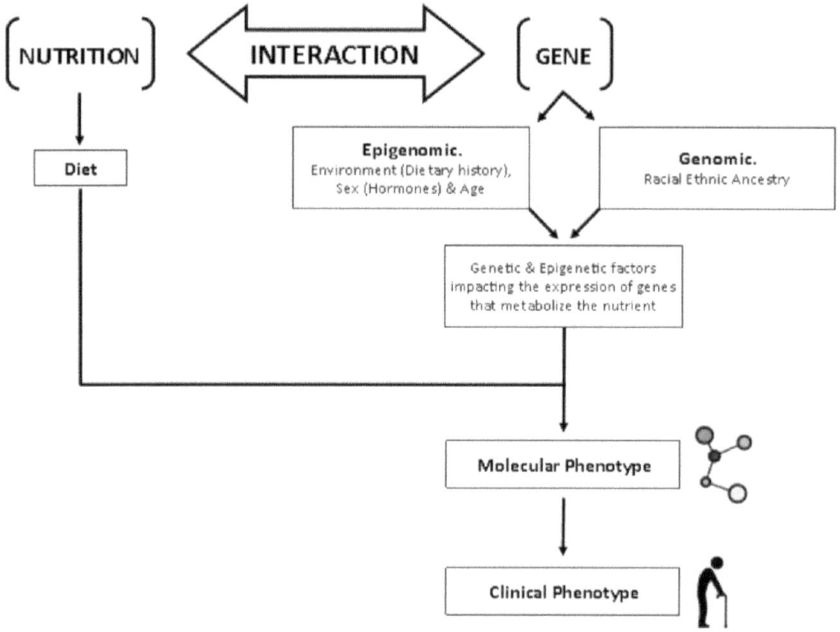


**FIGURE 6.4** Diet and gene interaction.

## 6.9 GENOMIC AND DIET STABILITY

Replication of DNA in eukaryotes begins at many locations across the genome and must be synchronized with transcription, sister chromatid cohesion, nucleosome assembly, and cell cycle progression. Besides the complex nature of the replication response, cells must also deal with DNA damage and blocked replication forks, which are usually caused by external and endogenous agents. DNA breakdowns, chromosome rearrangements, and mutations that can lead to cell viability loss result from the inability to cope with DNA lesions during replication or to protect or restart stalled forks, but DNA replication errors also cause a range of human disorders, which include premature aging, various cancer predispositions, and genetic problems. Cells use repair and detoxification pathways as well as monitoring systems called checkpoints to detect faults and coordinate repair with chromosomal separation and cell cycle advancement to address or lessen these issues. A large and broad variety of damage-reactive proteins are necessary to maintain the genomic stabilization of replication and recombination. To avoid tumor formation, the ability of a cell to decide where and when to release its DNA repair kit is crucial. There is scientific evidence that insufficient nutrition delivery can result in significant genome mutation levels and change gene expression. For the preservation of the genome, inadequacies in many micronutrients have been related to genetic damage and can possibly be linked to cancer. Due to chromosome abnormalities and impaired DNA repair, folic acid, niacin, vitamin B6, and vitamin B12 deficiency may raise the possibility of developing colon cancer, heart disease, and mental (Keservani et al., 2016b) retardation.

Traditional definitions of RDAs for micronutrients have been those levels needed to avoid symptoms of deficiency disorders. Higher amounts of many of these micronutrients may be needed for certain DNA maintenance activities, and existing RDAs for a few micronutrients could be insufficient to guard against genomic instability, according to mounting data. Similar to radiation exposure, mutagens, carcinogens, and nutritional imbalances may promote chromosomal abnormalities and genetic mutations in human populations. Because it affects all important mechanisms, such as exposure to dietetic carcinogens, DNA synthesis and repair, and death, nutrition may be a crucial determinant in determining genomic stability. Many micronutrients, such as vitamins and minerals, operate as complexes and/or intermediates in important DNA maintenance events, and their proper quantity in the cell could be crucial. Low amounts of crucial micronutrients needed for DNA

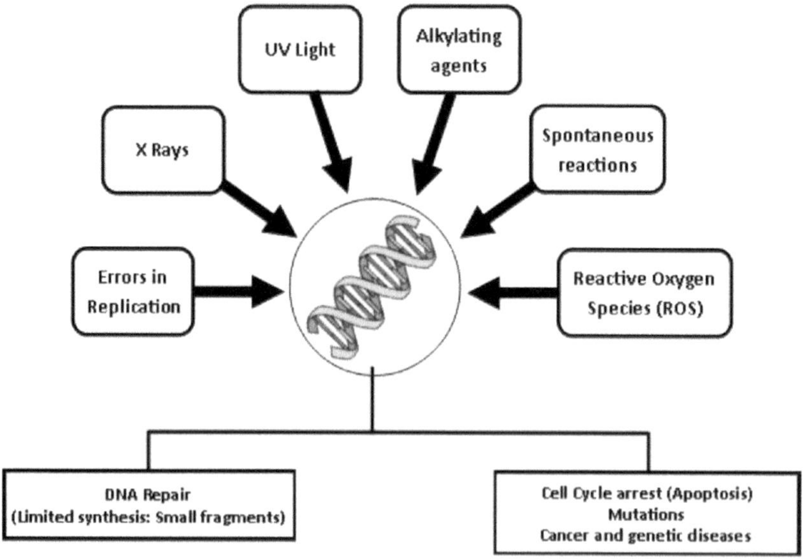


**FIGURE 6.5** A broad overview of DNA damages and their effects on the cell cycle and DNA repair.

repair will decrease genomic stability, similar to the effects of inherited hereditary defects or carcinogen exposure.

## 6.10 CONTROVERSIES

For both patients and their families, obtaining a cancer diagnosis is a severe blow. For people who are treated for, living with, or living beyond cancer, it can present a range of psychological, emotional, spiritual, and physical issues. As a result, it's understandable that a lot of people would try to change their diet to help fight cancer or reduce treatment side effects.

## 6.11 DIETS IN CANCER: MYTHS AND FACTS

### *6.11.1 DAIRY*

Hormones involved in milk production stimulate the development of hormone-related malignant tumors.

**Fact:** There is no scientific proof that dairy-based diets raise cancer risk or promote cancer growth due to hormones.

There is proof that higher calcium, vitamin D, butyrate, conjugated linoleic acid, and lactoferrin intake in dairy products protects against breast cancer development.

### *6.11.2 SOYA*

Isoflavones, which are contained in soy products, have a composition that is chemically comparable to that of the hormone estrogen. Because estrogen has been found to promote various malignancies, it was anticipated that meals containing isoflavones might have a similar effect.

**Fact:** According to current knowledge, a naturally occurring isoflavone-rich diet is safe.

Soy foods can be utilized as a nonmeat source of protein and fiber in a healthy, balanced diet.

### *6.11.3 SUGAR*

Sugar deficiency "starves" cancer cells.

**Fact:** Sugar nourishes all of the body's cells, including cancer cells. Even if carbohydrates (Singh et al., 2017b) or sugar aren't available, your body will produce sugar from other sources, such as protein and fat, to keep your brain and body operating normally.

So removing sugar from our diet is not only difficult, but it also won't get removed from our bodies. The most important thing is to keep sugar to a minimum to avoid weight gain, which increases cancer risk. Minimize added sugars and go for high-quality carbs derived from whole foods and plant-based sources (Heid, 2022).

### *6.11.4 LOW-CARBOHYDRATE DIET*

The keto diet reduces the size and spread of tumors.

**Fact:** A low-carbohydrate diet is known as the ketogenic diet. There are differences, with different macronutrient amounts depending on the type of ketogenic diet taken. Animals and brain tumors are the focus of current research. Human data are based on single cases and weight reduction, which could be problematic. When the protein consumed exceeds the usual nonstarvation protein turnover, a diet rich in protein can help offset it.

## 6.12 CONCLUSION

Cancer incidence is predicted to grow in the future, needing an effective prevention approach to solve this problem. Modifying a person's eating habits could be an effective method of reducing one's cancer risk. The biological effects of specific foods or bioactive compounds linked with cancer, as well as the prediction of individual sensitivity as a nutrient-nutrient interaction function and genetics, are critical factors in determining who will receive dietary interventions. A precise knowledge of how diet interacts with the interactions between cancer, the microenvironment, and systemic metabolism will enable us to provide patients with rational and individualized nutritional advice that will help them get the most out of their treatment.

## ACKNOWLEDGMENT

We thank the MET's Institute of Pharmacy, Bhujbal Knowledge City, Adgaon, Nashik, Principal Dr. Sanjay J. Kshirsagar Sir and our guide Dr. Neelam L. Dashputre Ma'am for their constant support and providing all facilities to complete this work. Also many thanks to all the known and unknowns who helped us right from the writing till the publishing of this chapter.

## KEYWORDS


- **cancer**
- **carcinogenesis**
- **chromosomes**
- **controversies**
- **de novo**
- **DNA methylation**
- **epigenetics**


## REFERENCES


Albans, D. Alpha-tocopherol and Beta-carotene Supplements and Lung Cancer Incidence in the Alpha-tocopherol, Beta-carotene Cancer Prevention Atudy: Effects of Baseline Characteristics and Study Compliance. *J. Natl. Cancer Inst.* **1996,** *88*, 1560–1570.

Ambrosone, C. B.; Freudenheim, J. L.; Sinha, R.; Graham, S.; Marshall, J. R.; Vena, J. E.; Laughlin, R.; Nemoto, T.; Shields, P. G. Breast Cancer Risk, Meat Consumption and N-acetyltransferase (NAT2) Genetic Polymorphisms. *Int. J. Cancer* **1998,** *75* (6), 825–830.

American Cancer Society 1996 Advisory Committee on Diet, Nutrition, and Cancer Prevention. Guidelines on Diet, Nutrition, and Cancer Prevention: Reducing the Risk of Cancer with Healthy Food Choices and Physical Activity. *CA Cancer J. Clin.* **1996,** *46* (6), 325–341.

Ames, B. N. DNA Damage from Micronutrient Deficiencies is Likely to be a Major Cause of Cancer. *Mutat. Res. Fundam. Mol. Mech. Mutagen.* **2001,** *475* (1-2), 7–20.

Amin, K.; Li, J.; Chao, W. R.; Dewhirst, M. W.; Haroon, Z. A. Dietary Glycine Inhibits Angiogenesis During Wound Healing and Tumor Growth. *Cancer Biol. Ther.* **2003,** *2* (2), 173–178.

Anand, P.; Kunnumakara, A. B.; Sundaram, C.; Harikumar, K. B.; Tharakan, S. T.; Lai, O. S.; Sung, B.; Aggarwal, B. B. Cancer is a Preventable Disease that Requires Major Lifestyle Changes. *Pharm. Res.* **2008,** *25* (9), 2097–2116.

Ananieva, E. A.; Wilkinson, A. C. Branched-chain Amino Acid Metabolism in Cancer. *Curr. Opin. Clin. Nutr. Metab. Care* **2018,** *21* (1), 64.

Anderson, K. E.; Pantuck, E. J.; Conney, A. H.; Kappas, A.. In *Nutrient Regulation of Chemical Metabolism in Humans*, Federation Proceedings, 1985; vol *44* (1) Pt 1, pp 130–133.

Arends, J.; Bachmann, P.; Baracos, V.; Barthelemy, N.; Bertz, H.; Bozzetti, F.; Fearon, K.; Hütterer, E.; Isenring, E.; Kaasa, S.; Krznaric, Z. ESPEN Guidelines on Nutrition in Cancer Patients. *Clin. Nutr.* **2017,** *36* (1), 11–48.

Argiles, J. M. Cancer-Associated Malnutrition. *Euro. J. Oncol. Nurs.* **2005,** *9*, S39–S50.

Austin, G. L.; Ogden, L. G.; Hill, J. O. Trends in Carbohydrate, Fat, and Protein Intakes and Association with Energy Intake in Normal-Weight, Overweight, and Obese Individuals: 1971–2006. *Am. J. Clin. Nutr.* **2011,** *93* (4), 836–843.

Backus, H. H. J.; Pinedo, H. M.; Wouters, D.; Padron, J. M.; Molders, N.; van Der Wilt, C. L.; Van Groeningen, C. J.; Jansen, G.; Peters, G. J. Folate Depletion Increases Sensitivity of Solid Tumor Cell Lines to 5-fluorouracil and Antifolates. *Int. J. Cancer* **2000,** *87* (6), 771–778.

Bray, F.; Møller, B. Predicting the Future Burden of Cancer. *Nat. Rev. Cancer* **2006,** *6* (1), 63–74.

Coates, P. M.; Nicastro, H.; Milner, J. A. Bioactive Components in Foods and Supplements for Health Promotion. In *Present Knowledge in Nutrition*, 2006; pp 1255–1267.

Committee on Diet, N. and National Research Council. In *Diet, Nutrition, and Cancer*, 1982.

Daniel, H. Genomics and Proteomics: Importance for the Future of Nutrition Research. *Br. J. Nutr.* **2002,** *87* (S2), S305–S311.

Davis, C. D. Nutritional Interactions: Credentialing of Molecular Targets for Cancer Prevention. *Exp. Biol. Med.* **2007,** *232* (2), 176–183.

Davis, C. D.; Milner, J. A. Biomarkers for Diet and Cancer Prevention Research: Potentials and Challenges. *Acta Pharm. Sinica* **2007,** *28* (9), 1262–1273.

Food, N. B. Institute of Medicine: Dietary Reference Intakes: Applications in Dietary Assessment; National Academy of Sciences: Washington, DC [Google Scholar], 2000.

German, J. B.; Roberts, M. A.; Watkins, S. M. Genomics and Metabolomics as Markers for the Interaction of Diet and Health: Lessons from Lipids. *J. Nutr.* 2003, *133* (6), 2078S–2083S.

Glade, M. J. Food, Nutrition, and the Prevention of Cancer: A Global Perspective. American Institute for Cancer Research/World Cancer Research Fund, American institute for Cancer Research, 1997. *Nutrition* 1999, *15*, 523–526.

Go, V. L. W.; Butrum, R. R.; Wong, D. A. Diet, Nutrition, and Cancer Prevention: The Postgenomic Era. *J. Nutr.* **2003,** *133* (11), 3830S–3836S.

Go, V. L. W.; Wong, D. A.; Butrum, R. Diet, Nutrition and Cancer Prevention: Where are we Going from Here?. *J. Nutr.* **2001,** *131* (11), 3121S–3126S.

Haenszel, W.; Kurihara, M. Studies of Japanese Migrants. I. Mortality from Cancer and Other Diseases among Japanese in the United States. *J. Natl. Cancer Inst.* **1968,** *40* (1), 43–68.

https://www.bda.uk.com/resource/cancer-diets-myths-and-more.html

https://www.cancer.gov/about-nci/overview/history/national-cancer-act-1971

https://www.cancerresearchuk.org/about-cancer/coping/physically/diet problems/managing/alternative-cancer-diets

https://www.mdanderson.org/publications/focused-on-health/food-truths-to-lower-cancer-risk.h18-1589046.html

https://www.who.int/dietphysicalactivity/publications/trs916/en/gsfao_cancer.pdf

Jarouliya, U.; Keservani, Raj K. Pathways Leading to Child Obesity, Chapter 12. In *Global Perspectives on Childhood Obesity: Current Status, Consequences and Prevention, Second Ed.*; Bagchi, D., Ed.; Academic Press, Elsevier, 2019a; pp 137–146, ISBN: 9780128128404.

Jarouliya, U.; Keservani, Raj K. Protein Function as Cell Surface and Nuclear Receptor in Human Diseases, Chapter 1. In *Medicinal Chemistry with Pharmaceutical Product Development*; Bharti, S.; Mahapatra, D. Kar, Ed.; Apple Academic Press, CRC Press, Taylor & Francis Group, 2019b; pp 1–32, ISBN: 9781771887106.

Jemal, A.; Siegel, R.; Ward, E.; Murray, T.; Xu, J.; Thun, M. J. Cancer Statistics, 2007. *CA Cancer J. Clin.* **2007,** *57* (1), 43–66.

Keservani, R. K.; Kesharwani, R. K.; Sharma, A. K.; Vyas, N.; Chadoker, A. Nutritional Supplements: An Overview. *Int. J. Curr. Pharm. Rev. Res.* **2010b,** *1* (1), 59–75.

Keservani, R. K.; Kesharwani, R. K.; Vyas, N.; Jain, S.; Raghuvanshi, R.; Sharma, A. K. Nutraceutical and Functional Food as Future Food: A Review. *Der Pharmacia Lettre* **2010a,** *2* (1), 106–116.

Keservani, R. K.; Sharma, A. K. Flavonoids: Emerging Trends and Potential Health Benefits. *J. Chinese Pharm. Sci.* **2014,** *23* (12), 815.

Keservani, R. K.; Sharma, A. K.; Jarouliya, U. Protein and Peptide in Drug Targeting and its Therapeutic Approach. *Ars Pharm.* **2015,** *56* (3), 165–177.

Keservani, R. K.; Sharma, A. K.; Kesharwani, R. K. Nutraceutical and Functional Foods for Cardiovascular Health. In *Food Process Engineering*, pp 291–312; Apple Academic Press, CRC Press, 2016a; pp 257–278, ISBN: 9781771884020.

Keservani, R. K.; Sharma, A. K.; Kesharwani, R. K. Medicinal Effect of Nutraceutical Fruits for the Cognition and Brain Health. *Scientifica* **2016b,** Article ID 3109254, 10. DOI: 10.1155/2016/3109254.

Keservani, R. K.; Sharma, A. K.; Kesharwani, R. K. An Overview and Therapeutic Applications of Nutraceutical and Functional Foods. In *Recent Advances in Drug Delivery Technology*, 2017; pp 160–201, ISBN: 9781522507543.

Keservani, R. K.; Sharma, A. K.; Kesharwani, R. K. Eds.; In *Nutraceuticals and Dietary Supplements: Applications in Health Improvement and Disease Management*; CRC Press, 2020, ISBN: 9781771888738.

Key, T. J.; Bradbury, K. E.; Perez-Cornago, A.; Sinha, R.; Tsilidis, K. K.; Tsugane, S. Diet, Nutrition, and Cancer Risk: What do we Know and what is the Way Forward?. *BMJ* **2020,** *368*.

Kolonel, L. N.; Altshuler, D.; Henderson, B. E. The Multiethnic Cohort Study: Exploring Genes, Lifestyle and Cancer Risk. *Nat. Rev. Cancer* **2004,** *4* (7), 519–527.

Lichtenstein, P.; Holm, N. V.; Verkasalo, P. K.; Iliadou, A.; Kaprio, J.; Koskenvuo, M.; Pukkala, E.; Skytthe, A.; Hemminki, K. Environmental and Heritable Factors in the Causation of Cancer—Analyses of Cohorts of Twins from Sweden, Denmark, and Finland. *New Engl. J. Med.* **2000,** *343* (2), 78–85.

Liu, R. H. Health Benefits of Fruit and Vegetables are from Additive and Synergistic Combinations of Phytochemicals. *Am. J. Clin. Nutr.* **2003,** *78* (3), 517S–520S.

Loeb, K. R.; Loeb, L. A. Significance of Multiple Mutations in Cancer. *Carcinogenesis* **2000,** *21* (3), 379–385.

Nepomuceno, J. C. Nutrigenomics and Cancer Prevention. *Cancer Treatment*; IntechOpen, 2013; pp 391–416.

O'Shaughnessy, J. A.; Kelloff, G. J.; Gordon, G. B.; Dannenberg, A. J.; Hong, W. K.; Fabian, C. J.; Sigman, C. C.; Bertagnolli, M. M.; Stratton, S. P.; Lam, S.; Nelson, W. G. Treatment and Prevention of Intraepithelial Neoplasia: An Important Target for Accelerated New Agent Development: Recommendations of the American Association for Cancer Research Task Force on the Treatment and Prevention of Intraepithelial Neoplasia. *Clin. Cancer Res.* **2002,** *8* (2), 314–346.

Rock, C. L.; Lampe, J. W.; Patterson, R. E. Nutrition, Genetics, and Risks of Cancer. *Ann. Rev. Public Health* **2000,** *21* (1), 47–64.

Singh, P.; Kesharwani, R. K.; Keservani, R. K. Antioxidants and Vitamins: Roles in Cellular Function and Metabolism. In *Sustained Energy for Enhanced Human Functions and Activity*; Academic Press, 2017a; pp 385–407, ISBN: 9780128054130.

Singh, P.; Kesharwani, R. K.; Keservani, R. K. Protein, Carbohydrates, and Fats: Energy Metabolism. In *Sustained Energy for Enhanced Human Functions and Activity*; Academic Press, 2017b; pp 103–115. ISBN: 9780128054130.

Surana, K. R.; Ahire, E. D.; Sonawane, V. N.; Talele, S. G.; Talele, G. S. Molecular Modeling: Novel Techniques in Food and Nutrition Development. In *Natural Food Products and Waste Recovery: Healthy Foods, Nutrition Design, and Extraction of Valuable Compounds*, 2021; p 17.

Surana, K. R.; Ahire, E. D.; Sonawane, V. N.; Talele, S. G.; Talele, G. S. Informatics and Methods in Nutrition Design and Development. In *Natural Food Products and Waste Recovery: Healthy Foods, Nutrition Design, and Extraction of Valuable Compounds*, 2021; p 33.

Tajan, M.; Vousden, K. H. Dietary Approaches to Cancer Therapy. *Cancer Cell* **2020,** *37* (6), 767–785.

Venter, J. C.; Adams, M. D.; Myers, E. W.; Li, P. W.; Mural, R. J.; Sutton, G. G.; Smith, H. O.; Yandell, M.; Evans, C. A.; Holt, R. A.; Gocayne, J. D. The Sequence of the Human Genome. *Science* **2001,** *291* (5507), 1304–1351.

World Health Organization. *Diet, Nutrition, and the Prevention of Chronic Diseases: Report of a Joint WHO/FAO Expert Consultation* (vol. 916), World Health Organization, 2003.

Young, V. R. 2001 WO Atwater Memorial Lecture and the 2001 ASNS President's Lecture: Human Nutrient Requirements: The Challenge of the Post-genome era. *J. Nutr.* **2002,** *132* (4), 621–629.

Ziegler, R. G.; Hoover, R. N.; Pike, M. C.; Hildesheim, A.; Nomura, A. M.; West, D. W.; Wu-Williams, A. H.; Kolonel, L. N.; Horn-Ross, P. L.; Rosenthal, J. F.; Hyer, M. B. Migration Patterns and Breast Cancer Risk in Asian-American Women. *J. Natl. Cancer Inst.* **1993,** *85* (22), 1819–1827.

## CHAPTER 7

# THE INFLUENCE OF DIETARY COMPONENTS ON GENE EXPRESSION

ANJALI B. TAJANPURE and PRANALI P. DESHMUKH

*Department of Pharmacology, MET's Institute of Pharmacy, Adgaon, Nashik, India*

## ABSTRACT

The capacity to control gene expression permits cells to supply a specific protein whenever required for everyday survival or function. The activity of a gene leads to the creation of the gene's protein product. The pattern of proteins produced in one tissue versus another is the easiest way to see how gene expression is controlled. Epigenetics is the study of semi-heritable genetics. Epigenetic processes are major determinants of gene expression activation or repression. The quantity and quality of food consumed have a significant impact on cellular activities at the molecular level. Dietary exposures can have long-term consequences for health. Nutrients have the ability to alter gene expression, controlling metabolic processes. This chapter gives an account of how dietary components affect gene expression, such as the influence of vitamin C, niacin, folate, vitamin $B_{12}$, vitamin D3, resveratrol, quercetin, fisetin, myricetin, gallic acid, EGCG, soybeans, fava beans, lupin, kudzu, *Psoralea*, curcumin, and grapes on gene expression.

## 7.1 INTRODUCTION

Gene expression is a basic biological concept that connects the information in a genome to a specific functional gene, such as a peptide or noncoding

Nutrigenomics and Nutraceuticals. Raj K. Keservani, Eknath D. Ahire, Shubham J. Khairnar, Sanjay J. Kshirsagar, & Rajesh K. Kesharwani (Eds.)

RNA (ncRNA). Transcription, mRNA splicing, translation, and post-translational protein modification are all part of the process of protein production. Altering this pathway at any stage can change the amount and spatiotemporal characteristics of functional protein manifestation. It's important for cellular structure and function, as well as formative processes like differentiation and morphogenesis. Controlling gene expression allows cells to produce a functioning protein even when it is not needed for adequate survival or function. This pathway is engaged in physiologic processes, including cellular adaptation to new environments, maintaining homeostasis, and repairing damage (Atianand and Fitzgerald, 2014).

A cell's activity is determined by the number and type of mRNA molecules it has. Many transcripts are generated in each cell every second. It's no surprise that the major area of regulation for gene expression is generally the commencement of transcription at the initiation of the protein synthesis process. Because a single mRNA molecule may produce numerous proteins, RNA transcription is an effective control point.

For eukaryotes, transcript processing offers an extra degree of control, which is made feasible by the presence of a nucleus. Although ribosomes are near embryonic mRNA molecules in prokaryotes, translation occurs before the transcript is completed. In eukaryotes, transcripts are altered in the middle before being released to the cytoplasm for translation.

Transcripts in eukaryotes are also far more complicated than those in prokaryotes. For example, RNA polymerase generates the first transcripts that are mostly not present in mature RNA. Introns are sections between the mature mRNA and the nucleus that are deleted. Exons are the spliced-together parts of the transcript that contain protein-coding sequences and are used to make mature mRNA. Transcripts in eukaryotes are likewise changed at the end, affecting their integrity and translation. In a variety of situations, cells can quickly adapt to changing environmental surroundings. In some circumstances, the regulatory operating system will occur following transcription. Because just a little amount of transcription occurs in the first cellular divisions after conception, most animals' early development is reliant on translational control. As a result, eggs contain a large number of maternally derived mRNA transcripts that can be translated following fertilization.

Cells may quickly change their protein levels by enzymatically breaking down RNA transcripts and existing protein molecules on the degradative side of the balance. Certain proteins are reduced as a result of both of these activities. This breakdown is frequently connected to certain cell events. Protein degradation is connected to biological activities in the eukaryotic cell cycle. This cycle is split into multiple stages, each of which has its own set of cyclin proteins that function as important regulators for that phase. A cell must first complete one stage of the cell cycle before moving on to the next cycle; it must break down the cyclin that defines that phase. The cycle cannot continue if a cyclin is not degraded.

Because the degree of gene expression leads to the synthesis of the gene's protein product, control must be defined as any mechanism that affects the protein's production. The pattern of proteins generated in one situation versus another can be used to visualize gene expression control (Raj et al., 2006).

## 7.2 DIETARY COMPONENTS AND GENE EXPRESSION

Genes may combine with diet and lifestyle variables and have a negative impact on health, predisposing to the development of noncommunicable diseases (NCDs) like heart diseases, diabetes mellitus, and carcinoma, for which avoidance is a top priority around the world, while other dietary supplements and lifestyle choices can have the opposite, protective effect (McKay and Mathers, 2011). These findings show that dietary exposures can result in long-term health consequences and raise issues concerning the processes by which early life exposures are "remembered" over extended periods of time and how they affect disease risk. Nutritional deficiencies during pregnancy can lead to improper development of particular cells and tissues, such as fewer nephrons being produced, which increases the risk of chronic kidney disease later in life (Vehaskari, 2010). Extracellular molecules have a crucial role in the growth of eukaryotic cells and organism growth since various distinct nutritional shortages can cause growth and development to be inhibited or severely disrupted at both levels. Despite the fact that some people still don't get all of the molecular processes that facilitate these effects, it is well-founded that nutrients can impact the expression of genes that control metabolic processes via signaling pathways via multiple transcriptional signal cascades and connections. The food offers nutrients to numerous bodily systems, including intestinal function, cellular development, and DNA repair, that are required for appropriate metabolism and optimal physiological processes. The amount and quality of the foods we eat have a big impact on our cellular functions at the molecular level. Epigenetics is a branch of biology that studies reversible, quasi-heritable changes in gene activity. The sequence of nucleotides in DNA molecules does not get altered; nevertheless, gene function is influenced by epigenetic events, as well as DNA sequence and coding. Recent research has revealed that environmental variables, such as nutrition and nutrients, can directly influence gene expression in cell signaling pathways that drive both normal and malignant cell growth. Epigenetic processes are major determinants of gene expression activation or repression (Davis and Ross, 2007). Epigenetic events, including the recruitment of a diverse set of chromatin remodeling complex coregulators, effectors, and transcription factors, result in chromatin remodeling in the right manner, that is, tightly or loosely wrapped chromatin. Important processes influencing gene expression include DNA methylation, histone modification, and the creation of noncoding microRNAs (miRNAs). Nutritional factors are capable of causing epigenetic alterations by connecting to or modifying membrane receptors upstream of signaling cascades, but ligands attaching to and activating nuclear receptors can also have a direct effect on gene expression. Upstream regulatory signal mechanisms that govern epigenetic modification and additional downstream signal processes in eukaryotic cells are therefore akin to upstream regulatory signal mechanisms that affect other gene expressions. Following are some examples of dietary components that lead to gene expression:

**TABLE 7.1** Dietary Components that Cause Gene Expression.

| Dietary component | Influence | Disorders |
|---|---|---|
| **Folic acid** (Ghoshal et al., 2006)<br>(Legumes, eggs, green leafy vegetables, citrus fruits, nuts, seeds, etc.) | DNA methyltransferases upregulation | Breast cancer and intestinal cancer |
| **Gallic acid** (Huang et al., 2019)<br>(Blueberry, blackberry, strawberry, plum, grape, mango, cashewnut, hazelnut, walnut, tea, wine, etc.) | DNA<br>methyltransferase<br>inhibition and histone deacetylase inhibition | Prostate, lung, and oral cancer and cardiovascular diseases |
| **Vitamin A** (Dadon and Reifen, 2015)<br>(Beef liver, cod liver oil, mango, carrot, spinach, broccoli, tomato, etc.) | DNA<br>methyltransferase<br>inhibition | Heart defects |
| **Vitamin D** (Lamblin et al., 2010)<br>(Oily fish, red meat, liver, egg yolk, cereals, oats, etc.) | Histone deacetylase<br>inhibition | Breast cancer |
| **Omega-3 fatty acid** (Ceccarelli et al., 2020)<br>(Sea foods, nuts, and seeds) | Upregulation of DNMT, DNA<br>methyltransferase<br>inhibition, and histone deacetylase inhibition | Intestinal and liver cancer |

**TABLE 7.1** *(Continued)*

| Dietary component | Influence | Disorders |
|---|---|---|
| **β-carotene** (Moreno et al., 2002) (Carrots, spinach, lettuce, tomatoes, sweet potatoes, broccoli, cantaloupe, and winter squash) | DNA methyltransferase inhibition | Colorectal cancer |
| **Cinnamic acid derivatives** (Anantharaju et al., 2017) (Cinnamon, balsams, and shea butter) | Histone deacetylase inhibition | Colon and cervical cancer |
| **Chrysin** (Ganai et al., 2021) (Mint, honey, passion fruit, and mushroom) | Histone deacetylase inhibition | Skin cancer |
| **Kaempherol** (Qiu et al., 2017) (Goose berry, blue berry, watermelon, kiwi, peach, cherry, etc.) | DNA methyltransferase inhibition | Bladder cancer |

## 7.3 MECHANISMS OF DIET COMPONENTS AND GENE EXPRESSION INTERACTION

Bioactive dietary components influence gene expression through changes in chromatin structure (including DNA methylation and histone modification), noncoding RNA, activation of transcription factors via signaling cascades, or direct ligand binding to nuclear receptors.

### *7.3.1 ALTERATION IN THE CHROMATIN STRUCTURE*

#### a. DNA methylation

The methylation of DNA is an enzymatic process. DNMT1, DNMT3a, and DNMT3b are three enzymes that attach a methyl group to a cytosine base at the CpG dinucleotide residue to create 5-methylcytosine (Meeran et al., 2010). The first, DNMT1, is a widely distributed enzyme that is thought to be the primary methyltransferase for maintenance methylation. It's vital because it catalyzes the movement of a methyl group from the methyl precursors to the methyl end product S-adenosyl-L-methionine (SAMe) at the 5-position of specific cytosines in CpG dinucleotides. The other two DNA methylating enzymes (DNMTs) act as de novo methyltransferases with distinct roles than DNMT1. In DNMT-mediated DNA methylation, as well as numerous other enzymatic methylation processes, SAMe is the methyl donor. DNA methylation is tissue- and cell-type-specific in higher eukaryotes (Bestor, 2000). The silencing of transcriptional genes may have a role in disease processes. Gene promoter CpG islands may become abnormally hypermethylated in malignancies due to alterations in DNMT expression and activity. Cancer may thus enhance DNA methylation, which has two important effects on gene transcription: First, sequence-specific transcriptional factors like AP-2, E2F, and c-Myc, which require the existence of unmethylated CpG inside the binding sites, may be blocked from binding to DNA (Tate and Bird, 1993). The alternative process involves methyl-CpG-binding domain (MBD) proteins binding to methylated DNA. Other proteins, including histone deacetylases and other chromatin-remodeling proteins, are then recruited by these proteins. They can change the acetylation and methylation states of histones bound to a specific genomic region, preventing proteins essential for transcription from reaching the genes in question. After that, a compact, inactive chromatin may develop. The process of DNA methylation appears to be reversible. As a result, its therapeutic implications may be significant, particularly because it causes stimulation through a foreign entity of certain genomes that are important in the advancement or regression of many disorders, including cancer (Laird, 2005).

**Effect of Diet and Nutrients on DNA Methylation:**

- Dietary deficiency of zinc and selenium: It has been found in rat liver to impede the consumption of methyl groups from SAMe. This will progress to hypomethylation of histones and hypomethylation of genomic DNA (Jiang et al., 2022).
- Vitamin C deficiency: In lung cancer cells, DNA is hypermethylated (Ou et al., 2022).
- Niacin: Niacin can assist preserve the unmethylated form of CpG dinucleotides by inhibiting enzymatic DNA methylation. Because it may engage in the production of various components involved, it appears that one of the main nutritional activities of niacin is to inhibit some enzymes participating in or regulating DNA methylation. However, niacin or tryptophan are required for the production of NAD+, which is required for proper DNA repair as a substrate for poly (ADP-ribose) polymerase (PARP). Niacin and tryptophan combination shortages may thus produce alterations in transcriptional epigenetic modifications as well as be directly mutagenic, which might be particularly problematic in nations in Sub-Saharan Africa, where maize is a key staple crop (Hageman and Stierum, 2001).
- Folate and vitamin $B_{12}$: Considering its role in one-carbon metabolism, folate provides the strongest evidence of nutritional involvement

in DNA methylation. One-carbon groups are found in folate and vitamin B12 in the form of methylcobalamin, which is a water-soluble vitamin B. Folate, as well as vitamin B12 and methionine, are required for the synthesis and resynthesis of SAMe (also known as AdoMet), the unique methyl donor for DNA methylation processes. Thus, dietary deficits of vitamin B12 and methionine should have comparable effects to folate insufficiency; if more than one of these flaws exists, they will interact in a synergistic way if present at the same time, as is the case in numerous impoverished nations, particularly in Sub-Saharan Africa. Methyl group donor foods like folate, methionine, choline, betaine, and vitamin B12, as well as omega-3 fatty acids, were demonstrated to affect DNA methylation. Folate is required for DNA methylation reprogramming during the early stages of development. Early-pregnancy folate insufficiency raises the chance of neural tube abnormalities. Low dietary folate has been proposed as a possible mechanism for abnormal DNA methylation reprogramming. The DNA methylation mechanism is, in reality, rather complicated (Steegers-Theunissen et al., 2000).

- Vitamin D3 and resveratrol: Resveratrol and vitamin D3 may be involved in the reduction of promoter methylation. By activating the promoters of PTEN and p21, these drugs suppress DNA methyltransferase. In MCF-7 cells, vitamin D3 and resveratrol were found to enhance the inhibitory effects of 2CdA and F-ara-A on PTEN methylation; however, only the joint action of vitamin D3 and 2CdA has been discovered to enhance the initiation of PTEN expression, implying that these compounds work together in other processes that drive changes in PTEN expression (Stefanska et al., 2012; Fang et al., 2007).
- Quercetin, fisetin, and myricetin: These have been shown to suppress DNMT and DNMT1-mediated signaling (Fang et al., 2007).

The methylation of DNA occurs in a concentration-dependent way.

- Gallic acid and EGCG: Green tea polyphenols, such as gallic acid and EGCG, are potential demethylating agents (Majid et al., 2008).
- Soybeans, fava beans, lupin, kudzu, and *Psoralea*: Genistein is found in soybeans, fava beans, lupin, kudzu, and *Psoralea* and is one of the isoflavones, which are a subclass of the flavonoid group of chemicals. In several types of cancers, they exhibit anticancer and antiangiogenic properties. There's evidence that genistein inhibits DNA methyltransferases while also promoting histone modification

in cancer cells. It is also a nonspecific tyrosine kinase inhibitor, which can affect signal pathways governing tumor cell proliferation and death at numerous loci, thus in different pathways, and is even more potentially essential for its anticancer properties. Genistein was found to promote cancerous cells expression of the tumor suppressor genes p21WAF1/CIP1 and p16INK4a through changing promoter methylation and histone modification (Majid et al., 2008).

- Curcumin: Curcumin was recently discovered to prevent the catalytic thiolate group irreversibly at the C1226 position of DNMT1, explaining its inhibitory impact through a proposed mechanism involving molecular docking. Nonetheless, the amplitude of this impact was smaller than that of other bioactive dietary components such as EGCG and genistein. Curcumin treatment of isolated nuclear DNA from a leukemia cell line was reported, as a result of hypomethylation, worldwide. Curcumin looks to be a powerful DNA hypomethylating agent with potential anti-inflammatory, anticancer, and activities against other chronic diseases (Liu et al., 2005).
- Grapes: Resveratrol may be found in grapes, other berries, peanuts, and other plants, and it has anticancer characteristics shown by modifying signal transduction pathways. Resveratrol has an impact on cell division and proliferation, as well as apoptosis, inflammation, angiogenesis, and metastasis. This even helps to avoid malignancy (Bishayee, 2009).

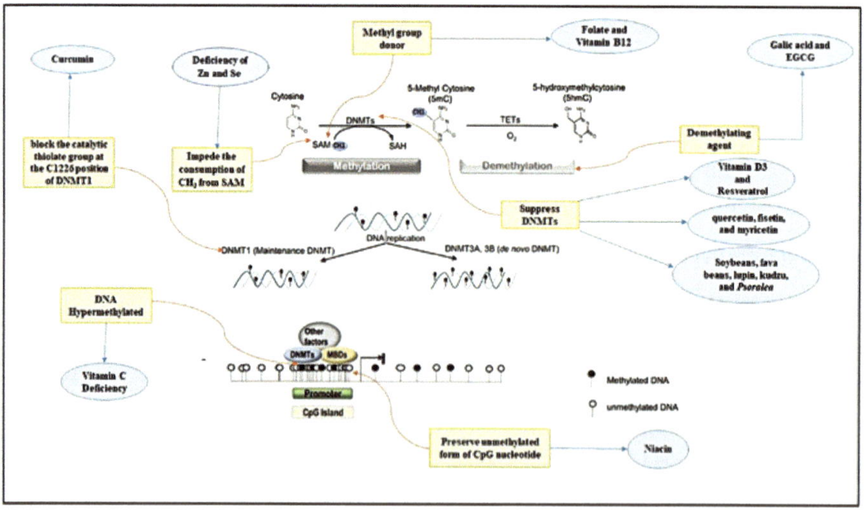


**FIGURE 7.1** Effect of dietary components on DNA methylation.

*Source*: Adapted from Wu et al, 2015.

**b. Histone Modifications**

DNA is wrapped within the nucleus by a complex wrapping around an octet of globular proteins called histones, which contains two copies of four core histones (H3, H4, H2A, and H2B) (Kouzarides, 2007). Post-translational modifications of certain amino acid residues, including acetylation and ubiquitination of lysine residues, phosphorylation of serines, and methylation of lysine and arginine residues, are found in the tails of these histones that protrude from the globular cores (Berger, 2007). Histones go through approximately a hundred different post-translational modifications. A histone code is assumed to be made up of individual histone modifications and/or patterns of changes, which is referred to as "histone decorating." It affects the expression of related genes in combination with the DNA methylation state and the presence or absence of particular miRNAs. This intricate superstructure, which sits on top of DNA's fundamental genetic information, allows for sophisticated gene expression regulation based on cell and tissue location, time, and environment, with substantial implications for cell destiny decisions, normal and pathological development, and aging (Mathers, 2006).

## 7.4 BIOACTIVE NUTRIENTS AND HISTONE MODIFICATION

HDAC inhibitors may be taken to prevent and cure cardiovascular diseases, cancer, and a variety of other physiopathological disorders. Both curcumin and EGCG are effective HAT inhibitors.

- EGCG: EGCG has been shown to prevent p300-driven p65 acetylation in vitro and in vivo, resulting in an increase in cytosolic Iκ-Bα levels and decreasing NF-κB activation produced by tumor necrosis factor alpha (TNF-α). EGCG causes p65 deacetylation, which results in a cytoplasmic increase of Iκ-Bα and subsequent cytosolic sequestration of NF-κB, which inhibits inflammatory genes including COX-2, IL-6, and NOS2 (Choi et al., 2009).
- Curcumin: Curcumin is a powerful HDAC inhibitor, even more potent than the well-known HDAC inhibitors valproic acid and sodium butyrate. It also works as a HAT inhibitor. In vitro and in vivo, curcumin inhibits p300/CREBbinding protein (CBP) HAT activity but not p300/CBP-associated factor (Karanam Balasubramanyam, 2004). Curcumin inhibits p300-mediated acetylation of RelA, an NF-κB component that is deacetylated by association with HDAC3. This deacetylation increases efficient binding to Iκ-Bα

and leads to the complex's nuclear export via a chromosomal region maintenance-1-dependent route (Chen, 2001). Curcumin treatment lowers NF-κB binding by lowering HAT activity, p300 levels, and acetylated CBP/p300 gene expression. Curcumin's ability to reduce p300/CBP HAT activity may, at least in part, be responsible for its potent NF-κB inhibitory impact (Yun et al., 2011).

- Genistein: In prostate cancer cells, genistein promotes the expression of the tumor suppressor genes p16INK4a and p21WAF1/CIP1 through altering promoter methylation and histone modification. Induction of HATs may also enhance acetylated histones 3 and 4, as well as H3 lysine 4, at the p16INK4a and p21WAF1/CIP1 transcription start sites. In human prostate cancer cells, genistein-mediated histone demethylation and acetylation activate tumor suppressor genes; this will be followed by a stoppage within the cell cycle and death driven by the cyclin and caspase pathways (Kikuno et al., 2008).

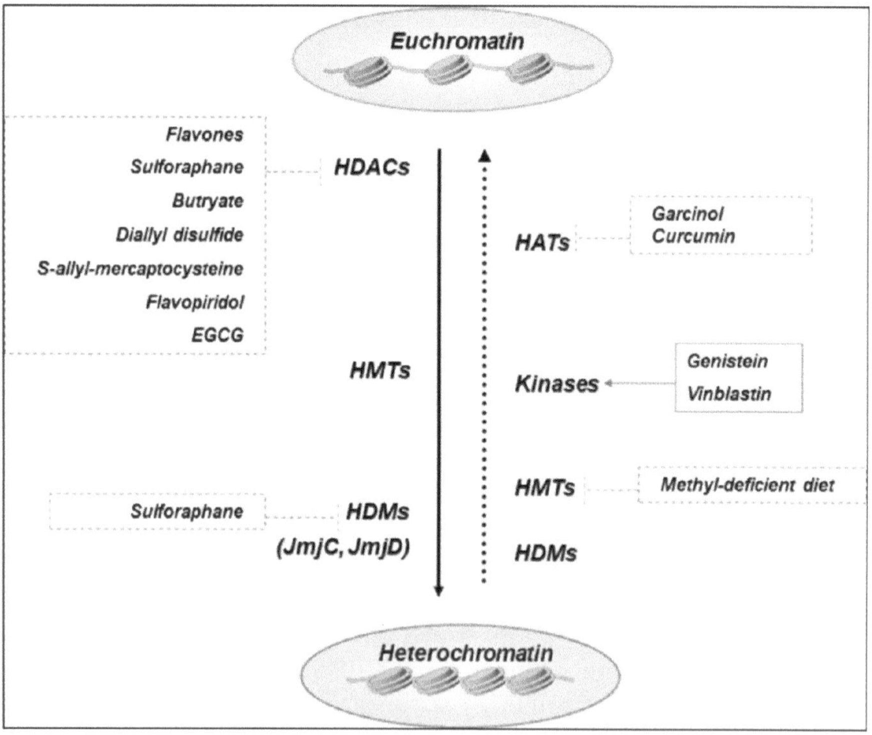


**FIGURE 7.2** Effect of bioactive compounds on histone modification.

*Source*: Adapted from Barghi et al., 2017.

### 7.4.1 NONCODING RNA

MicroRNA (miRNA) is a vast class of tiny noncoding RNAs (22 nucleotides long). The human genome contains at least 1046 miRNAs, which are encoded on all human chromosomes and range from 94 miRNAs on chromosome X to 12 miRNAs on chromosome 21. These miRNAs control the transcription of 30% of all protein-coding genes by binding to RNA in particular sequences. Each miRNA binds to the target mRNA's 3'-untranslated region and inhibits expression by altering mRNA stability and/or directing the mRNA for destruction. MiRNA can regulate practically all genetic pathways by targeting transcription factors, secreted factors, receptors, and transporters and controls the expression of around one-third of human mRNAs. Various dietary variables, ranging from macronutrients (fat, protein, and alcohol) to micronutrients (vitamin E), are increasingly being shown to affect the expression of many miRNAs in rats and humans. The genetic sequences encoding miRNA are now known to be polymorphic. The relationships between the genome and the epigenetic apparatus, that control gene expression, are now far more complicated.

## 7.5 MIRNA-RELATED EPIGENETIC REGULATION MEDIATED BY DIET

The discovery of microRNAs marked a turning point in gene regulation research and understanding. A miRNA is a 22-nucleotide noncoding RNA molecule that controls gene expression after transcription. Two epigenetic mechanisms that influence the expression of protein-coding genes and miRNAs are histone modification and DNA methylation. Epigenetic regulators include DNA methyltransferases, histone deacetylases, and polycomb group genes. Some might be affected by miRNA, which is part of the epigenetic-miRNA regulatory circuit. Damage to this regulatory circuit can induce a number of disorders by disrupting normal physiological functions. MiRNAs have been linked to cancer at every stage of its development, from its onset through its development. They impact cell proliferation, differentiation, death, angiogenesis, and metastasis. According to recent discoveries from human and animal models, nutrition (e.g., vitamin E, fat, protein, and alcohol) appears to change the expression of several miRNAs. Diet can influence miRNAs involved in different stages of cancer development, including early chemoprevention versus delayed therapeutic effects. Treatment with vitamin A has been demonstrated to increase the activity of multiple tumor suppressor miRNAs in human APL cells. Garzon et al. detected putative NF-κB consensus elements in the upstream genomic region of the let-7 cluster after all-trans-retinoic acid (ATRA) treatment. Because of their capacity to attach to their receptors, vitamin D and its metabolites have anticancer capabilities (VDR). MiR-125b may have a sequence match in the 3'-UTR regions of human VDR mRNA. This shows that downregulation of VDR might be utilized to target cancer therapy in humans. Vitamin E influences miR-122 and miR-125b expression, which monitor lipid metabolism, inflammation, and other cancer-related pathways. Curcumin, resveratrol, catechins, ellagitannins, and indoles are bioactive substances that will have considerable down-regulation of tumor marker miRNAs and significant upregulation of tumor suppressor miRNAs, indicating anticancer effects. Nutrients can also influence epigenetic inheritance from generation to generation (Chango and Pogribny, 2015).

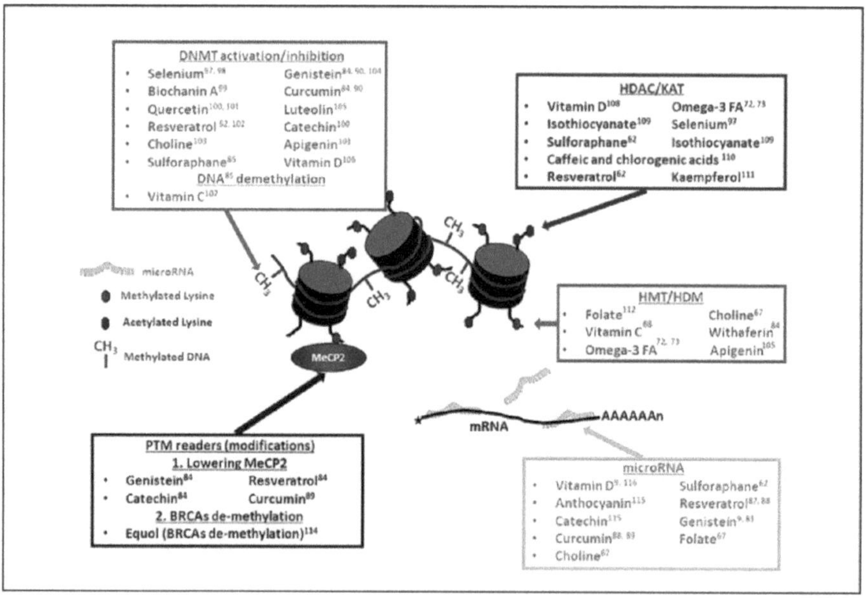


**FIGURE 7.3** Effect of dietary components on epigenetics.

*Source*: Adapted from Georgel et al., 2021.

### 7.5.1 ACTIVATION OF TRANSCRIPTION FACTOR

Hormones or the nervous system are among the mechanisms used by organisms to keep metabolic and energy balances in check. Numerous studies have shown that direct and indirect nutrients, as well as their derivatives, provide health advantages and regulate gene expression in a hormone-independent manner by interacting directly with cell components.

The peroxisome proliferator-activated receptors (PPARs) are a nuclear receptor family that affects lipid and glucose metabolism. Nutritional lipids (FAs) and their metabolites activate these receptors, which cause a reaction when redirected. Sterol regulatory element-binding proteins regulate sterol and fatty acid metabolism (SREBPs). SREBP-1c transcription can be stimulated by activation of the liver X receptor (LXR), a steroid nuclear receptor induced by oxysterols (cholesterol derivatives).

Proteins are the third most important macromolecule that must be consumed through food. The amino acid response is triggered when the amount of amino acids in the body diminishes. A healthy and balanced diet must meet all of an organism's amino acid requirements. The AAR regulates gene expression at several levels, including chromatin structure and transcription start sites. By deacetylating matching tRNAs, it activates the general control nonderepressible 2 (GCN2) kinase. Eukaryotic initiation factor 2 alpha (eIF2) is then activated by the kinase. When asparagine is depleted, the quantity of asparagine synthetase mRNA is increased. Activating the stress response pathway in the endoplasmic reticulum (ER) has the same effect as glucose deprivation. By attaching to nutrient-sensing response elements, both strategies promote asparaginase transcription (Barrat et al., 2019).

## 7.6 THE EFFECT OF DIETARY COMPONENTS ON DISEASES

According to epidemiological research, those who eat a high diet of fruits and vegetables are less likely to get cancer. Fruits and vegetables are high in fiber, vitamins, and minerals, and they also include health-promoting chemicals, including terpenes, alkaloids, and phenolics (Block, et al., 1992). Over 500 chemicals originating from food have been discovered as potential cancer modifiers. They include substances originating from plants, animals, and fungi, as well as metabolites produced by human microflora (Działo, et al., 2016). Modifications in chromatin structure, epigenetic changes, and transcription factor activation are all ways that these substances control gene expression. Allergies are one of the most significant issues that can arise as a result of poor food selection. Although the mechanism of allergic responses to foods is reasonably well understood, several areas, particularly those involving epigenetic control, deserve more investigation. For example, a mother's food (during pregnancy and

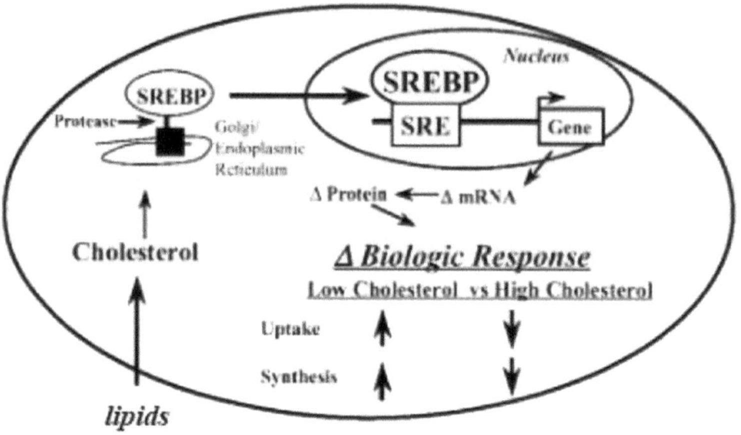


**FIGURE 7.4** Involvement of lipid in gene transcription.

*Source*: Adapted from Jump, 2001.

lactation) has a significant and long-term influence on the development of her offspring's immune system, and as a result, on the prevalence of allergies. The activity of the PKCζ (protein kinase C zeta) gene in T cells throughout newborn life has been linked to the severity of allergy illness because of increased PKC gene expression, which is likely regulated epigenetically by changes in DNA methylation or histone acetylation (Harb et al., 2017). Pregnant women who take a 3-polyunsaturated fatty acid (PUFA)-rich fish oil supplement lessen the prevalence of allergies in their children. Supplementing with fish oil influences epigenetic alterations in genes related to the immune system, such as IL13 and T-box 21 (Alhamwe et al., 2018). Another study found that adding olive oil to pregnant women's diets influenced histone acetylation in genes that regulate the immune system throughout fetal life. For example, increased acetylation of H3 histone at the promoter of the FOXP3 gene (encoding one of the regulators of immunological homeostasis) was reported. FOXP3, the primary transcription factor in regulatory T cells (Tregs), is essential for the development of tolerance to self and environmental antigens (including food) (Acevedo et al., 2019). However, because PUFAs stimulate proinflammatory pathways by affecting DNA methylation and histone acetylation and also modulate RNA transcription (miRNAs and lncRNAs), they can be allergenic, and only a well-balanced and appropriately selected diet composition should be considered. Adequate healthy nutrition is important at any age because it helps to prevent not only cancer but also noninfectious diseases like obesity, type-2 diabetes, heart disease, neurodegenerative disease, and allergic disease. Vitamins, minerals, polyphenols, carotenoids, and isothiocyanates, among other bioactive dietary components, influence gene expression, resulting in a decrease in the start and course of illnesses. DNA methylation, histone changes, telomerase suppression, and the regulatory influence of RNA or transcription factors all contribute to this.

### *7.6.1 DIETARY FIBER*

The application of molecular biologic tools has moved the field of dietary gene regulation to the forefront of nutrition research. Using these methods, researchers were able to evaluate the influence of high-fiber meals on the expression of particular genes in the small intestine and colon. Dietary fiber's effects on gene expression are difficult to understand since they are likely indirect and entail a complex interaction of several components (e.g., hormones, enzymes, and cellular secondary messengers). However, research comparing the impact of high-fiber versus fiber-free diets on certain gene products has yielded useful information. The effects of cholesterol-free, high-fiber diets on apolipoprotein expression in the ileum, jejunum, and liver of rats were investigated. Apo A-I is a key protein component of high-density lipoprotein, the predominant transporter of plasma cholesterol in rats, and is generated in the liver and small intestine. The cholesterol-free diets were fed for 14 days with either beet fiber (15 g/100 g of food) or no fiber supplement. Plasma cholesterol concentrations of apo A-I were lower in rats fed a fiber-supplemented diet than in rats fed a fiber-free diet. According to a northern blot examination of total RNA obtained from the animals' liver, ileum, and jejunum, the expression of liver and ileum apo A-I mRNA was significantly lower in the group provided with beet fiber than in the group fed fiber-free diets. There were no significant variations in the expression of apo A-I mRNA in the jejunum between the groups. Reduced hepatic apo A-I mRNA expression was linked to lower plasma cholesterol levels in beet fiber-fed rats when compared to controls. A similar pattern of alterations in apo A-IV expression was seen in response to the beet-fiber diet. The levels of apo E mRNA in the liver were unaffected. However, these findings showed that a dietary component has a tissue-specific influence on gene expression, though the exact mechanism remains unknown (Luo, 2021). In one study, the effect of a high-fiber diet versus a fiber-free diet on the expression of intestinal proglucagon mRNA in rats was studied. The precursor of glucagon-like protein-1 (GLP-1), which promotes insulin production and modifies glycemic responses, is intestinal proglucagon 7. For 14 days, rats were fed either an elemental diet or one enriched with 30% fiber. Rats fed a high-fiber diet had considerably higher levels of ileal proglucagon mRNA than those fed a low-fiber diet. When rats on a 30% fiber diet were compared to those on a fiber-free diet, there was a tendency toward higher expression of colonic proglucagon mRNA. When compared to controls, the rise in proglucagon mRNA was associated with a substantial increase in short-chain fatty acid synthesis in the cecum and colon of rats on a high-fiber diet. These findings support the theory that dietary fiber controls GLP-1 gene expression and secretion (Reimer and Mc Burney, 1996).

### *7.6.2 DIETARY ZINC*

Dietary zinc has been demonstrated to alter metallothionein gene expression by binding to proteins that are connected to metal response elements in the promoter (King, 2000). This association is one of the best-studied examples of nutrient-induced gene regulation. It's used as an example of a rather well-understood gene-nutrient relationship. The effects of zinc on gene regulation, on the other hand, have not been fully studied. We discovered mRNAs in the rat small intestine that may be influenced by dietary zinc deficiency. The rats were kept on one of three diets for 18 days. One diet was zinc-sufficient, whereas the other was zinc-insufficient. Zinc-deficient diets are linked to lower food intake; to account for this, a third group—the zinc-adequate, pair-fed group—was included. These rats were provided a diet that was zinc-sufficient but had the same total amount of food as the zinc-deficient rats. Total RNA was extracted from the intestinal mucosa of each group of animals at the conclusion of this time period and evaluated using mRNA differential display. Differential display is a method for selecting, amplifying, and displaying subpopulations of mRNAs on polyacrylamide gel electrophoresis (PAGE), eventually screening all genes expressed at any given moment. To locate genes that are controlled differently, we used the gel display to examine three or more conditions side by side. Upregulated genes appear as bands that are absent or more abundant in the zinc-adequate or pair-fed samples than in the zinc-deficient sample. Genes that are upregulated by zinc (or downregulated by zinc deficiency) appear as bands in the zinc-adequate or pair-fed samples but are reduced or absent in the zinc-deficient sample. We discovered 47 zinc-regulated candidate genes, 30 of which were upregulated and 17 of which were downregulated (underexpressed) in zinc shortage. Several of these genes have been cloned so far, and their zinc regulation has been validated by northern blot analysis. Many have been sequenced and found to be expressed sequence tags (Wu et al., 2022). One of the best-studied examples of nutrient-induced gene regulation is this relationship. It's used to illustrate a very well-known gene-nutrient link. The impact of zinc on the expression of other genes has not been well investigated. Zinc has been demonstrated to control the metallothionein gene directly. Young males took part in an 18-day placebo-controlled study in which they were given 50 mg of zinc each day. The quantity of metallothionein mRNA in peripheral monocytes in the zinc-supplemented group was considerably higher by day 2 compared to the placebo controls and remained elevated throughout the 18-day supplementation period. The levels of metallothionein mRNA reverted to baseline within days after the zinc supplement was removed, suggesting the quick gene-regulating action of this dietary component (King, 2000).

## 7.7 CONCLUSION

Dietary food components regulate gene expression through a variety of processes, including chromatin structural changes, noncoding RNA, signaling cascade activation of transcription factors, and direct ligand binding to the nuclear receptor. The identification of these substances and the understanding of their mechanisms of action will allow for more effective diet recommendations for groups of people or individuals. Many food-related disorders, including malignancies such as heart disease, type-2 diabetes, and obesity, are prevented and treated with bioactive diet components. It appears that further research into nutrigenetics and nutrigenomics is needed, as well as encouragement for dieticians to use this information to make more successful dietary recommendations.

## KEYWORDS

- **dietary supplement**
- **gene expression**
- **gene**
- **genetics**
- **nutrients**

## REFERENCES

Alhamwe, B. A.; Khalaila, R.; Wolf, J.; von Bülow, V.; Harb, H.; Alhamdan, F.; Hii, C. S.; Prescott, S. L.; Ferrante, A.; Renz, H.; Garn, H.; Potaczek, D. P. Histone Modifications and their Role in Epigenetics of Atopy and Allergic Diseases. *Allergy Asthma Clin. Immunol.* **2018,** *14*, 39.

Acevedo, N.; Frumento, P.; Harb, H.; Alhamwe, B. A.; Johansson, C.; Eick, L.; Alm, J.; Renz, H.; Scheynius, A.; Potaczek, D. P. Histone Acetylation of Immune Regulatory Genes in Human Placenta in Association with Maternal Intake of Olive Oil and Fish Consumption. *Int. J. Mol. Sci.* **2019,** *20*, 1060.

Anantharaju, P. G.; Reddy, D. B.; Padukudru, M. A.; Chitturi, Ch. M. K.; Vimalambike, M. G.; Madhunapantula, S. V. Induction of Colon and Cervical Cancer Cell Death by Cinnamic Acid Derivatives is Mediated Through the Inhibition of Histone Deacetylases (HDAC). *PLoS One* **2017,** *12* (11).

Atianand, M. K.; Fitzgerald, K. A. Long Non-coding RNAs and Control of Gene Expression in the Immune System. *Trends Mol. Med.* **2014,** *20* (11), 623–631.

Balasubramanyam, K.; Varier, R. A.; Altaf, M.; Swaminathan, V.; Siddappa, N. B.; Ranga, U.; Kundu, T. K. Curcumin, a Novel p300/CREB-binding Protein-specific Inhibitor of Acetyltransferase, Represses the Acetylation of Histone/Nonhistone Proteins and Histone Acetyltransferase-dependent Chromatin Transcription. *J. Biol. Chem.* **2004,** *279* (49), 51163–51171.

Barghi, S.; Amiri, M.; Hajipour, H.; Namaki, S. The Effect of Dietary Constituents on Regulation of Epigenetic Changes in Cancer. *J. Babol Univ. Med. Sci.* **2017,** *19* (7), 63–71.

Barrat, F. J.; Crow, M. K.; Ivashkiv, L. B.; Interferon Target-gene Expression and Epigenomic Signatures in Health and Disease. *Nat. Immunol.* **2019,** *20*, 1574–1583.

Berger, S. L. The Complex Language of Chromatin Regulation During Transcription. *Nature* **2007,** *447*.

Bishayee, A. Cancer Prevention and Treatment with Resveratrol: From Rodent Studies to Clinical Trials. *Cancer Prev. Res.* **2009,** *2*, 409–418.

Block, G.; Patterson, B.; Subar, A. Fruit, Vegetables, and Cancer Prevention: A Review of the Epidemiological Evidence. *Nutr. Cancer* **1992,** *18* (1), 1–29.

Ceccarelli, V.; Ronchetti, S.; Marchetti, M. C.; Calvitti, M.; Riccardi, C.; Grignani, F.; Vecchini, A. Molecular Mechanisms Underlying Eicosapentaenoic Acid Inhibition of HDAC1 and DNMT Expression and Activity Incarcinoma Cells. *Biochim. Biophys. Acta Gene. Regul. Mech.* **2020,** *1863* (2), 194481.

Chen, L. F.; Fischle, W.; Verdin, E.; Greene, W. C. Duration of Nuclear NF-kB Action Regulated by Reversible Acetylation. *Science* **2001,** *293* (5535), 1653–1657.

Choi, K. C.; Jung, M. G.; Lee, Y. H.; Yoon, J. C.; Kwon, S. H.; Kang, H. B.; Kim, M. J.; Cha, J. H.; Kim, Y. J.; Woo Jin Jun, W. J.; Jae Myun Lee, Ho-Geun Yoon Kyung-Chul Choi, M. G. J. Y.-H. L. J. C. Y. S. H. K., 2009. Epigallocatechin-3-Gallate, a Histone Acetyltransferase Inhibitor Inhibits EBV-Induced B Lymphocyte Transformation via Suppression of RelA Acetylation. *Cancer Res.* **2009,** *69* (2), 583–592.

Clouaire, T.; Stancheva, I. Methyl-CpG Binding Proteins: Specialized Transcriptional Repressors or Structural Components of Chromatin?. *Cell. Mol. Life Sci.* **2008,** *65*, 1509–1522.

Cousins, R. J. Metral Element and Gene Expression. *Ann. Rev. Nutr.* **1994,** *14*, 449–69.

Dadon, S. B.; Reifen, R. Vitamin A and the Epigenome. *Crit. Rev. Food Sci. Nutr.* **2015,** 2404–2411.

Davis, C. D.; Ross, S. A. Dietary Components Impact Histone Modifications and Cancer Risk. *Nutr. Rev.* **2007,** *65* (2), 88–94.

Działo, M.; Mierziak, J.; Korzun, U.; Preisner, M.; Szopa, J.; Kulma, A.; Andrade, P. The Potential of Plant Phenolics in Prevention and Therapy of Skin Disorders. *Int. J. Mol. Sci.* **2016,** *17*, 160.

Fang, M.; Chen, D.; Yang, C. S. Dietary Polyphenols May Affect DNA Methylation. *J. Nutr.* **2007,** *137*, 223S–228S.

Epigenet, C. The DNA Methyltransferase of Mammals. *Human Mol. Genet.* **2000,** *9* (16), 2395–2402.

Georgel, P. T.; Georgel, P. Where Epigenetics Meets Food Intake: Their Interaction in the Development/Severity of Gout and Therapeutic Perspectives. *Front. Immunol.* **2021,** *12*, 752359.

Ganai, S. A.; Sheikh, F. A.; Baba, Z. A. Plant Flavone Chrysin as an Emerging Histone Deacetylase Inhibitor for Prosperous Epigenetic-Based Anticancer Therapy. *Phytother. Res.* **2021,** *35*, 823–834.

Ghoshal, K.; Li, X.; Datta, J.; Bai, S.; Pogribny, I.; Pogribny, M.; Huang, Y.; Young, D.; Jacob, S. T. A Folate- and Methyl-Deficient Diet Alters the Expression of DNA Methyltransferases and Methyl CpG Binding Proteins Involved in Epigenetic Gene Silencing in Livers of F344 Rats. *Nutr. Dis.* **2006,** *136*, 1522–1527.

Hageman, G. J.; Stierum, R. H. Niacin, poly(ADP-ribose) Polymerase-1 and Genomic Stability. *Mutat. Res.* **2001,** *475*, 45–56.

Harb, H.; Irvine, J.; Amarasekera, M.; Hii, C. S.; Kesper, D. A.; Ma, Y. F.; D'Vaz, N.; Renz, H.; Potaczek, D. P.; Prescott, S. L.; Ferrante, A. The Role of PKC in Cord Blood T-cell Maturation Towards Th1 Cytokine Profile and its Epigenetic Regulation by Fish Oil. *Biosci. Rep.* **2017,** *37* (2), BSR20160485.

Huang, D.; Cui, L.; Ahmed, S.; Zainab, F.; Wu, Q.; Wang, X.; Yuan, Z. An Overview of Epigenetic Agents and Natural Nutrition Products Targeting DNA Methyltransferase, Histone Deacetylases and microRNAs. *Food Chem. Toxicol.* **2018,** *18*, S0278–6915.

Jenuwein, T.; Allis, C. D. Translating the Histone Code. *Science* **2001,** *293* (5532), 1074–1080.

Jump, D. B. Nutrients as Regulators of Gene Expression, 2001; p 32.

Kikuno, N.; Shiina, H.; Urakami, S.; Kawamoto, K.; Hirata, H.; Tanaka, Y.; Majid, S.; Igawa, M.; Dahiya, R. Genistein Mediated Histone Acetylation and Demethylation Activates Tumor Suppressor Genes in Prostate Cancer Cells. *Int. J. Cancer* **2008,** *123*, 552–560.

Kouzarides, T. Chromatin Modifications and Their Function. *Cell* **2007,** *128*, 693–705.

Laird, P. W. Cancer Epigenetics. *Human Mol. Genet.* **2005,** *14* (1).

Lamblin, M.; Dabbas, B.; Spingarn, R.; Mendoza-Sanchez, R.; Wang, T. T.; An, B. S.; Huang, D. C.; Kremer, R.; White, J. H.; Gleason, J. L. Vitamin D Receptor Agonist/Histone Deacetylase Inhibitor Molecular Hybrids. *Bioorg. Med. Chem.* **2010,** *18*, 4119–4137.

Liu, H. L.; Chen, Y.; Cui, G. H.; Zhou, J. F. Curcumin, a Potent Anti-tumor Reagent, is a Novel Histone Deacetylase Inhibitor Regulating B-NHL Cell Line Raji Proliferation. *Acta Pharm. Sinica* **2005,** *26* (5), 603–609.

Majid, S.; Kikuno, N.; Nelles, J.; Noonan, E.; Tanaka, Y.; Kawamoto, K.; Hirata, H.; Li, L. C.; Zhao, H.; Okino, S. T.; Place, R. F.; Pookot, D.; Dahiya, R. Genistein Induces the p21WAF1/CIP1 and p16INK4a Tumor Suppressor Genes in Prostate Cancer Cells by

Epigenetic Mechanisms Involving Active Chromatin Modification. *Cancer Res.* **2008,** *68*, 2736–2744.
Mathers, J. C. Nutritional Modulation of Ageing: Genomic and Epigenetic Approaches. *Mech. Ageing Dev.* **2006,** *127*, 584–589.
McKay, J. A.; Mathers, J. C. Diet Induced Epigenetic Changes and their Implications for Health. *Acta Physiol.* **2011,** *202*, 103–118.
Meeran, S. M.; Ahmed, A.; Tollefsbol, T. O. Epigenetic targets of bioactive dietary components for cancer prevention and therapy. *Clin. Epigenet.* **2010,** *1*, 101–116.
Moreno, F. S.; Wu, S. T.; Naves, M. M. V.; Silveira, E. R.; Oloris, S. C.; da Costa, M. A. L.; Dagli, M. L. Z.; Ong, T. P. Inhibitory Effects of β-Carotene and Vitamin A During the Progression Phase of Hepatocarcinogenesis Involve Inhibition of Cell Proliferation but Not Alterations in DNA Methylation. *Nutr. Cancer* **2014,** *44* (1), 80–88.
Nie, Y.; **Luo, F.** Dietary Fiber: An Opportunity for a Global Control of Hyperlipidemia. *Oxid. Med. Cell. Longev.* **2021,** *2021*, 20.
Ou, J.; Liao, Q.; Du, Y.; Xi, W.; Meng, Q.; Li, K.; Cai, Q.; Pang, C. L. SERPINB7, TMEM158, and SERPINE1 Signature as Predictors of Non-Small-Cell Lung Cancer Prognosis after Intravenous Vitamin C Treatment; Research Square, 2022.
Qiu, W.; Lin, J.; Zhu, Y.; Zhang, J.; Zeng, L.; Su, M.; Tian, Y. Kaempferol Modulates DNA Methylation and Downregulates DNMT3B in Bladder Cancer. *Cell. Physiol. Biochem.* **2017,** *41*, 1325–1335.
Raj, A.; Peskin, C. S.; Tranchina, D.; Vargas, D. Y.; Tyagi, S. Stochastic mRNA Synthesis in Mammalian Cells. *PLoS Biol.* **2006,** *4* (10), e309.
Reimer, R. A.; McBurney, M. I. Dietary Fiber Modulates Intestinal Proglucagon Messenger Ribonucleic Acid and Postprandial Secretion of Glucagon-Like Peptide-l and Insulin in Rats. *Endocrinology* **1996,** *137* (9).
Steegers-Theunissen, R. P.; Smith, S. C.; Steegers, E. A.; Guilbert, L. J.; Baker, P. N. Folate Affects Apoptosis in Human Trophoblastic Cells. *Br. J. Obstet. Gynaecol.* **2000,** *107*, 1513–1515.
Stefanska, B.; Salamé, P.; Bednarek, A.; Fabianowska-Majewska, K. Comparative Effects of Retinoic Acid, Vitamin D and Resveratrol Alone and in Combination with Adenosine Analogues on Methylation and Expression of Phosphatase and Tensin Homologue Tumour Suppressor Gene in Breast Cancer Cells. *Br. J. Nutr.* **2012,** *107*, 781–790.
Tata, P. H.; Bird, A. P. Effects of DNA Methylation on DNA-binding Proteins and Gene Expression. *Curr. Opin. Genet. Dev.* **1993,** *3*, 226–231.
Vehaskari, V. M. Prenatal Programming of Kidney Disease. *Curr. Opin. Pediatr.* **2010,** *22*, 176–182.
Wu, J.; Zhao, Y.; Wang, X.; Kong, L.; Johnston, L. J.; Lu, L.; Ma, X. Dietary Nutrients Shape Gut Microbes and Intestinal Mucosa via Epigenetic Modifications. *Crit. Rev. Food Sci. Nutr.* **2020,** *62* (3), 783–797.
Wu Y, Sarkissyan M, Vadgama JV. Epigenetics in breast and prostate cancer. Cancer Epigenetics: Risk Assessment, Diagnosis, Treatment, and Prognosis. 2015:425-66.
Yiang, Y. G.; Wang, Y. H.; Zhang, H.; Wang, Z. Y., Liu, Y. Q. Effects of Early-Life Zinc Deficiency on Learning and Memory in Offspring and the Changes in DNA Methylation Patterns. *Nutr. Neurosci.* 2020.
Yun, J. M.; Jialal, I.; Devaraj, S. Epigenetic Regulation of High Glucose-Induced Proinflammatory Cytokine Production in Monocytes by Curcumin. *J. Nutr. Biochem.* **2011,** *22*, 450–458.

Zhang, X.; Hu, M.; Lyu, X.; Li, C.; Thannickal, V. J.; Sanders, Y. Y. DNA methylation regulated gene expression in organ fibrosis. *Biochim. Biophys. Acta Mol. Basis Dis.* **2017,** *1863* (9), 2389–2397.

# CHAPTER 8

# THE ROLE OF NUTRITIONAL FACTORS IN PATHOGENESIS OF DISEASES

RAHUL R. SABLE,[1] SHUBHAM J. KHAIRNAR,[1] EKNATH D. AHIRE,[2] JAYESH D. KADAM,[1] and SURABHI B. PATIL[1]

[1]*Department of Pharmacology, MET's Institute of Pharmacy, Adgaon, Nashik, India*

[2]*Department of Pharmaceutics, MET's Institute of Pharmacy, Adgaon, Nashik, India*

## ABSTRACT

Diet and nutrition are critical aspects in the development and maintenance of overall health throughout one's life. It has long been assumed that nutrition, environment, and hereditary factors all play a role in our susceptibility to infection and disease. Nutrition is gaining traction as a key modifiable predictor of chronic illness, with scientific evidence increasingly supporting the notion that dietary changes have significant implications on health throughout life, both good and negative. The majority of scientific research on mental health focuses on depression, cognition, and dementia, with less information on other psychiatric disorders like schizophrenia. The longer a person lives, the more mental problems become prevalent, and the more attention is paid to them. Many foods have been involved in inflammatory bowel disease aetiology ideas. While data has accumulated that nutritional factors as part of overall lifestyle variations may play a key role in the increasing incidence, no specific dietary recommendations can currently be made to lower the likelihood of acquiring Crohn's disease or ulcerative colitis, with the exception of promoting breastfeeding. Diet consists of a

Nutrigenomics and Nutraceuticals. Raj K. Keservani, Eknath D. Ahire, Shubham J. Khairnar, Sanjay J. Kshirsagar, & Rajesh K. Kesharwani (Eds.)

mixture of protective, mutagenic, and carcinogenic substances, the majority of which are processed by biotransformation enzymes.

## 8.1 INTRODUCTION

Individuals' dietary consumption and nutritional status have a tremendous impact on mental health and the development of different pathological diseases. Nutrition is a complex and interesting topic, with a well-documented function in the development, exacerbation, and amelioration of a wide range of human disorders. The majority of scientific evidence on mental health focuses on dementia, cognitive function, and depression and there is less evidence on other psychiatric diseases, such as schizophrenia (Lim et al., 2016). The longer a person lives, the more mental problems are prevalent, and the more attention is paid to them. Omega-3 fatty acids, phospholipids, cholesterol, vitamin B6, and vitamin B12 are among the dietary factors that may be advantageous to mental health. Unhealthy fats and simple sugar are both thought to be bad for your brain. The evidence regarding cholesterol's impact is mixed; although, blood cholesterol levels are generally linked to a lower incidence of depression. The aims of this topic are to elaborate known nutritional factors (Keservani et al., 2017; Keservani et al., 2020) that have an impact on cognitive function involve in pathogenesis of diseases (Diotallevi et al., 2022). Nutritional factors are assumed to be responsible for roughly 30% of cancers in affluent nations, placing them second only to tobacco as a treatable cause of cancer (Doll et al., 1981). Diet consists of a combination of protective, mutagenic, and carcinogenic substances, the majority of which are processed by biotransformation enzymes.

## 8.2 HISTORY

Experimental research done in the 1940s initially discovered the association between dietary elements and cancer pathogenesis, which was marked by the sequential confinement of nutritional factors. In mice, the incidence of cancer was lowered (Tannenbaum, 1940). Other advancements, such as the development of cancer registries and research focused on the impact of environmental variables on human health, have painted a larger picture of the geographical variables in cancer incidence, which could be because of differences in diet and lifestyle, after two decades. (Doll, 1966; Doll, 1970; Armstrong, 1975). In 2006, epidemiology research looked at the Crohn's disease growth in Japan and found that higher dietary consumption of n-6 polyunsaturated fatty acids and animal protein, as well as a shift toward a more westernized diet, were related to an increased risk of the disease (Shoda et al., 1996). Other surveys and case-control studies have linked refined sugar, fast meals, margarine, and dairy products to heart disease, whereas vegetables, fruits, seafood, and dietary fibre tend to protect against it (Amre et al., 2016).

## 8.3 FACTORS AFFECTING MENTAL HEALTH

Mental well-being is a critical component of optimal health, and it is a state in which people can handle stress from daily life and achieve positive goals in pursuit of public good and community service (Manwell, 2015). Genetic factors, stress, food, physical inactivity, medicines, and other environmental variables all have a role in the progression of mental illnesses (Pinto, 2015). Although dietary variables are not primary etiologies, they may exacerbate or alleviate symptoms and the course of the disorders. Polyunsaturated fatty acids (PUFAs), particularly omega-3 PUFAs, phospholipids, niacin, folate, cholesterol, vitamin B6, vitamin B12, and vitamin D are all nutritional factors that are advantageous to mental health (Lauritzen, 2016; Lakhan, 2008; Smith, 2008). Among them few are discussed below.

### *8.3.1 OMEGA-3 FAs*

Omega-3 fatty acids have been studied widely in direct relation to brain health. The function of omega-3 FAs in the retina is primarily as a structural and functional part of membrane phospholipids (Sinclair, 2007; Haag, 2003; Lauritzen, 2001). Flaxseed oil and soybean oil include alpha-linolenic acid, a plant-derived omega-3 FA, whereas fish oil is the primary nutritive source of eicosapentaenoic acid (EPA) and docosahexaenoic acid (DHA). These omega-3 fatty acids are powerful transcription factor activators and inflammatory modulators (Wang, 2015; Zúñiga, 2011). Omega-3 FAs' anti-inflammatory properties are often connected to the inhibition of excessive pro-inflammatory effects of omega-6 fatty acids. Dementia is one type of mental illness; its prevalence is clearly increasing as the elderly population grows. Low blood levels of omega-3 fatty acids were found in persons with dementia, which in first stages is age-related memory impairment and cognitive and functional impairment. Mild-to-severe protein-energy malnutrition (PEM) commonly occurs when dementia advances to stages 2 (early dementia) and 3 (dementia with behavioral symptoms), necessitating energy supplementation and oral nutritional supplements with protein rather than omega-3 FAs supplementation

(Kang, 2014). The intervention of omega-3 fatty acids (FAs) reduced the symptoms of depression, Alzheimer's disease (AD), and schizophrenia. Regulation of membrane integrity and fluidity, neurite formation, neurotransmitters, endothelium, neuronal survival, neurodegeneration, transcription, and inflammation are among the mechanisms behind omega-3 FAs' positive effects on symptoms of mental disorders (Ahire et al., 2020).

### 8.3.2 VITAMIN B: NIACIN, FOLATE, VITAMIN B6, VITAMIN B12

Vitamin B, in the form of cofactors such as flavin adenine dinucleotide (FAD) and nicotinamide adenine dinucleotide (NAD), is important in energy metabolism (FAD). Mental health is influenced by B vitamins, including niacin, vitamin B6, vitamin B12, and folate. The "homocysteine hypothesis," which claims that too much homocysteine produces psychiatric symptoms, is a well-known mental disease hypothesis. High levels of homocysteine and low levels of the B vitamins were reported in patients with MCI, dementia, and depression, while low levels of the B vitamins and high levels of homocysteine were observed in persons with MCI, dementia, and depression (Smith, 2018). Homocysteine's effect on brain function is thought to be mediated via variations in cerebral vasculature and neurotransmitter function, as well as an increase in neurotoxicity and oxidative stress. In comparison to other vitamin B supplements, niacin's effect on brain function has received less attention. According to a case study, a person with pellagra, a niacin deficiency condition, displayed mental symptoms, namely behavior deterioration and dementia, which were reduced by niacin supplementation (Wang, 2012). Research has been conducted in recent times on the effects of dietary niacin, particularly nicotinamide riboside (NR), on beta-secretase 1 degradation, and on the expression of mitochondrial metabolism-related genes (aconitase, citrate synthase, glucose phosphate isomerase 1, phosphoglycerate kinase, and pyruvate dehydrogenase kinase) linked to the action of proliferator-activated receptor-$\gamma$ coactivator 1$\alpha$ (Gong, 2013; Chi, 2013). So far, vitamin B nutritional interventions for brain function have yielded mixed findings in terms of their effectiveness on cognitive performance (Ford, 2012; Clarke, 2014).

### 8.3.3 ANTIOXIDANTS

Because it possesses a lipid-rich region, particularly in the neuronal membrane, and is active metabolically, the brain is susceptible to oxidative stress. To keep the brain's structural integrity and overall functions, a tight balance between the antioxidant system and oxidative stress is essential

(Birben, 2012). Vitamins (Keservani et al., 2010a; Keservani et al., 2010b; Singh et al., 2017a) A, C, and E are significant non-enzymatic antioxidants found in foods, and new evidence suggests that they may protect against mental disorders such as anxiety disorders, attention deficit hyperactivity disorder, autism, bipolar disorder, depression, schizophrenia, and substance abuse (Ng, 2008; Brown, 2014). Antioxidant vitamin levels in the blood are shown to be low in people suffering from a variety of mental illnesses. In the prenatal determinants of schizophrenia study, perinatal retinol deficiency, as measured by low serum retinol concentrations, was found to be strongly related with an elevated risk (more than threefold) of schizophrenia and other schizophrenia spectrum disorders (Bao, 2012). In the Rotterdam study, people who consumed a high tertile of vitamins C and E had a reduced risk of Alzheimer's disease than those who consumed lower tertiles of these antioxidant vitamins (Engelhart, 2002). Low levels of vitamins C and E in blood and/or cerebrospinal fluid have been seen in AD patients, and amyloid-beta accumulation in the brain linked to increase in oxidative stress is one of the key causes of AD (Butterfield, 2002; Lopes da Silva, 2014). Vitamin E therapy lowers amyloid beta deposition, reactive oxygen species, and nitric oxide generation, as well as cognitive impairment and the course of Alzheimer's disease (Yang, 2010). A novel technique has recently been utilized to determine the underlying processes of neurodegenerative illnesses and to assess the efficacy of interventions to treat the symptoms of mental disorders. The use of a redox proteomics technique allows for the identification of illness stage-specific alterations in oxidative stress-related molecules as well as the demonstration of a cluster of protein changes (Singh et al., 2017b; Keservani et al., 2015) oxidative modification as a result of a specific dietary (Keservani et al., 2020) intervention. In general, existing intervention studies demonstrate that antioxidants improve general symptoms of mental illnesses, and it is necessary to research vitamin combinations, the best antioxidant, and the recommended length of antioxidant vitamin consumption.

Challenges to researching nutrition-mental scientific studies:

1. Subject recruitment, the collecting of accurate and trustworthy data on memory-based surveys from subjects, the identification of the cause-effect relation, and the acquisition of intervention compliance are all challenges.
2. Intervention responses are complicated and varied because of diversity.
3. The study necessitates the participation of highly qualified professionals.

### 8.3.4 MAGNESIUM

Magnesium is essential for immune cell differentiation, proliferation, function, and mobility, as well as for balanced immune cell activities. Several research has examined serum magnesium levels in patients with various mental disorders; nevertheless, a pathophysiological connection between psychiatric conditions and magnesium in patients has yet to be established. Magnesium has been shown to decrease neuroinflammation in rodent models of illness in several investigations. Because of magnesium's well-known role in several basic mechanisms of depressed pathophysiology, including oxidative stress and inflammation, the vast majority of research conducted thus far in patients has focused on depression. Magnesium, for example, has been claimed to treat depression by inhibiting the NMDA receptor, whose malfunction is a key cause of depression pathophysiology. To determine the involvement of Mg in the pathogenesis or therapy of bipolar illness and OCD, more research is needed.

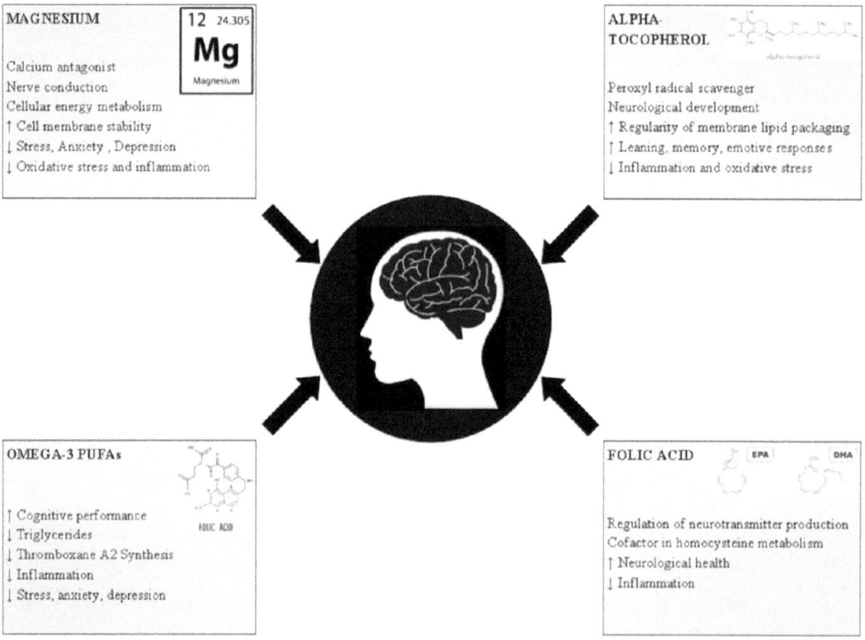


**FIGURE 8.1** Summary of the main molecular and physiological effects exerted by omega-3 PUFAs, alpha-tocopherol, magnesium, and folic acid in mental health and well-being. (↓) indicates downregulation; (↑) indicates upregulation. PUFAs, polyunsaturated fatty acids; SPMs, specialized pro-resolving mediators.

*Source*: Modified from Muscaritoli et al. (2021).

## 8.4 FACTORS AFFECTING INFLAMMATORY BOWEL DISEASE

Many foods have been involved in inflammatory bowel disease aetiology ideas. While evidence has accumulated that nutritional factors as part of overall lifestyle alterations may play a role in the rising incidence, no specific dietary recommendations can currently be made to reduce the risk of developing Crohn's disease or ulcerative colitis, with the exception of promoting breastfeeding. However, enteral feeding with a semi-elemental diet appears to be equally beneficial as corticosteroids in generating and sustaining remission in children and adolescents with Crohn's disease.

### *8.4.1 HISTORY*

Several foods have been identified as a potential danger for the growth of inflammatory bowel disease. In 2006, epidemiology research looked at the growth of Crohn's disease in Japan and defined a link between higher dietary intake of n-6 polyunsaturated fatty acids and animal protein, as well as a shift toward a more westernized diet, and an increased risk of the disease.

**TABLE 8.1** Nutritional Risk Factors for Selected Cancers.

| Site for Cancer | High risk factors | Low risk factors |
|---|---|---|
| **Stomach** | Salty food | Milk and dairy products |
| | Salted or dried fish | Raw vegetables |
| | Large amount of rice | Fruit |
| | Hot drinks and food | |
| | Irregular meals | |
| **Colorectum** | High-fat diet<br>Low-fibre diet | Fibre-rich diet (grains and pulses) |
| | Beer (rectal cancer)<br>Cholesterol | Protein-rich diet (cheese, meat) |
| **Oesophagus** | Hot drinks and food<br>Diet poor in protein, Vitamins and minerals | Diet rich in good protein, vitamins and minerals |
| **Breast** | High fat/calorie diet | - |
| **Lung** | Cholesterol | Green-yellow vegetables carotene, vitamin A |

*Source*: Adapted from Patel et al. (2018).

### 8.4.2 INFLAMMATORY BOWEL DISEASE'S EFFECT ON NUTRITION STATUS

Growth failure, delayed puberty, anaemia, osteopenia, and osteoporosis are all symptoms of inflammatory bowel disease in children. Chronic inflammation, proinflammatory cytokines such as TNF- and IL-1, hormonal abnormalities, reduced IGF-1, and exogenous steroids all contribute to decreased linear growth. (1) Weight loss was seen in 70% of children with Crohn's disease and just 34% of those with ulcerative colitis in a 12-year population study of paediatric inflammatory bowel illness in France. Short height was found solely in Crohn's disease patients in research of 739 new IBD cases in children in the United Kingdom and Ireland (Sawczenko, 2003). Iron, folate, and vitamin B12 deficiency are common micronutrient deficiencies, although vitamin E, vitamin A, beta-carotene, magnesium, selenium, and zinc may also be deficient. In an intervention trial involving 154 Crohn's disease patients with low bone mineral density, the beneficial effects of vitamin D, and calcium supplementation on bone health were revealed.

### 8.4.3 NUTRITIONAL THERAPY IN TREATMENT OF CROHN'S DISEASE (PAEDIATRICS)

Precursors of thromboxane A3 (a vasoconstrictor and a weak platelet aggregator), prostacyclin PGI3 (which leads to a rise in total prostacyclins, active vasodilators, and inhibitors of platelet aggregation), and leukotriene B5 are

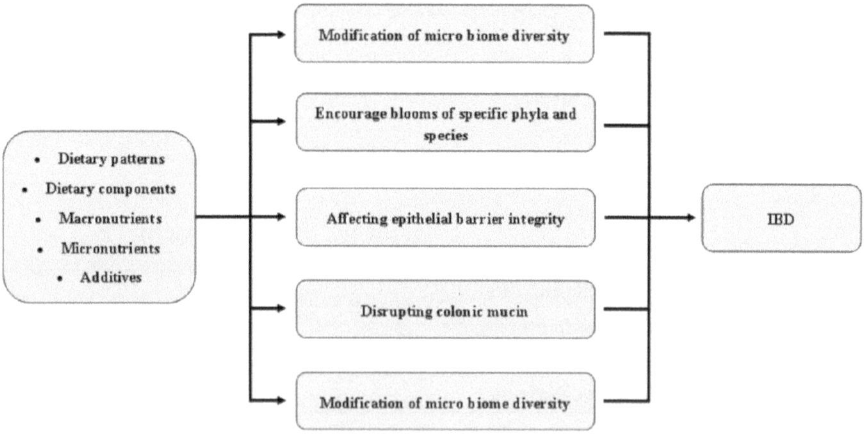


**FIGURE 8.2** Potential mechanisms of influence of dietary components on the pathogenesis of inflammatory bowel diseases (IBDs).

*Source*: Adapted from Sasson et al. (2021).

eicosapentaenoic acid and docosahexaenoic acid (a chemotactic agent and a weak inducer of inflammation). Proinflammatory cytokines derived from n-6 polyunsaturated fatty acids, such as prostaglandin E2 metabolites (potent vasoconstrictors), thromboxane A2 (a vasoconstrictor and a potent platelet aggregator that participates in the inflammatory process) by increasing vascular permeability, and others, are reduced in response to a diet rich in fish oil. Butyrate is created by bacteria fermenting primarily undigested food carbohydrates, and it appears to have anti-inflammatory properties in the colon.

## 8.5 ROLE OF NUTRITIONAL FACTORS AFFECTING ON CANCER

### *8.5.1 CANCER OF ORAL CAVITY AND PHARYNX*

The inclusion of one serving of vegetables and fruit per day appeared to lower the risk of oral cancer by around 50%, with fruit and vegetables providing protection mostly due to selected micronutrients, such as $\beta$ -carotene and different other carotenoids (La Vecchia et al., 1991; Zheng et al., 1993). Oral cancer's nutritional epidemiology is also characterized by two risk variables that appear to be significantly stronger than nutrition: using cigarettes and drinking alcohol (Marshall et al., 1992). Overall, a diet high in fruits and vegetables decreases the risk of oral cancer, but consumption of extremely hot drinks and foods, which is common in some cultures, likely raises the risk of oral cavity and pharyngeal malignancies. Identified a relationship between meat, saturated fats, cholesterol consumption, and the risk of mouth cancer, as well as a correlation between retinol, a meat intake indication, and the risk of oral cancer in research done in Northern Italy (La Vecchia et al., 1991).

### *8.5.2 CANCER OF GASTROINTESTINAL*

Malignancies of the GI tract and accessory organs of digestion, such as the oesophagus, stomach, pancreas, small intestine, rectum, and anus, are referred to as gastrointestinal (GI) cancer. Oesophageal cancer is one of the very common malignancies of the upper digestive system, and it is more common in developing and poor nations. There is a lot of research that suggests that eating raw fruits and vegetables, especially citrus fruits, can reduce the oesophageal cancer risk (Ziegler et al., 1981). The cooking method might have a role in oesophageal cancer development. Several studies have shown that vitamin A plays a key role in the defence of damaged epithelial cells against carcinogens, and that oesophageal epithelial cells are particularly sensitive to vitamin A shortage. According to a study done in Spain, an increased intake of salt and smoked and also pickled foods may be linked to an increased risk of gastric cancer, and this association may result in the creation of nitrosamines in the stomach (Ramón et al., 1993). Salt is not a direct carcinogen, but it is hypothesized to enhance the risk of gastric cancer by causing gastritis, increased deoxyribonucleic acid (DNA) production, and cell proliferation through direct injury to the stomach mucosa (World Cancer Research Fund, 1997). Dietary nutrients that prevent DNA damage or mutations by lowering oxidative stress and inflammation, such as -carotene and total carotenoids, high serum -tocopherol concentration, or vitamin C, have been related to a lower risk of pancreatic cancer (Stolzenberg-Solomon et al., 2009). In the Western world, colorectal cancer is the biggest cause of mortality. It has a complex cause that includes both hereditary and environmental factors. Certain lifestyle variables, such as diet, have been linked to an increased risk of cancer (Newmark et al., 1984).

### *8.5.3 SKIN CANCER AND DIET*

The most frequent type of cancer in the United States is skin cancer, affecting roughly one in every five people (Skin Cancer Facts, 2014). Skin cancer, comprising basal cell carcinoma (BCC), squamous cell carcinoma (SCC), and cancer of the melanocytes, has been diagnosed more than all other forms of cancer combined. There are various established risk factors for skin cancer, one of which is nutritional. Nutritional therapies may also help persons who are at high danger of skin cancer, according to the evidence. Researchers identified that persons who ate more vegetables and fruits had a 54% lower risk of SCC (Ibiebele et al., 2007). Consumption of green leafy vegetables was linked to a lower risk. High consumption of vitamins A, C, D, - and -carotene, cryptoxanthin, lutein, and lycopene has been linked to a decreased incidence of melanoma than reduced intake (Millen et al., 2004). Researchers discovered that the photoprotective effect exhibited in the combination of carotenoids and -tocopherol significantly reduced UV light–induced erythema (Stahl et al., 2000). A number of epidemiological studies have linked excess consumption of red meats like beef, hog, and lamb, as well as processed meats like bacon, sausage, and hot dogs, to an increased risk of cancer and all-cause death.

### 8.5.4 BREAST CANCER

Hormone receptor status and menopausal state can affect the biological and clinical characteristics of breast tumors. Dietary variables have long been assumed to have a vital role in the development of breast cancer; it stands out among the modifiable risk factors, and as a result, it has been explored in numerous researches, primarily focusing on individual nutrients or components. In a group of premenopausal women, red meat consumption was connected to an increased incidence of hormone receptor–positive breast cancer but not of hormone receptor–negative cancer.

### 8.5.5 CERVICAL CANCER

The second most common cancer is cervical cancer in women between the ages of 15 and 44 worldwide, and the most common in developing nations (Parkin et al., 2001). Human papillomaviruses (HPV) are responsible for up to 83% of all cervical cancer cases, and they are acknowledged as a necessary but insufficient cause of cervical cancer. Endogenous and exogenous co-factors may be influenced by the risk of cervical cancer when HPV is present (Bosch et al., 2002).

## 8.6 ROLE OF DIETARY FACTORS IN THE PATHOGENESIS OF ASTHMA AND CHRONIC OBSTRUCTIVE PULMONARY DISEASE

Decreased pulmonary function and respiratory blockage markers are substantial predictors of death (Schünemann et al., 2001), and thus a relationship between food and disease pathogenesis would have major public health consequences. Inflammation of the airways is a feature of both asthma and COPD. Asthma is a long-term inflammatory condition that affects a variety of cells and cell membranes. Reversible airway constriction is a feature of this condition. It is the most frequent chronic condition in children, with an increasing frequency (Mutius et al., 2000). COPD is a term used to describe a group of diseases that affect the lungs, including emphysema and chronic bronchitis. The presence of airflow restriction and a slow, progressive reduction in forced expiratory volume in one sec (FEV1) (Schünemann et al., 2001) explain it.

COPD is a prominent source of illness and mortality around the world. It is estimated to become the fifth main cause of global mortality and disability by 2020 (Anto et al., 2001). Several dietary variables have been linked to the development of asthma and COPD, owing to their possible influence on inflammatory responses, airway smooth muscle activity, and enzymatic reactions that impact neuromuscular signaling (Romieu et al., 1998).

### 8.6.1 ANTIOXIDANT VITAMINS

The interaction of oxidizing free radicals in asthma and COPD is assumed to be considerable. The balance between the detrimental effects of oxidant exposure and antioxidant defense has been hypothesized to be significant in the etiology of the two diseases, and there is a theory that dietary antioxidant deficiencies increase asthma and COPD sensitivity (Smit et al., 2001). Vitamins C and E are two of the most crucial vitamins for lung antioxidant protection. The most prevalent antioxidant in the ECF lining of the lungs is vitamin C, a water-soluble vitamin. Through numerous processes, including free-radical scavenging, it contributes to antioxidant activity. It also aids in the regeneration of vitamin E and aids immunological function (Romieu et al., 1998). The body's main defense against oxidant-induced membrane harm is vitamin E, a lipid-soluble vitamin (Romieu et al., 1998). It may also influence other activities in the cell by affecting membrane integrity and penetration, as well as the production of inflammatory mediators (Fogarty et al., 2000). Vitamin A, retinol, and carotenoids may also help to protect the lungs from free radicals.

### 8.6.2 MINERALS

Several minerals have been hypothesized to play protective functions in asthma and COPD. Selenium serves as a cofactor for the antioxidant enzyme glutathione peroxidase and has overlapping functions with vitamin E in antioxidant defense (Romieu et al., 1998). Magnesium is vital in bronchodilation because it has an anti-inflammatory impact and a massive impact on the electromotive force between cell membranes (Greene et al., 1999). Sodium has an important part in neuron and muscle function. It has a key role in the aetiology of hypertension, another disease characterized by smooth muscle dysfunction, and hence has a biologically relevant role in asthma (Fogarty et al., 2000).

### *8.6.3 FATTY ACIDS*

The intake of omega-6 and omega-3 fatty acids has been linked to the onset of airway inflammation. Through the synthesis of inflammatory mediators, changing dietary fatty acids can influence inflammatory activity. Inflammatory mediator synthesis is competitively inhibited by omega-6 and omega-3 fatty acids. There are fewer inflammatory mediators produced when omega-3 fatty acids are the enzymes for this activity than when omega-6 FAs are the enzymes (Schwartz et al., 2000). Since the 1960s, the amount of omega-6 fatty acids consumed in the dietetic pattern has increased dramatically, which may have resulted in an increased omega-6 fatty acid used as substrate and the synthesis of inflammatory cytokines (Calder et al., 2002).

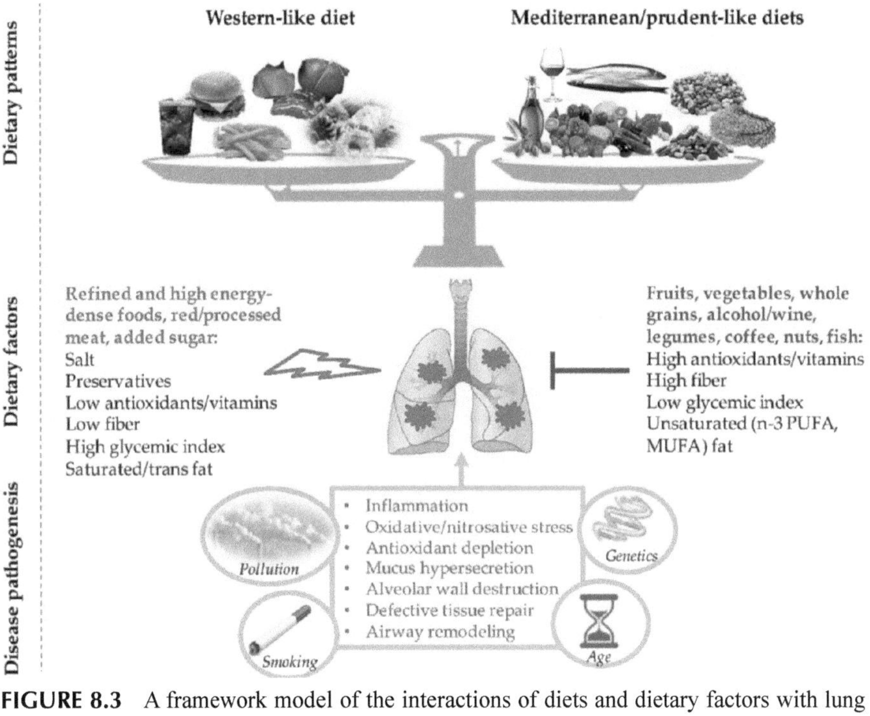


**FIGURE 8.3** A framework model of the interactions of diets and dietary factors with lung function and COPD development and progression

*Source*: Adapted from Scoditti et al. (2019).

## 8.7 ROLE OF DIETARY VARIABLES IN THE DEVELOPMENT OF CHRONIC KIDNEY DISEASE

It appears that once a crucial point of renal functional degradation is reached in illnesses of the kidney defined by irreversible injury, the development of end-stage disease happens invariably, even if the starting event or condition is cured or eliminated. Most kinds of primary renal diseases in humans are thought to be caused by immunopathogenesis processes. Although immunologic activities and the agents of glomerular harm they did may be responsible for the occurrence of most glomerular disorders, clinical and experimental evidence suggests that the rate at which these diseases advance is affected by several nonimmunologic factors (Ambard, 1920).

The cause of progressive loss of renal function in some diseases, such as cell proliferation of glomerulonephritis, is very likely the preliminary underlying process; however, the cause of progressive renal dysfunction in other diseases, such as interstitial scarring and glomerular sclerosis, is less noticeable. Recent research reveals that dietary changes, such as protein and/or phosphate consumption decreases, can slow or stop the course of chronic renal insufficiency in experimental animal models. As a result, the exciting prospect arises that dietary alterations could modify the natural history of several renal illnesses (Ambard, 1920). Protein intake has been shown to exacerbate the clinical status of people with renal dysfunction since the turn of the century. Uremic individuals are frequently depleted and perform badly, especially when they consume animal products. The medical background prompted efforts to reduce protein consumption in the diet to reduce the "neurotoxicity of uremia." "In situations of end-stage renal disease, a decrease in nitrogen intake to levels of grams per day avoids the increase in urine concentration for extended periods of time; markedly greater urea, indican, and xanthoprotein values decline, and first indicators of uremic intoxication vanish," Volhard observed. These studies, and also a slew of others thereafter, have demonstrated that lowering protein intake alleviates many symptoms of uremic syndrome (Ambard, 1920).

## 8.8 ROLE OF NUTRITIONAL FACTORS IN THE PATHOGENESIS OF TYPE 1 DIABETES

Type 1 diabetes (T1D) is caused by insulin depletion produced by autoimmune-mediated death of insulin-producing pancreatic islet beta cells (insulitis) in genetically vulnerable individuals. It is caused by poorly understood interactions between beta cells, the immune system, and the environment.

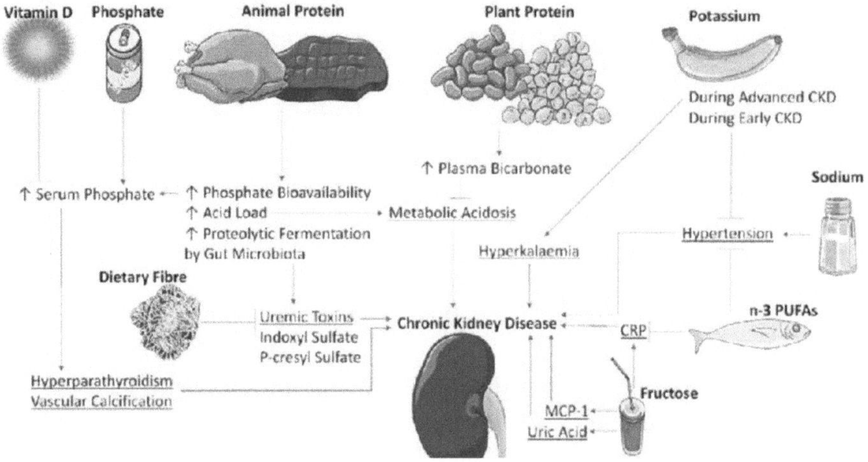


**FIGURE 8.4** Mechanisms of dietary factors impact on chronic kidney disease.
CRP, C-reactive protein; MCP-1, monocyte chemoattractant protein-1; N-3 PUFAs, omega 3 polyunsaturated fatty acids.

*Source*: Adapted from Snelson et al. (2017).

T1D concordance between monozygotic twin pairs has been shown to be around 30%, implying that environmental factors have a substantial role in the pathophysiology and diagnostic course of T1D. Viruses, pollutants, and dietary elements are examples of environmental influences. Fulminant T1D, which accounts for 15–20% of T1D cases in Japan with ketosis or ketoacidosis start, is linked to symptoms of the colds, such as fever and coughing, and pathogen has been proposed as the cause. A variety of viruses have indeed been proposed as potential environmental contributors to T1D. It has been observed that the Th1 type immune response against coxackie B4 virus, that is, interferon-gamma production and Tbet gene expression, is reduced. T-bet is a type-1 transcription factor that controls the growth of type-1 T cells and antitumor immunity. Several nutrients, including cow's milk protein, activated vitamin D, vitamin C, and polyunsaturated fatty acids, have been proposed as environmental factors that affect or are involved in the aetiology of T1D, which has lately increased in occurrence. Breastfeeding has been shown to lessen the risk of T1D in humans, while other studies have found no significant difference in breastfeeding rates between T1D and control children. Predicated on its molecular homology with pancreatic autoantigens, bovine serum albumin has been proposed as the environmental element affecting the possible relation between cow's milk consumption and T1D presence to induce autoimmunity against islet cells; however, the causative relationship remains controversial. Vitamin C is thought to protect islet cells from free radical damage. In foods, multiple chemical compounds such as oxidizers, antioxidants, and synthetic coloring agents are present, and they may be cytotoxic to islet cells individually or combined, or they may predispose islet cells to be significantly injured by other factors (Kagohashi and Otani, 2014).

## 8.9 CONCLUSION

Polymorphisms in the DNA that affect protein production or the activity of these enzymes can influence the risk of cancer. What you consume, as well as how you live, can all contribute to slow the progression of cancer. So, this description explains about role of Nutritional Factors in Pathogenesis of different Diseases. Many foods have been involved in inflammatory bowel disease aetiology ideas. While evidence has accumulated that nutritional factors as part of overall lifestyle alterations may play a key role in the rising incidence, no specific dietary recommendations can currently be made to reduce the risk of developing Crohn's disease or ulcerative colitis, with the exception of promoting breastfeeding.

## ACKNOWLEDGMENT

We thank the Principal Dr. Sanjay J. Kshirsagar Sir, MET's Institute of Pharmacy, Bhujbal Knowledge City, Adgaon, Nashik, for their constant support and providing all facilities to complete this work. Also, many thanks to all the known and unknowns who helped us right from the documentation till the publishing of this chapter.

## KEYWORDS

- **alpha-tocopherol**
- **antioxidants**
- **cancer**
- **COPD**
- **diabetes**
- **pathogenesis**
- **PUFAs**

## REFERENCES


Ahire, E. D.; Surana, K. R.; Patil, C. D.; Shah, H. S.; Sonwane, G. B.; Talele, S. G. Role of Omega-3 Fatty Acids in Different Neurodegenerative Disorders. In *Applied Pharmaceutical Science and Microbiology: Novel Green Chemistry Methods and Natural Products*, 2020; p 173.

Ambard, L. In *Physiologie normale et pathologique des reins*; Masson, 1920.

Anto, J. M.; Vermeire, P.; Vestbo, J.; Sunyer, J. Epidemiology of Chronic Obstructive Pulmonary Disease. *Eur. Respir. J.* **2001,** *17* (5), 982–994.

Atkinson, M. A.; Maclaren, N. K. The Pathogenesis of Insulin-Dependent Diabetes Mellitus. *N. Engl. J. Med.*,**1994,** *331* (21), 1428–1436.

Bao, Y.; Ibram, G.; Blaner, W. S.; Quesenberry, C. P.; Shen, L.; McKeague, I. W.; Schaefer, C. A.; Susser, E. S.; Brown, A. S. Low Maternal Retinol as a Risk Factor for Schizophrenia in Adult Offspring. *Schizophr. Res.* **2012,** *137*, 159–165.

Berlyne, G. M.; Shaw, A. B.; Nilwarangkur, S. Dietary Treatment of Chronic Renal Failure. *Nephron* **1965,** *2* (3), 129–147.

Birben, E.; Sahiner, U. M.; Sackesen, C.; Erzurum, S.; Kalayci, O. Oxidative Stress and Antioxidant Defense. *World Allergy Organ J.* **2012,** –, 9–19.

Bosch, F. X.; Lorincz, A.; Muñoz, N.; Meijer, C. J.; Shah, K. V. The Causal Relation between Human Papillomavirus and Cervical Cancer. *J. Clin. Pathol.* **2002,** *55*, 244–265.

Brown, H. E.; Roffman, J. L. Vitamin Supplementation in the Treatment of Schizophrenia. *CNS Drugs* **2014,** *28*, 611–622.

Butterfield, D.; Castegna, A.; Pocernich, C.; Drake, J.; Scapagnini, G.; Calabrese, V. Nutritional Approaches to Combat Oxidative Stress in Alzheimer's Disease. *J. Nutr. Biochem.* **2002,** *13*, 444–461.

Calder, P. C.; Yaqoob, P.; Thies, F.; Wallace, F. A.; Miles, E. A. Fatty Acids and Lymphocyte Functions. *Br. J. Nutr.* **2002,** *87* (S1), S31–S48.

Ceriello, A.; Novials, A.; Ortega, E.; Canivell, S.; La Sala, L.; Pujadas, G.; Bucciarelli, L.; Rondinelli, M.; Genovese, S. Vitamin C Further Improves the Protective Effect of Glucagon-like Peptide-1 on Acute Hypoglycemia-induced Oxidative Stress, Inflammation, and Endothelial Dysfunction in Type 1 Diabetes. *Diabetes Care* **2013,** *36* (12), 4104–4108.

Chi, Y.; Sauve, A. A. Nicotinamide Riboside, A Trace Nutrient in Foods, is a Vitamin B3 with Effects on Energy Metabolism and Neuroprotection. *Curr. Opin. Clin. Nutr. Metab. Care* **2013,** *16*, 657–661.

Clarke, R.; Bennett, D.; Parish, S.; Lewington, S.; Skeaff, M.; Eussen, S. J.; Lewerin, C.; Stott, D. J.; Armitage, J.; Hankey, G. J., Lonn, E.; Spence, J. D.; Galan, P.; de Groot, L. C.; Halsey, J.; Dangour, A. D.; Collins, R. Grodstein FB-Vitamin Treatment Trialists' Collaboration. Effects of Homocysteine Lowering with B Vitamins on Cognitive Aging: Meta-analysis of 11 Trials with Cognitive Data on 22,000 Individuals. *Am. J. Clin. Nutr.* **2014,** *100*, 657–666.

Diotallevi, F.; Campanati, A.; Martina, E.; Radi, G.; Paolinelli, M.; Marani, A.; Molinelli, E.; Candelora, M.; Taus, M.; Galeazzi, T.; Nicolai, A.; Offidani, A. The Role of Nutrition in Immune-Mediated, Inflammatory Skin Disease: A Narrative Review. *Nutrients* **2022,** *14* (3), 591. https://doi.org/10.3390/nu14030591.

Eisenbarth, G. S. Type I Diabetes Mellitus. *N Engl. J. Med.* **1986,** *314* (21), 1360–1368.

Engelhart, M. J.; Geerlings, M. I.; Ruitenberg, A.; van Swieten, J. C.; Hofman, A.; Witteman, J. C.; Breteler, M. M. Dietary Intake of Antioxidants and Risk of Alzheimer Disease. *JAMA* **2002,** *287*, 3223–3229.

Fogarty, A. The Role of Diet in the Aetiology of Asthma. *Clin. Exper. Allergy* **2000,** *30*, 615.
Ford, A. H.; Almeida, O. P. Effect of Homocysteine Lowering Treatment on Cognitive Function: A Systematic Review and Meta-analysis of Randomized Controlled Trials. *J. Alzheimers Dis.* **2012,** *29*, 133–149.
Giordano, C.; Esposito, R.; De Pascale, C.; De Santo, N. G. Dietary Treatment in Renal Failure1. In *Clinical Nephrology*; Karger Publishers, 1967; vol 3, pp 214–229.
Giovannetti, S.; Maggiore, Q. A Low-Nitrogen Diet with Proteins of High Biological Value for Severe Chronic Uraemia. *Lancet* **1964,** *1*, 1000–1003.
Gong, B.; Pan, Y.; Vempati, P.; Zhao, W.; Knable, L.; Ho, L.; Wang, J.; Sastre, M.; Ono, K.; Sauve, A. A.; Pasinetti, G. M. Nicotinamide Riboside Restores Cognition Through an Upregulation of Proliferator-Activated Receptor-γ Coactivator 1α Regulated β-secretase 1 Degradation and Mitochondrial Gene Expression in Alzheimer's Mouse Models. *Neurobiol. Aging* **2013,** *34*, 1581–1588.
Greene, L. S. Asthma, Oxidant Stress, and Diet. *Nutrition* 1999, *15* (11-12), 899–907.
Haag, M. Essential Fatty Acids and the Brain. *Can. J. Psychiat.* **2003,** *48*, 195–203.
Haan, M. N.; Miller, J. W.; Aiello, A. E.; Whitmer, R. A.; Jagust, W. J.; Mungas, D. M.; Allen, L. H.; Green, R. Homocysteine, B Vitamins, and the Incidence of Dementia and Cognitive Impairment: Results from the Sacramento Area Latino Study on Aging. *Am. J. Clin. Nutr.* **2007,** *85*, 511–517.
Hanafusa, T.; Imagawa, A. Fulminant Type 1 Diabetes: A Novel Clinical Entity Requiring Special Attention by all Medical Practitioners. *Nat. Clin. Pract. Endocrinol. Metab.* **2007,** *3* (1), 36–45.
Ibiebele, T. I.; van der Pols, J. C.; Hughes, M. C.; Marks, G. C.; Williams, G. M.; Green, A. C. Dietary Pattern in Association with Squamous Cell Carcinoma of the Skin: A Prospective Study. *Am. J. Clin. Nutr.* **2007,** *85*, 1401–1408.
Jomova, K.; Vondrakova, D.; Lawson, M.; Valko, M. Metals, Oxidative Stress and Neurodegenerative Disorders. *Mol. Cell Biochem.* **2010,** 345, 91–104.
Kagohashi, Y.; Otani, H. Role of Nutritional Factors at the Early Life Stages in the Pathogenesis and Clinical Course of Type 1 Diabetes. *BioMed. Res. Int.* **2015,** *2015*.
Kang, H. J.; Hong, J. W.; Han, J. W.; Yang, S. J.; Kim, S. W.; Shin, I. S.; Kim, K. W.; Yoon, J. S.; Kim, J. M. Nutritional Biomaker in Alzheimer Disease. *J. Korean Soc. Biol. Ther. Psychiatry* **2014,** *20*, 187–200.
Keservani, R. K.; Kesharwani, R. K.; Sharma, A. K.; Vyas, N.; Chadoker, A. Nutritional Supplements: An Overview. *Int. J. Curr. Pharm. Rev. Res.* **2010b,** *1* (1), 59–75.
Keservani, R. K.; Kesharwani, R. K.; Vyas, N.; Jain, S.; Raghuvanshi, R.; Sharma, A. K. Nutraceutical and Functional Food as Future Food: A Review. *Der Pharmacia Lettre* **2010a,** *2* (1), 106–116.
Keservani, R. K.; Sharma, A. K.; Jarouliya, U. Protein and Peptide in Drug Targeting and its Therapeutic Approach. *Ars Pharm.* **2015,** *56* (3), 165–177.
Keservani, R. K.; Sharma, A. K.; Kesharwani, R. K. An Overview and Therapeutic Applications of Nutraceutical and Functional Foods. In *Recent Advances in Drug Delivery Technology*, 2017; pp 160–201, ISBN: 9781522507543.
Keservani, R. K.; Sharma, A. K.; Kesharwani, R. K. Eds.; *Nutraceuticals and Dietary Supplements: Applications in Health Improvement and Disease Management*; CRC Press, 2020; ISBN: 9781771888738.
Kopple, J. D.; Coburn, J. W. Metabolic Studies of Low Protein Diets in Uremia: i. Nlteogen and Potassium. *Medicine* **1973,** *52* (6), 583–595.

Lakhan, S. E.; Vieira, K. F. Nutritional Therapies for Mental Disorders. *Nutr. J.* **2008,** *7*, 2.

Lauritzen, L.; Hansen, H. S.; Jørgensen, M. H.; Michaelsen, K. F. The Essentiality of Long Chain n-3 Fatty Acids in Relation to Development and Function of the Brain and Retina. *Prog. Lipid Res.* **2001,** *40*, 1–94.

La Vecchia, C.; Negri, E.; D'Avanzo, B.; Boyle, P.; Franceschi, S. Dietary Indicators of Oral and Pharyngeal Cancer. *Int. J. Epidemiol.* **1991,** *20*, 39–44.

Lim, S. Y.; Kim, E. J.; Kim, A.; Lee, H. J.; Choi, H. J.; Yang, S. J. Nutritional Factors Affecting Mental Health. *Clin. Nutr. Res.* **2016,** *5* (3), 143–152. DOI: 10.7762/cnr.2016.5.3.143

Lopes da Silva, S.; Vellas, B.; Elemans, S.; Luchsinger, J.; Kamphuis, P.; Yaffe, K.; Sijben, J.; Groenendijk, M.; Stijnen, T. Plasma Nutrient Status of Patients with Alzheimer's Disease: Systematic Review and Meta-Analysis. *Alzheimers Dement.* **2014,** *10*, 485–502.

Mehers, K. L.; Gillespie, K. M. The Genetic Basis for Type 1 Diabetes. *Br. Med. Bull.* **2008,** *88* (1), 115–129.

Millen, A. E.; et al. Diet and Melanoma in a Case-Control Study. *Cancer Epidemiol. Biomarkers Prev* **2004,** *13*, 1042–1051.

Muscaritoli, M. The Impact of Nutrients on Mental Health and Well-being: Insights from the literature. *Front. Nutr.* **2021,** *8*, 97.

Newmark, H. L.; Wargovich, M. J.; Bruce, W. R. Colon Cancer and Dietary Fat, Phosphate, and Calcium: A Hypothesis. *J. Natl. Cancer Inst.* **1984,** *72*, 1323–1325.

Ng, F.; Berk, M.; Dean, O.; Bush, A. I. Oxidative Stress in Psychiatric Disorders: Evidence Base and Therapeutic Implications. *Int. J. Neuropsychopharmacol.* **2008,** *11*, 851–876.

Nokoff, N.; Rewers, M. Pathogenesis of Type 1 Diabetes: Lessons from Natural History Studies of High-Risk Individuals. *Ann. N York Acad. Sci.* **2013,** *1281* (1), 1–15.

Parkin, D. M. Global Cancer Statistics in the Year 2000. *Lancet Oncol.* **2001,** *2*, 533–543.

Parletta, N.; Milte, C. M.; Meyer, B. J. Nutritional Modulation of Cognitive Function and Mental Health. *J. Nutr. Biochem.* **2013,** *24*, 725–743.

Patel, A.; Pathak, Y.; Patel, J.; Sutariya, V. Role of Nutritional Factors in Pathogenesis of Cancer. Food Qual. Saf. **2018,** *2* (1), 27–36.

Pinto, R. Q.; Soares, I.; Carvalho-Correia, E.; Mesquita, A. R. Gene-environment Interactions in Psychopathology Throughout Early Childhood: A Systematic Review. *Psychiatr. Genet.* **2015,** *25*, 223–233.

Ram6n, J. M.; Serra, L.; Cerd6, C.; Oromí, J. Dietary Factors and Gastric Cancer Risk. A Case-Control Study in Spain. *Cancer* **1993,** *71*, 1731–1735.

Romieu, I.; Meneses, F.; Ramirez, M.; Ruiz, S.; Padilla, R. P.; Sienra, J. J.; Gerber, M.; Grievink, L.; Dekker, R.; Walda, I.; Brunekreef, B. Antioxidant Supplementation and Respiratory Functions Among Workers Exposed to High Levels of Ozone. *Am. J. Respir. Crit. Care Med.* **1998,** *158* (1), 226–232.

Sawczenko, A.; Sandhu, B. K. Presenting Features of Inflammatory Bowel Disease in Great Britain and Ireland. *Arch. Dis. Child* **2003,** *88*, 995–1000.

Schünemann, H. J.; Grant, B. J.; Freudenheim, J. L.; Muti, P.; Browne, R. W.; Drake, J. A.; Klocke, R. A.; Trevisan, M. The Relation of Serum Levels of Antioxidant Vitamins C and E, Retinol and Carotenoids with Pulmonary Function in the General Population. *Am. J. Respir. Crit. Care Med.* **2001,** *163* (5), 1246–1255.

Schwartz, J. Role of Polyunsaturated Fatty Acids in Lung Disease. *Am. J. Clin. Nutr.* **2000,** *71* (1), 393s–396s.

Shoda, R.; Matsueda, K.; Yamato, S.; Umeda, N. Epidemiologic Analysis of Crohn Disease in Japan: Increased Dietary Intake of n-6 Polyunsaturated Fatty Acids and Animal Protein

Relates to the Increased Incidence of Crohn Disease in Japan. *Am. J. Clin. Nutr.* **1996,** *63*, 741–745.

Sinclair, A. J.; Begg, D.; Mathai, M.; Weisinger, R. S. Omega 3 Fatty Acids and the Brain: Review of Studies in Depression. *Asia Pac. J. Clin. Nutr.* **2007,** *16* Suppl 1, 391–397.

Singh, P.; Kesharwani, R. K.; Keservani, R. K. Antioxidants and Vitamins: Roles in Cellular Function and Metabolism. In *Sustained Energy for Enhanced Human Functions and Activity*; Academic Press, 2017a; pp 385–407, ISBN: 9780128054130.

Singh, P.; Kesharwani, R. K.; Keservani, R. K. Protein, Carbohydrates, and Fats: Energy metabolism. In *Sustained Energy for Enhanced Human Functions and Activity*; Academic Press, 2017b, pp 103–115, ISBN: 9780128054130.

Skarsvik, S.; Puranen, J.; Honkanen, J.; Roivainen, M.; Ilonen, J.; Holmberg, H.; Ludvigsson, J.; Vaarala, O. Decreased in vitro Type 1 Immune Response Against Coxsackie Virus B4 in Children with Type 1 Diabetes. *Diabetes*, **2006,** *55* (4), 996–1003.

Skin Cancer Facts. Skin Cancer Foundation, 2014 [Online]. http://www.skincancer.org/ skin-cancer-information/skin-cancer-facts. (accessed Jan 3, 2017).

Smit, H. A. Chronic Obstructive Pulmonary Disease, Asthma and Protective Effects of Food Intake: From Hypothesis to Evidence?. *Respir. Res.* **2001,** *2* (5), 1–4.

Smith, A. D. The Worldwide Challenge of the Dementias: A Role for B Vitamins and Homocysteine? *Food Nutr. Bull.* **2008,** *29*, S143–S172.

Snelson, M.; Clarke, R. E.; Coughlan, M. T. Stirring the Pot: Can Dietary Modification Alleviate the Burden of CKD?. *Nutrients* **2017,** *9* (3), 265.

Spagnuolo, AI; Patti, A.; Sebastiani, G.; Nigi, L.; Dotta, F. The Case for Virus-Induced Type 1 Diabetes. *Curr. Opin. Endocrinol. Diabetes Obes.* **2013,** *20* (4), 292–298.

Stahl, W.; Heinrich, U.; Jungmann, H.; Sies, H.; Tronnier, H. Carotenoids and Carotenoids Plus Vitamin E Protect Against Ultraviolet Light-induced Erythema in Humans. *Am. J. Clin. Nutr.* **2000,** *71*, 795–798.

Stolzenberg-Solomon, R. Z.; et al. Vitamin E Intake, Alpha-tocopherol Status, and Pancreatic Cancer in a Cohort of Male Smokers. *Am. J. Clin. Nutr.* **2009,** *89*, 584–591.

Tisch, R.; McDevitt, H. Insulin-Dependent Diabetes Mellitus. *Cell* **1996,** *85* (3), 291–297.

Volhard, F. Die Behandlung der echten Uramie. In *Hanbach der Innersen Medizin*, 1977; p 806.

von Mutius, E. The Burden of Childhood Asthma. *Arch. Dis. Childhood* **2000,** *82* (suppl 2), ii2–ii5.

Wang, Y.; Huang, F. N-3 Polyunsaturated Fatty Acids and Inflammation in Obesity: Local Effect and Systemic Benefit. *Biomed. Res. Int.* **2015,** *2015*, 581469.

Wang, W.; Liang, B. Case Report of Mental Disorder Induced by Niacin Deficiency. *Shanghai Arch. Psychiatry* **2012,** *24*, 352–354.

World Cancer Research Fund in Association with the American Institute of Cancer Research. In *Food, Nutrition, and Cancer Prevention: A Global Prospective*; American Institute of Cancer Research, Washington, DC, 1997.

Yang, S. G.; Wang, W. Y.; Ling, T. J.; Feng, Y.; Du, X. T.; Zhang, X.; Sun, X. X.; Zhao, M.; Xue, D.; Yang, Y.; Liu, R. T. α-Tocopherol Quinone Inhibits β-amyloid Aggregation and Cytotoxicity, Disaggregates Preformed Fibrils and Decreases the Production of Reactive Oxygen Species, NO and Inflammatory Cytokines. *Neurochem. Int.* **2010,** *57*, 914–922.

Ziegler, R. G.; Morris, L. E.; Blot, W. J.; Pottern, L. M.; Hoover, R.; Fraumeni, J. F. Jr. Esophageal Cancer Among Black Men in Washington, D.C. II. Role of Nutrition. *J. Natl. Cancer Inst.* **1981,** *67*, 1199–1206.

Zúñiga, J.; Cancino, M.; Medina, F.; Varela, P.; Vargas, R.; Tapia, G.; Videla, L. A.; Fernández, V. N-3 PUFA Supplementation Triggers PPAR-α Activation and PPAR-α/NF-κB Interaction: Anti-inflammatory Implications in Liver Ischemia-reperfusion Injury. *PLoS One* **2011,** *6*, e28502.

## CHAPTER 9

# GENE-BASED DIETARY ADVICE AND EATING BEHAVIOR

SHUBHAM J. KHAIRNAR, RAHUL R. SABLE, SIMONA E. D'SOUZA, APURVA P. TORANE, and PAVAN B. UDAVANT

*Department of Pharmacology, MET's Institute of Pharmacy, Adgaon, Nashik, India*

### ABSTRACT

Nutrition is a modifiable important element that can influence human health by interacting with the genome as well as the epigenome. Specific genetic polymorphisms can influence how the body reacts to nutritional requirements and food material, and the diet can modulate gene expression as well. We've discussed regarding how proper nutrition is vital for staying healthy in this chapter. The chapter also emphasises which foods should be ingested and which should be avoided in order to avoid their contact with our genes, which might produce gene variation, which can further lead to numerous health problems. Furthermore, how can diversity in gene expression be passed down from generation to generation? The chapter also discusses how genetic variations can cause a variety of ailments and what dietary changes are required in such circumstances; how direct-to-consumer gene testing can be used to test gene variation; and how tailored nutrition uses genetic, phenotypic, clinical, dietary, and other data to deliver more precise and effective personalized healthy eating advice and to drive appropriate dietary adjustments in order to maintain and improve health. This chapter studies relationships between epigenetic variables and eating disorders (EDs) and involves investigation on how gene expression and epigenetic programming can explain for gene–environment interactions in EDs. Finally, a case study

Nutrigenomics and Nutraceuticals. Raj K. Keservani, Eknath D. Ahire, Shubham J. Khairnar, Sanjay J. Kshirsagar, & Rajesh K. Kesharwani (Eds.)

is explained on how gene expression profiles are different with different dietary patterns, which, in turn, can result in long-term illness.

## 9.1 INTRODUCTION

Human nutrition is a biochemically based science that is still relatively new. The scientific study of food and human requirements began less than 100 years ago, despite its roots in millennia of medical understanding. The field of human nutrition is constantly evolving. It helps to give a clearer picture of persons and their bioanalytical profiles based on geographic epidemiologic investigations, and it has attained integration with pharmacology. The knowledge and comprehension of the mechanisms causing specific diseases, as well as their association with dietary components, exploded in the latter part of the 20th century. This has created the groundwork for nutrition counseling and food education (Di Renzo et al., 2019). "Man is what he eats," wrote German philosopher Ludwig Feuerbach in 1848. Indeed, 160 years later, we can demonstrate that nutrition is a significant element influencing human quality of life. Dietary habits including high fat and sugar consumption, alcohol abuse, and a lack of vegetables and fruits are critical risk factors for noncommunicable diseases (NCDs), which are characterized by high blood cholesterol and glucose, hypertension, and obesity. As a result, a healthy eating pattern is advocated as a preventative measure against heart disease, cancer, diabetes, and other NCDs, which is the primary cause of a disease eventually leading to death (Franzago et al., 2020). Individual dietary recommendations are possible thanks to the application of new technology and the development of new fields of study such as nutrigenetics and nutrigenomics. Nutrigenomics is indeed the science of the relationships between nutrients and gene expression, as well as the discovery of systems that control how food ingredients affect human health. Nutrigenetics focuses on genetic differences in individual patients' genomes and employs analysis of these differences to develop dietary recommendations for individualized nutrition (Mierziak et al., 2021).

## 9.2 HEALTHY DIET

All forms of malnutrition, various types of NCDs, and illnesses can all be prevented with a well-balanced diet. However, as a result of greater processing of packaged foods, rising population, and living standards, dietary trends have evolved. People are consuming foods which are high in calories, fats, free sugars, and salt, and many aren't consuming enough fruit, vegetables, and other types of dietary fiber like whole grains. Individual characteristics (such as age, gender, lifestyle, and physical activity), ethnic values, regionally available foods, and dietary habits all have an impact on the composition of a varied, balanced, and nutritious diet. However, the core conceptions of what a healthy diet should constitute has evolved.

**A healthy diet's components and its advantages:**

- A nutritious (Keservani et al., 2020) diet is one in which macronutrients are consumed in proper amounts to support energetic and physiologic demands while also supplying sufficient micronutrients and fluids to fulfil the body's physiologic needs.
- Macronutrients (carbohydrates, proteins, and lipids) supply the energy required for everyday biological functions.
- Micronutrients (vitamins and minerals) are needed in minute amounts for appropriate growth, development, metabolism, and physiologic function (Skerrett et al., 2010).

**Two healthy eating patterns are represented by their constituent elements.**

1. Diet of the Mediterranean type
    - The large percentage of the food consumed on a daily basis consists of vegetables, fruits, grains, nuts, beans, and seeds.
    - Fat accounts for approximately 40% of total daily calories, most of it coming from olive oil.
    - Every day, a little piece of cheese or yoghurt is consumed, along with a meal of fish, chicken, or eggs.
    - Red meat is occasionally consumed.
    - With meals, little quantity of red wine is usually consumed.
2. DASH diet
    - 7–8 servings of grains and grain-related products, with greater than half of those being whole-grain items.
    - 4–5 servings fruits.
    - 4–5 servings vegetables.
    - 2–3 servings low-fat or fat-free dairy products.
    - 2 servings or less of lean meats, fish, or poultry.
    - 4–5 servings of legumes, nuts, and seeds per week.
    - 2–3 meals per day of added fats.
    - Sweets are in short supply.

**The following foods consists of a healthy diet:**

- Fruit, vegetables, nuts, legumes, seeds, and grains are considered to be the best sources of dietary fibre (e.g., unrefined maize, wheat, oats, millet, and brown rice) (Keservani et al., 2010a; Keservani et al., 2010b; Keservani et al., 2017).
- Each day minimum of about 400 g (five servings) of fruits and vegetables should be consumed whereas starchy roots like potatoes, sweet potatoes, and cassava should be avoided.
- Free sugars should account for less than 10% on an average of daily energy intake, which in itself is 50 g for a healthier weight, eating approximately 2000 calories every day. However, for potential health benefits, consumption of energy intake should be below 5%. Sugars that have been added by the manufacturer, chef, or customer to foods or beverages and also sugar that is found naturally in honey, fruit juices, syrups, and fruit juice concentrates, are all considered free sugars.
- Fats account for about 30% less of the total energy consumption. Saturated fats (mainly found in meat containing fat, palm, butter, cream, ghee, cheese, lard, and coconut oil) and trans-fats (mainly present in food that is baked and fried, as well as snacks and foods that are processed like frozen pizza, cookies, pies, wafers, biscuits, spreads, and cooking oil) are favorable to fats which are unsaturated (present in avocado, fish, and nuts, as well as sunflower, canola, soybean, and olive oils; in dairy foods and meat from ruminant animals, such as cows, sheep, goats, and camels). Saturated fats should account for about 10% less of total caloric intake, while trans fats should account for less than 1%.
- Not more than 5 g of salt each day (about one teaspoon). Iodized salt should be used.

## 9.3 DIET, HEREDITARY, AND GENE RELATION

In light of recent research on diet and gene expression control, the ancient adage that pregnant women eat for two may be more accurate than ever. Epidemiological and molecular research provides a more complicated and sophisticated picture of how a pregnant woman's dietary habits influence the expression of her own genes—and presumably those of her foetus—and hence her children's long-term health. Type 2 diabetes, obesity-related heart disease, insulin resistance, and hypertension are the disorders most closely linked to maternal diet during pregnancy, according to Karen Lillycrop, a perinatal nutrition expert at the University of Southampton in the United Kingdom. Furthermore, evidence shows that dietary impacts, which can modify gene expression and regulation, may have had a role in human evolution. According to a new study, the most genetic diversity exists among both chimpanzees and humans in genes that have control over metabolism and are strongly linked to nutrition (Somel et al., 2008).

Epidemiologists have accumulated a rapidly increasing body of evidence demonstrating heritable epigenetic changes induced by nutrition over the last decade. Several studies have found that overweight or diabetic mothers during pregnancy—which is influenced in part by diet—are more likely to give birth to children who are obese (Mayer-Davis et al., 2006). Other research has urgent public health implications since it helps to give an idea on some of the human health issues that occurred in the process of modern nutrition, particularly in industrialized countries. Although people in industrialized countries are living longer than they have in the past, modern living has also created plenty of issues, many of which, as is becoming obvious, involve epigenetic impacts. The most serious of them is a growing obesity epidemic—particularly among children—that is afflicting every country in the industrialized world and wreaking havoc on public health and health-care systems. Obviously, the genetic material of a population cannot change in one single generation, leaving two possible explanations for obesity caused during childhood: one of which is variations in diet, particularly the consumption of addictive junk foods which is quite high in calories; and the other being lack of exercise in children; or because of both reasons combined. However, this does not account for the entire scope of the obesity pandemic; it appears that the mother's diet and health have a significant role as well. Diet-induced epigenetic changes in animals are also said to be transgenerational, meaning that they can be passed down to offspring (Hunter et al., 2008).

## 9.4 DIET AFFECTING GENE EXPRESSION

The study of nutrient–gene interactions and their impact on health is known as nutritional genomics. It looks at how diet impacts gene expression, how genes affect the way nutrients are being used by the body, and how the processes which occur at molecular level in the body are affected by nutrients. A diet that supports one person's health and wellbeing may be detrimental to another's health and wellbeing, and the same diet may lead to obesity and subsequent metabolic disorders that undermine quality of life (Walker et al., 2004). Several cardiovascular risk variables, including obesity, serum lipids, blood pressure, serum lipids, and inflammatory indicators like C-reactive protein, are linked to dietary patterns (CRP). The risk of dying from heart disease and cancer is known to increase due to these reasons. Data from microarrays can be utilized to investigate variations in gene expression for thousands of genes at the same time. Studies of gene expression have found links to disorders including cancer and cardiovascular disease. Gene expression can be influenced by dietary patterns by a variety of methods. For example, PUFA with PPARs binding to transcription factors which modulates their activity. The distribution of macronutrients appears to influence gene expression regulation as well. Diet containing high saturated fats resulted in profile of proinflammatory gene expression than a diet which is high in monounsaturated fats. The Mediterranean diet has been linked to lower gene expression involved in response to inflammation (Bouchard-Mercier et al., 2013). Another research has found that a high-sugar diet regulates apoprotein gene expression, that nutrition regulates expression of gene in lipogenesis, and that PUFA suppresses fatty acid synthase transcription. According to reports, hormones and the nervous system were assumed to be the principal determinants of gene expression regulation in response to various changes in dietary environment until recently. However, prior research indicates that main (glucose, amino acids, fatty acids) and marginal (vitamin, iron) nutrients, or their metabolites, influence the expression of gene in a hormone-independent sort of way (Walker et al., 2004). Epigenetic changes are those that permits the genome to adapt to environmental factors like diet (Hunter et al., 2008).

## 9.5 EPIGENETICS

The term "epigenome" refers to a group of epigenetic mechanisms that control when (during a person's lifetime) and where (in the body) genes are expressed, chiefly through chemical alterations to DNA, histone proteins, and chromatin structure. DNA methylation (DNAm) is the most commonly studied epigenetic function in human research stress on early life and mental risk since it is more easily quantifiable and more stable than the other epigenetic processes (Cecil et al., 2020). DNAm includes adding methyl groups to

CpG (Cytosine and Guanine separated by only one phosphate group) sites. Certain CpGs can be methylated to silence or decrease gene expression, allowing for environmental programming of genetic expression in theory. The available evidence strongly shows that methylation of DNA is known to regulate gene expression sensitive to environmental exposures. However, there are certain limitations to these conclusions, both methodologically and conceptually (Steiger et al., 2020).

### 9.5.1 EATING BEHAVIOR

Traditional views of Anorexia nervosa, Bulimia nervosa, and other eating disorders have claimed that they are "sociocultural inventions" (Steiger et al., 2020). EDs are marked by obsessive thoughts about food, weight, and body image, as well as maladaptive eating habits such calorie restriction, binge eating, self-induced vomiting, and compulsive exercise. AN, BN, BED, RD, ARFID, and pica are the six subtypes of EDs recognized by the most recent edition of the *DSM*-5. Additional identified eating or feeding disorder and undefined eating or feeding disorder are residual categories that capture ED variants that have clinical relevance but do not meet full-threshold syndrome criteria. AN is characterized by calorie restriction and a prolonged avoidance of weight growth, culminating in a subnormal body weight. People who binge eat regularly and engage in behaviors to compensate the same are also observed in people with BN, usually without significant weight loss. BED patients, almost 40% of whom are obese, have recurring binge-eating episodes with no compensatory behaviors. ARFID is a new diagnosis in the *DSM*-5 that is defined by a restriction of food consumption for reasons other than body image and weight, such as compulsive avoidance of foods deemed impure or unhealthy, or foods that provoke disgust or aversion. Currently, neither genetic nor epigenetic research on RD, ARFID, or pica has been conducted (Steiger et al., 2020).

### 9.5.2 DNA METHYLATION SEEN IN PEOPLE WITH EATING DISORDERS

Studies have shown methylation changes in genes responsible for the production of alpha-synuclein (a protein involved in neurotransmitter release), dopamine (linked to mood, impulse control, reward sensitivity, and binge eating), oxytocin (related to social attachment) (Kim et al., 2014), histone deacetylase

**TABLE 9.1** Eating Disorder Types and Characteristics of Obesity.

| Disorder | Diagnostic characteristics |
|---|---|
| Anorexia nervosa (AN) | 1. Refusal to sustain weight of the body at or above the minimum normal weight for one's age and height.<br>2. Insecurity regarding gaining weight.<br>3. Misperception of one's own body weight or shape.<br>4. Amenorrhea being more common in postmenarchal women. |
| Restricting type | All of the above, plus no bingeing or purging. |
| Binge-eating/ purging type | The above, as well as binge eating or purging behavior on a routine basis. |
| Bulimia nervosa (BN) | 1. Binge eating episodes that occur on a regular basis.<br>2. Recurrent behavior with the goal of compensating bingeing in order to avoid weight gain.<br>3. For three months, characteristics 1 and 2 recur twice a week.<br>4. Misperception of one's own body weight or shape.<br>5. Point 4 does not only occur during anorexia nervosa episodes. |
| Purging type | All of the above, plus the person induced puking or misused purgatives, diuretics, or enemas during the episode of bulimia. |
| Nonpurging type | All of the above, plus the individual used methods other than purging to avoid excessive weight gain during the bulimia incident (e.g., fasting or excessive exercise). |
| Eating disorder not otherwise specified | 1. Except for characteristic 4, all AN criteria are met.<br>2. Except for the individual's normal weight, all AN criteria are met.<br>3. Except for the frequency of bingeing and corrective behavior occurring not more than twice a week over a 3-month period, all BN criteria's are met.<br>4. Individual of normal weight who exhibits compensatory behavior after consumption of small amounts of food.<br>5. Chewing, spitting out (large amounts of food) but not swallowing.<br>6. Binge eating disorder; recurrent binge eating incidents without BN compensatory behavior. |
| Obesity | 1. BMI: 30.0–34.9 class I<br>BMI: 35.0–39.9 class II<br>BMI: >40 class III<br>2. Women's waist circumference should not be more than 35 inches.<br>3. Waist circumference for men should not be more than 40 inches. |

*Source*: Adapted from de Krom et al. (2009).

(broadly influencing expression of gene), and the hormone leptin has been discovered in AN study (which inhibits hunger) (Neyazi et al., 2019). Using relevant information from resting-state functional connectivity and measurements of methylation of the SLC6A4 gene. By trying to combine measures of SLC6A4 gene methylation with resting-state functional connectivity data, Boehm et al., linked variation in epigenetics in the SLC6A4 gene with experiment revealed in the salience network—main brain circuit for mood regulation (Boehm et al., 2020). Another study discovered that people with BN had more methylation of the ANP gene promoter (which is associated with cardiovascular homeostasis) as compared to people who did not have the condition. Similarly, the methylation of potential genes that could be linked to bulimic ED variations, including as the receptor genes of glucocorticoid and dopamine DRD2, has been identified in previous investigations. However, available evidence suggests that the epigenetic alterations seen are more relevant to comorbidity (e.g., personality disorders or suicidality) in comparison to BN itself. There were no studies that looked at methylation differences among patients with BED and healthy eaters. According to one study, individuals with bipolar illness and binge eating showed hypomethylation seen in the SLC1A2 gene which is involved in clearing glutamate present in the synapse especially in comparison to patients suffering from bipolar disorder who do not binge eat (Jia et al., 2017; Steiger et al., 2020).

## 9.6 DISEASE, DIET, AND GENE RELATION

A single environmental exposure can have various impacts on the likelihood of acquiring a disease in people with different genotypes, due to gene–environment interaction. As a result, the diet exhibits feature of effect on general well-being as a long-term determinant. Obesity and related disorders are on the rise, having increased by more than 100-fold in the last three decades. Obesity is a sign of more than 40 illnesses or chromosomal disorders. According to the GWAS, DNA sequences and about 600 genes are related to human obesity, but with mixed results. Many genes and polymorphisms, such as TNF, (IL)-6, MTHFR, and FTO gene, have been implicated in the pathophysiology of overweight/obesity in recent decades. Personalized diets are recommended to study the relationships between genes and nutrients and to improve people's health (Di Renzo et al., 2019). Examples of some gene variations are listed below:

### 9.6.1 APOE

Apolipoprotein gene abbreviated as APOE has three major alleles: Apolipoprotein 2 (APOE2), Apolipoprotein3 (APOE3), and Apolipoprotein4 (APOE4). In all populations tested, APOE3 being the most common form (>60%). Hepatic binding, take-up, and chylomicron catabolism, chylomicron residues, VLDL, and high-density lipoprotein subtypes are mediated by the polymorphism, which has functional consequences on lipoprotein metabolism (Eichner et al., 2002). Defects in the APOE protein may reduce its capacity to bind to receptors, resulting in an increase in blood cholesterol levels (Zhang et al., 2015). The prevention of cardiovascular disease necessitates maintaining normal cholesterol levels. The APOE allele variant is determined by SNPs rs429358 and rs7412. Individuals who seem to be heterozygous transmitters (C/T) at rs429358 and homozygous transmitters (C/C) at rs4712 have one APOE4 allele and one APOE3 allele. The existence of the APOE4 allele has long been associated with higher levels of total and LDL serum cholesterol, but the existence of the APOE2 allele is linked to a reduced effect when it comes to cholesterol effects from the three allele which increases their chances for both AD to three-fold and heart disease (Mullins et al., 2020) (Zhang et al., 2015). Significant lifestyle improvements reduce cardiovascular disease incidence and total mortality, according to a review of APOE variations and disease risk, and these lifestyle changes play a higher impact in decreasing cardiovascular disease and Alzheimer's disease incidence in APOE4 phenotypic groups than in APOE2 communities. Reduced saturated fats and consumption of cholesterol, increased PUFA intake, and increased exercise were among the lifestyle changes (Mullins et al., 2020).

### 9.6.2 MTHFR

5,10-Methylenetetrahydrofolate reductase (MTHFR), which is an enzyme necessary for homocysteine to methionine conversion in the FA cycle with the help of Vitamin B12. MTHFR alleles are produced by three frequent SNPs, one of which (rs1801133) has been associated to enzyme deficits (Mullins et al., 2020). MTHFR's catalytic activity is reduced by a missense mutation in exon 4 that changes cytosine (C) to thymine (T) (Jin et al., 2018). The rs1801133 MTHFR SNP causes an alanine-to-valine amino acid change in the enzyme's catalytic domain. Because of the gene variant, MTHFR's thermolability is increased, which results in decreased folate binding and the activity of MTHFR enzyme (Wu et al., 2015). The fact that rs1801133 raises total homocysteine levels in the blood is well known which is linked to a number of medical problems, including severe birth outcomes (neural tube abnormalities, premature birth, and congenital heart disease), pregnancy issues, malignancies, adult cardiovascular diseases, and neurodegenerative disorders. In order to overcome such disorders, Vitamin B12 (Singh et al., 2017a) and folic acid supplements should be taken as they provide the substrate required for proper metabolism by lowering plasma homocysteine levels (Mullins et al., 2020).

### 9.6.3 *TNF∝*

Obesity, dyslipidaemia, insulin resistance, and cardiometabolic illness all have inflammatory processes that TNF∝ plays a role in. The level of TNF∝ mRNA in the adipose tissue and skeletal muscle are favorably correlated with BMI, fat mass, insulin levels, and inversely correlated with the action of lipoprotein lipase. The polymorphism of G/A 308 TNF is linked to fasting glucose concentration, postprandial triacylglycerol levels, and BMI in reaction to a fat-rich meal (Di Renzo et al., 2019). A study was conducted on obese rats which demonstrated a higher PUFA consumption is linked to higher HDL cholesterol and Apo A-I levels in persons with TNF∝- genotypes linked to decreased TNF-production (Fontaine-Bisson et al., 2007). According to Fernández-Real et al., a mutation in the (TNFR)-2 gene tumor necrosis factor receptor gene can explain the correlation between obesity, insulin resistance, and elevated leptin levels. TNF∝-disrupted leptin subjects have a higher BMI and higher leptin levels, as well as disproportionately higher levels of soluble TNFR2. Dietary requirements in such type of cases are Mediterranean diet, which include plant-based food like whole grains, vegetables, fruits, seeds, and herbs (Sharma et al., 2020; Rane et al., 2020a; Rane et al., 2020b; Rane et al., 2020c) also polyphenols (Keservani and Sharma, 2014) and dried fruits (Di Renzo et al., 2019).

### 9.6.4 *IL-6*

Interleukin 6 (IL-6) is a cytokine that controls inflammation. VAT deposition and obesity, lipid metabolism, and an increased chance of CVD have all been linked to higher circulating levels of IL-6. Furthermore, significant evidence that polymorphisms of the IL-6 gene are likely to cause risk of dyslipidaemia and obesity (Joffe et al., 2014). The transcription of the human gene IL-6 is regulated by a frequent polymorphism of G174C (rs603573) in the promoter. Thin people have a higher prevalence of the G allele, and people who have the genotype of G174C or G174G burn much more energy than people who have the C174C genotype. IL-6 has the ability to alter energy expenditure by enhancing sympathetic neuron stimulation. Loss of weight by LAGB in pathologically obese people may not be as good as expected. This is believed to be caused due to a failure to adhere to the nutritional treatment, but actually, this could also be due to each individual's genetic background. Patients with the genotype G174G or G174C who undergo LAGB lose weight faster than others, having followed the same nutritional regimen. Furthermore, the significance of pathways associated with the regulation of appetite, fat storage, and modification of gut microbiota to the obesity GPS distribution cannot be overlooked. Nutrition to be provided during such type of variation is Mediterranean diet, polyphenols, dried fruits, etc. (Joffe et al., 2014). Mediterranean diet contributes significantly to weight loss, especially when combined with caloric restriction and regular physical activity (Di Renzo et al., 2018).

### *9.6.5 FTO*

Obesity and diabetes are strongly linked to a SNP within the FTO gene in a variety of ethnic groups, according to studies (Olszewski et al., 2009). Fasting and restricted food access control FTO expression in the hypothalamus, whereas diet containing high fat regulates FTO expression in the hypothalamus. In vitro investigations have revealed that depriving cells of amino acids and glucose reduces FTO expression. Polymorphisms in the very first intron of the FTO gene is associated to obesity and a higher BMI. When compared to patients homozygous for the T allele, homozygosity for the A allele of FTO rs9939609 phenotypically corresponds with an increase in BMI and adiposity and also raises the risk of obesity by 1.7 times. Subjects with a genotype of AA had altered appetite and greater food consumption, which is likely due to changes in ghrelin production and secretion. Children and adolescents who carry one or both FTO rs9939609 alleles having risk of obesity show more frequent LOC eating episodes and prefer higher-fat foods. Excess body weight can be caused by both LOC eating and far more frequent consumption of energy-dense, enticing foods (Tanofsky-Kraff et al., 2009). Mediterranean diet is preferred in case of such variation in the FTO gene (Di Renzo et al., 2019).

### 9.6.6 *MC4R*

Melanocortin-4 receptor (MC4R) regulates food intake and energy balance in the body. Obesity and resistance to insulin were linked to frequent polymorphisms near MC4R, according to recent genome-wide scans. Binge eating, extreme hunger, hyperphagia, and food-seeking behavior were previously linked to infrequent genetic changes in the gene of MC4R (Qi et al., 2008). The MC4R is a downstream target of the adipocyte signaling molecule leptin, which functions through the activation and inhibition of proopiomelanocortin and agouti-related protein (Singh et al., 2017b; Jarouliya et al., 2019b), respectively (Larsen et al., 2005). The common genetic variations surrounding the gene MC4R (rs17782313 and rs17700633) are linked to greater risk of obesity (Jarouliya et al., 2019a) and resistance to insulin (Qi et al., 2008). According to studies, because MC4R rs17782313 was linked to childhood and adult height, it had a stronger relationship with body weight than BMI (Hardy et al., 2010). In the case of such MC4R gene variation, a Mediterranean diet is recommended (Di Renzo et al., 2019).

### 9.6.7 *GSTM1*

GST enzymes are phase II enzymes that help to detoxify endogenous and exogenous electrophilic chemicals, as well as the metabolism of molecules generated under oxidative stress. In the genes1 coding for GST family enzymes, polymorphisms have been discovered. Interindividual differences in response to increased vulnerability to oxidative stress-related disorders is related to the loss of enzyme function produced by gene deletion (Yuan et al., 2014). The GSTM1-null genotype (complete gene deletion and lack of enzyme in homozygotes) has now been linked to poor lung function and growth abnormalities in children (Tujague et al., 2006). Furthermore, when exposed to cruciferous-rich diets, persons with GSTM1 and GSTT1 null genotypes were found to be more protected from lung cancer than those with GSTM1/GSTT1 positive genotypes. As a result, it was hypothesized that people with an unfavorable polymorphism would be more prone to oxidative damage and would benefit more from dietary antioxidants (Yuan et al., 2014).

### 9.6.8 *MNSOD*

Manganese superoxide dismutase (MnSOD) protects mitochondria from oxidative damage caused by ROS. It's made in the cytosol and then changed for transport into the mitochondria via post-transcriptional modifications. Cancer cells are known to reduce overall antioxidant enzymes than their normal counterparts, particularly MnSOD (Mitrunen et al., 2001). Oxidative stress and mitochondrial DNA damage are thought to play essential roles in breast cancer development. MnSOD is a main enzyme in the mitochondria which helps to detoxify the reactive oxygen species. A T → C mutation in the MnSOD which results in a change of Val Ala at the 9th place of the mitochondrial designed to target sequence (Val-9Ala), affecting MnSOD delivery into the mitochondria. It is recommended that those with a greater degree of oxidative stress or who are deficient in other antioxidative protection take an antioxidant-rich diet (Cai et al., 2004).

**TABLE 9.2** Gene Variation and Nutrients to be Given in Such Cases.

| Gene | Activity | Disease caused | Nutrients to be consumed |
|---|---|---|---|
| APOE | Lipoprotein metabolism | Cardiovascular disease | Polyunsaturated fatty acids |
| MTHRF | Homocysteine metabolism | Cardiovascular disease<br>Chronic-degenerative disease<br>Obesity | Folate<br>Vitamin B12<br>Protein |
| TNF$^{\propto}$ | Inflammation | Cardiovascular disease<br>Chronic-degenerative disease<br>Obesity | Mediterranean diet<br>Dried fruits<br>Polyphenols |
| IL-6 | Inflammation | Cardiovascular disease<br>Chronic-degenerative disease<br>Obesity | Mediterranean diet<br>Dried fruits<br>Polyphenols |
| FTO | Hunger keeper<br>Fat mass storage<br>Energy homeostasis | Obesity | Mediterranean diet |
| MC4R | Leptin pathways | Obesity | Mediterranean diet |
| GSTM1 | Oxidative stress | Lung cancer | Compounds containing antioxidants |
| MnSOD | Oxidative stress | Breast cancer | Compounds containing antioxidants |
| HMGCR | Lipid and carbohydrate metabolism | Chronic-degenerative disease | Mediterranean diet |
| IL- β | Lipid and carbohydrate metabolism | Chronic-degenerative disease | ώ 3 long chain PUFAs |
| NFkB | Proliferation angiogenesis, inflammation | Cancer<br>Chronic-degenerative disease | ώ 3, ώ 6 long chain PUFAs |

*Source*: Adapted from Di Renzo et al. (2018).

### 9.6.9 DISEASE STATE TO HEALTHY STATE

Dietary chemical components can affect the balance between health and sickness. In reality, nutrition and lifestyle are significant risk factors for a variety of CDDs. A healthy lifestyle, based on adequate nutrient and bioactive food intake, leads to a lesser risk of suffering from a disease, also a substantially higher life expectancy. MD in particular is linked to a delay in ageing or death as a result of the requirement of antioxidant compounds and the reduction of inflammatory cytokines linked to a variety of disorders, including CVD.

In various epidemiological studies, the diet rich in fruits and vegetables have been proven to have cardio-protective effects. Many nutrients and active compounds present in fruits and vegetables, such as fiber, potassium, and folate, may help to lower the risk of CVD. In order to avoid CVD, many studies propose a healthy lifestyle focused on physical activity and a diet that is fiber-rich and includes antioxidants, ω-3 PUFAs, nutraceuticals, vitamins, and minerals.

The danger being related to high triglyceride level (e.g., accelerated apoptotic cell death) may be reduced by eating a Mediterranean diet which is rich in olive oil, fish oil, and vegetable lipids (e.g., 4 parts olive oil, 1 part fish oil, and 1 part vegetable oil). This diet has the potential to delay the process of type 2 diabetes.

The beneficial effects of ώ-3 PUFA on lipid metabolism, insulin sensitivity, and nonoxidative glucose metabolism are all dependent on the quantity of PUFA consumed. Polyphenols, and other vitamins (such as vitamin E and C), have found to have beneficial benefits on human health. Polyphenols can have a direct or indirect pharmacological effect. Direct scavenging of ROS and free radicals is one of their antioxidant effects, while indirect antioxidant activity requires enzyme inhibition. Polyphenols may also have anti-inflammatory, anti-proliferative, anti-diabetic, and gene-editing properties.

Several studies have discovered a link between the consumption of dietary fruit and vegetable and plasma antioxidant vitamin levels (ascorbic acid, tocopherol, and carotenoids) and the chances of dying from coronary heart disease or cancer.

By lowering serum triglyceride levels and improving endothelial dysfunction, ώ-3 fatty acids have the ability to minimize the risk of developing CHD. ALA, a form of ώ-3 fatty acid found in canola, flaxseed, and soybean oils, can all be transformed to EPA and DHA, which help to prevent heart disease by lowering the risk of thrombosis.

Providing antioxidants, anti-inflammatory agents, vitamin B6, B12, folic acid, and decreasing saturated fats and cholesterol may be especially beneficial in persons prone to neurodegenerative illnesses with higher levels of homocysteine and the 4 allele of apolipoprotein E (Di Renzo et al., 2019).

## 9.7 DIRECT-TO-CONSUMER GENETIC TESTING

DTC-GT is a procedure of directly providing genetic heritage and health-related genetic tests to consumers without the involvement or supervision of a healthcare provider. In the recent decade, the number of manufacturers offering tests that include individualized nutritional or dietary advice based on one's genetic data has grown. DTC-GT companies may not provide clinical diagnosis, instead focusing on the threat of monogenic disorders like intolerance and sensitivity panels (for example, alcohol and caffeine), macronutrient and energy metabolism (for example, NAFLD), management of weight and obesity (for example, FTO), and vitamin and mineral prerequisites (e.g., metabolism of vitamin D) (Mullins et al., 2020). Nutrigenomic testing, which involves genetic testing for numerous genes linked to more prevalent illnesses like heart disease, diabetes, and osteoporosis, has attracted both scientific and journalistic attention. The results of genetic testing, as well as information provided by the tested individual on diet and lifestyle behaviors, are utilized to assess potential health concerns in this scenario (Goddard et al., 2006).

In this scenario, saliva samples are commonly used in genetic tests. Some DTC genetic test firms simply look at a single genetic variant, while others look into hundreds of SNPs. The very first step is to register over the internet. After that, the buyer receives a box containing instruments for analyzing the genetics of each individual using saliva samples. The company gives a customized dietary advise based on the customer's profile and genetic background. Customers then receive their particular genetic test results, as well as the nutritional prescription, by email or by downloading the information from the company's website's online account. Customers may also receive a recommendation for supplements that they may buy directly from the company to help them follow the gene-based dietary guidelines (Drabsch et al., 2019).

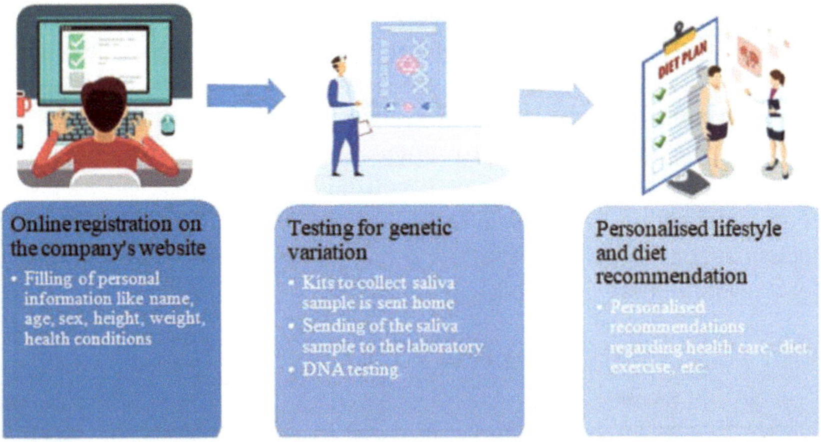


**FIGURE 9.1** Gene-based dietary recommendation schematic representation. The step-by-step procedure for using DCT-GT.

*Source*: Adapted from Drabsch and Holzapfel (2019).

## 9.8 PERSONALIZED NUTRITION

Personalized nutrition is a novel idea that aims to detect individual nutritional needs based on health state, genotype, and/or phenotype, and then provide healthy eating recommendations that are customized to the individual. D-T-C genetic tests are becoming more popular among consumers. Those with a stake in the delivery of personalized nutrition see enormous potential to improve preventative and therapeutic nutrition, benefiting public health and lowering healthcare costs in the process (Grimaldi et al., 2017). The additional benefit of nutrigenetic dietary guidance developed within such a dietary composition is that it may help offset the negative aspects of Western dietary patterns, such as excessive consumption of high-energy, processed foods. Their primary goal is figuring out how people's health is affected by interactions with both gene and nutrition. Nutrigenomics investigates the connection between our dietary and lifestyle habits and how they interact with our genetic material through terms of gene regulation involved in metabolism control. Type 2 diabetes, for example, remains among the most serious public health issues in Western Europe. Dietary changes that promote less consumption of cereals and milk products and a higher consumption of meat and vegetables have said to improve blood glucose control in type-2 diabetes patients (Thunders et al., 2013; Surana et al., 2021).

The short-term and long-term impacts of a gene-based personalized nutrition on dietary consumption were explored in a double-blinded randomized controlled study. Nielsen and El-Sohemy discovered no noticeable difference in daily diet between the control and intervention groups after three months of intervention. The intervention group was given information about their genetic makeup and also a dietary recommendation based on their genetic profile. Some substantial improvements in food intake, such as a lower sodium intake in the group of personalized nutrition, were observed at 12 months, indicating a long-term shift in dietary habits. However, the precise mechanisms and factors impacting the long-term impact of individualised diet are still unknown (Drabsch et al., 2019).

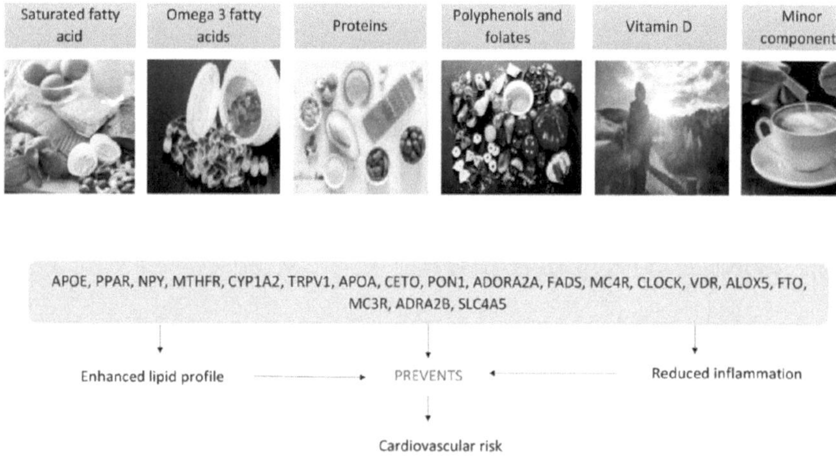


**FIGURE 9.2** Personalized nutrition based on nutrigenetics in obesity and cardiovascular disease.
*Source*: Adapted from Barrea et al., 2020

A schematic illustration of personalized initiatives for lowering Cardiovascular risk in obese patients based on individual sensitivity to key nutrients or dietary plans. A nutrigenomics-based personalized diet improves both lipid profiles and inflammation, and may be useful in the prevention of cardiovascular disease in obese patients (Barrea et al., 2020).

## 9.9 NUTRIGENOMIC IMPACT

**A cross-sectional study of healthy men's and women's dietary habits and gene expression:**

- **Objective of the study**: Several mechanisms regulate gene expression profiles in response to diet. The goal of this research was to look at gene expression in connection to eating habits.

- **Method followed:** An average of 254 people had all been recruited from the greater Quebec City metropolitan area. The study protocol was completed by 210 participants. Factor analysis was used to extract dietary patterns from a FFQRNA was extracted from 30 fasting subjects' PBMCs, and the higher expression of 47,231 messenger RNA transcripts was determined using Illumina Human-6 v3 Expression BeadChipsW. Flexarray software was used before the microarray data, which was then analyzed using IPA.

### 9.9.1 *RESULT DRAWN FROM THE STUDY*

There were two distinct eating trends discovered. The Prudent diet consists of a high intake of fruits, vegetables, whole grain products, and a lower consumption of refined grain products, whereas the Western meal consisted of a high intake of refined grain products, desserts, sweets, and processed meats. When people with high Prudent dietary pattern scores were compared to people with low Prudent dietary pattern scores, there were 2083 transcripts differently expressed in males, 1136 transcripts in women, and 59 transcripts that were overlapped in men and women. There were 1021 transcripts that differed in expression between men with high vs. low Western food pattern results, 1163 transcripts in women, and 23 transcripts that overlapped throughout men and women for the Western diet plan. According to IPA, genes mainly expressed for both patterns were discovered in networks associated with the immunological and/or inflammatory response, cancer, and cardiovascular disorders. Dietary habits have different gene expression profiles, which may influence the risk of many chronic diseases (Bouchard-Mercier et al., 2013).

## 9.10 CONCLUSION

Healthy diets share many common elements and generally fit with the WHO Global Action Plan for the Prevention and Control of Noncommunicable Diseases, whether they are based on tradition or design. When especially in comparison to a Western diet, these more nutritious options are high in plant-based foods such as fruits and vegetables, seeds, legumes, and nuts and lesser in animal-based foods like fatty and highly processed foods. Epidemiological studies of patients show that certain eating habits reduce the chances of noncommunicable diseases ranging from heart disease to cancer.

More work is needed to incorporate this good food and living choices into day-to-day life in communities around the world, as well as making healthy eating achievable, accessible, and long-term. Personalized nutrition should choose the best foods for each person, depending on their effects on expression of gene and/or the composition of the gut microbiota, for example. It is feasible to establish dietary regimes that are generally useful and successful by just considering human needs and features, rather than the individual's genetic profile.

## ACKNOWLEDGMENT

We would like to thank METs Institute of Pharmacy, Bhujbal Knowledge City, Nashik for their support and providing us the facilities to complete this work.

## KEYWORDS

- **dietary interaction**
- **direct-to-consumer gene testing**
- **disease state**
- **eating disorders**
- **epigenetics**
- **gene expression**
- **gene variation**

## REFERENCES

Barrea, L.; Annunziata, G.; Bordoni, L.; Muscogiuri, G.; Colao, A.; Savastano, S. Nutrigenetics—Personalized Nutrition in Obesity and Cardiovascular Diseases. *Int. J. Obes. Suppl.* **2020,** *10* (1), 1–13.

Boehm, I.; Walton, E.; Alexander, N.; Batury, V. L.; Seidel, M.; Geisler, D.; King, J. A.; Weidner, K.; Roessner, V.; Ehrlich, S. Peripheral Serotonin Transporter DNA Methylation is Linked to Increased Salience Network Connectivity in Females with Anorexia Nervosa. *J. Psychiatry Neurosci.* **2020,** *45* (3), 206–213.

Bouchard-Mercier, A.; Paradis, A. M.; Rudkowska, I.; Lemieux, S.; Couture, P.; Vohl, M. C. Associations between Dietary Patterns and Gene Expression Profiles of Healthy Men and Women: A Cross-Sectional Study. *Nutr. J.* **2013,** *12* (1), 1–13.

Cai, Q.; Shu, X. O.; Wen, W.; Cheng, J. R.; Dai, Q.; Gao, Y. T.; Zheng, W. Genetic Polymorphism in the Manganese Superoxide Dismutase Gene, Antioxidant Intake, and Breast Cancer Risk: Results from the Shanghai Breast Cancer Study. *Breast Cancer Res.* **2004,** *6* (6), 1–9.

Cecil, C. A.; Zhang, Y.; Nolte, T. Childhood Maltreatment and DNA Methylation: A Systematic Review. *Neurosci. Biobehav. Rev.* **2020,** *112*, 392–409.

Cena, H.; Calder, P. C. Defining a Healthy Diet: Evidence for the Role of Contemporary Dietary Patterns in Health and Disease. *Nutrients* **2020,** *12* (2), 334.

de Krom, M.; Bauer, F.; Collier, D.; Adan, R. A. H.; La Fleur, S. E. Genetic Variation and Effects on Human Eating Behavior. *Ann. Rev. Nutr.* **2009,** *29*, 283–304.

Di Daniele, N.; Noce, A.; Vidiri, M. F.; Moriconi, E.; Marrone, G.; Annicchiarico-Petruzzelli, M.; D'Urso, G.; Tesauro, M.; Rovella, V.; De Lorenzo, A. Impact of Mediterranean Diet on Metabolic Syndrome, Cancer and Longevity. *Oncotarget* **2017,** *8* (5), 8947.

Di Renzo, L.; Cioccoloni, G.; Falco, S.; Abenavoli, L.; Moia, A.; Sinibaldi Salimei, P.; De Lorenzo, A. Influence of FTO rs9939609 and Mediterranean Diet on Body Composition and Weight Loss: A Randomized Clinical Trial. *J. Transl. Med.* **2018,** *16* (1), 1–12.

Di Renzo, L.; Gualtieri, P.; Romano, L.; Marrone, G.; Noce, A.; Pujia, A.; Perrone, M. A.; Aiello, V.; Colica, C.; De Lorenzo, A. Role of Personalized Nutrition in Chronic-Degenerative Diseases. *Nutrients* **2019,** *11* (8), 1707.

Drabsch, T.; Holzapfel, C. A Scientific Perspective of Personalised Gene-based Dietary Recommendations for Weight Management. *Nutrients* **2019,** *11* (3), 617.

Eichner, J. E.; Dunn, S. T.; Perveen, G.; Thompson, D. M.; Stewart, K. E.; Stroehla, B. C. Apolipoprotein E Polymorphism and Cardiovascular Disease: A HuGE Review. *Am. J. Epidemiol.* **2002,** *155* (6), 487–495.

Fontaine-Bisson, B.; Wolever, T. M.; Chiasson, J. L.; Rabasa-Lhoret, R.; Maheux, P.; Josse, R. G.; Leiter, L. A.; Rodger, N. W.; Ryan, E. A.; Connelly, P. W.; Corey, P. N. Genetic Polymorphisms of Tumor Necrosis Factor-α Modify the Association Between Dietary Polyunsaturated Fatty Acids and Fasting HDL-Cholesterol and apo AI Concentrations. *Am. J. Clin. Nutr.* **2007,** *86* (3), 768–774.

Franzago, M.; Santurbano, D.; Vitacolonna, E.; Stuppia, L. Genes and Diet in the Prevention of Chronic Diseases in Future Generations. *Int. J. Mol. Sci.* **2020,** *21* (7), 2633.

Goddard, K. A.; Moore, C.; Ottman, D.; Szegda, K. L.; Bradley, L.; Khoury, M. J. Awareness and Use of Direct-to-consumer Nutrigenomic Tests, United States. *Genet. Med.* **2006,** *9* (8), 510–517.

Grimaldi, K. A.; van Ommen, B.; Ordovas, J. M.; Parnell, L. D.; Mathers, J. C.; Bendik, I.; Brennan, L.; Celis-Morales, C.; Cirillo, E.; Daniel, H.; de Kok, B. Proposed Guidelines to Evaluate Scientific Validity and Evidence for Genotype-Based Dietary Advice. *Genes Nutr.* **2017,** *12* (1), 1–12.

Hardy, R.; Wills, A. K.; Wong, A.; Elks, C. E.; Wareham, N. J.; Loos, R. J.; Kuh, D.; Ong, K. K. Life Course Variations in the Associations Between FTO and MC4R Gene Variants and Body Size. *Hum. Mol. Genet.* **2010,** *19* (3), 545–552.

Hunter, P. We are what we Eat: The Link Between Diet, Evolution and Non-genetic Inheritance. *EMBO Rep.* **2008,** *9* (5), 413–415.

Jarouliya, U.; Keservani, R. K. Pathways Leading to Child Obesity, Chapter 12. *Global Perspectives on Childhood Obesity: Current Status, Consequences and Prevention*, Second Ed.; Bagchi, D., Ed.; Academic Press, Elsevier, 2019a; pp 137–146, ISBN: 9780128128404.

Jarouliya, U.; Keservani, R. K. Protein Function as Cell Surface and Nuclear Receptor in Human Diseases, Chapter 1. In *Medicinal Chemistry with Pharmaceutical Product Development*; Bharti, S.; Mahapatra, D. kar, Ed.; Apple Academic Press, CRC Press, Taylor & Francis Group, 2019b, pp 1–32, ISBN: 9781771887106.

Jia, Y. F.; Choi, Y.; Ayers-Ringler, J. R.; Biernacka, J. M.; Geske, J. R.; Lindberg, D. R.; McElroy, S. L.; Frye, M. A.; Choi, D. S.; Veldic, M. Differential SLC1A2 Promoter Methylation in Bipolar Disorder with or Without Addiction. *Front. Cell. Neurosci.* **2017,** *11*, 217.

Jin, H.; Cheng, H.; Chen, W.; Sheng, X.; Levy, M. A.; Brown, M. J.; Tian, J. An Evidence-Based Approach to Globally Assess the Covariate-Dependent Effect of the MTHFR Single Nucleotide Polymorphism rs1801133 on Blood Homocysteine: A Systematic Review and Meta-Analysis. *Am. J. Clin. Nutr.* **2018,** *107* (5), 817–825.

Joffe, Y. T.; Van der Merwe, L.; Evans, J.; Collins, M.; Lambert, E. V.; September, A. V.; Goedecke, J. H. Interleukin-6 Gene Polymorphisms, Dietary Fat Intake, Obesity and Serum Lipid Concentrations in Black and White South African Women. *Nutrients* 2014, *6* (6), 2436–2465.

Keservani, R. K.; Kesharwani, R. K.; Sharma, A. K.; Vyas, N.; Chadoker, A. Nutritional Supplements: An Overview. *Int. J. Curr. Pharm. Rev. Res.* **2010b,** *1* (1), 59–75.

Keservani, R. K.; Kesharwani, R. K.; Vyas, N.; Jain, S.; Raghuvanshi, R.; Sharma, A. K. Nutraceutical and Functional Food as Future Food: A Review. *Der Pharmacia Lettre* **2010a,** *2* (1), 106–116.

Keservani, R. K.; Sharma, A. K. Flavonoids: Emerging Trends and Potential Health Benefits. *J. Chinese Pharm. Sci.* **2014,** *23* (12), 815.

Keservani, R. K.; Sharma, A. K.; Kesharwani, R. K. An Overview and Therapeutic Applications of Nutraceutical and Functional Foods. In *Recent Advances in Drug Delivery Technology*, 2017; pp. 160–201, ISBN:9781522507543.

Keservani, R. K.; Sharma, A. K.; Kesharwani, R. K. Eds.; In *Nutraceuticals and Dietary Supplements: Applications in Health Improvement and Disease Management*; CRC Press, 2020; ISBN: 9781771888738.

Kim, Y. R.; Kim, J. H.; Kim, M. J.; Treasure, J. Differential Methylation of the Oxytocin Receptor Gene in Patients with Anorexia Nervosa: A Pilot Study. *PloS One* **2014,** *9* (2), e88673.

Larsen, L. H.; Echwald, S. M.; Sørensen, T. I.; Andersen, T.; Wulff, B. S.; Pedersen, O. Prevalence of Mutations and Functional Analyses of Melanocortin 4 Receptor Variants Identified Among 750 Men with Juvenile-Onset Obesity. *J. Clin. Endocrinol. Metab.* **2005,** *90* (1), 219–224.

Mierziak, J.; Kostyn, K.; Boba, A.; Czemplik, M.; Kulma, A.; Wojtasik, W. Influence of the Bioactive Diet Components on the Gene Expression Regulation. *Nutrients* **2021,** *13* (11), 3673.

Mitrunen, K.; Sillanpää, P.; Kataja, V.; Eskelinen, M.; Kosma, V. M.; Benhamou, S.; Uusitupa, M.; Hirvonen, A. Association Between Manganese Superoxide Dismutase (MnSOD) Gene Polymorphism and Breast Cancer Risk. *Carcinogenesis* **2001,** *22* (5), 827–829.

Mullins, V. A.; Bresette, W.; Johnstone, L.; Hallmark, B.; Chilton, F. H. Genomics in Personalized Nutrition: Can you "Eat for your Genes"?. *Nutrients* **2020,** *12* (10), 3118.

Neyazi, A.; Buchholz, V.; Burkert, A.; Hillemacher, T.; De Zwaan, M.; Herzog, W.; Jahn, K.; Giel, K.; Herpertz, S.; Buchholz, C. A.; Dinkel, A. Association of Leptin Gene DNA Methylation with Diagnosis and Treatment Outcome of Anorexia Nervosa. *Front. Psychiatry* **2019,** *10*, 197.

Olszewski, P. K.; Fredriksson, R.; Olszewska, A. M.; Stephansson, O.; Alsiö, J.; Radomska, K. J.; Levine, A. S.; Schiöth, H. B. Hypothalamic FTO is Associated with the Regulation of Energy Intake not Feeding Reward. *BMC Neurosci.* **2009,** *10* (1), 1–12.

Qi, L.; Kraft, P.; Hunter, D. J.; Hu, F. B. The Common Obesity Variant Near MC4R Gene is Associated with Higher Intakes of Total Energy and Dietary Fat, Weight Change and Diabetes Risk in Women. *Hum. Mol. Genet.* **2008,** *17* (22), 3502–3508.

Rane, B. R.; Bharath, M. S.; Patil, R. R.; Keservani, R. K.; Jain, A. S. Novel Approaches in Nutraceuticals, Chapter 11. In *Enhancing the Therapeutic Efficacy of Herbal Formulations through Novel Drug Delivery Systems,* Kesharwani, R. K., Keservani, R. K., Sharma, A. K., Ed.; IGI Global International Publisher: Pennsylvania, USA, 2020b; pp 241–266, ISBN13: 9781799844532.

Rane, B. R.; Patil, A. S.; Keservani, R. K.; Jain, A. S. Novel Approaches in Herbal Formulation, Chapter 2. In *Enhancing the Therapeutic Efficacy of Herbal Formulations Through Novel Drug Delivery Systems*; Kesharwani, R. K., Keservani, R. K., Sharma, A. K., Ed.; IGI Global International Publisher: Pennsylvania, USA, 2020a; pp 43–68, ISBN13: 9781799844532.

Rane, B. R.; Tadavi, S. A.; Keservani, R. K. Naturopathy, Chapter 12. In *Herbal Product Development*; Sharma, A. K., Keservani, R. K., Gautam, S. P., Ed.; Apple Academic Press, CRC Press, Taylor & Francis Group, 2020c; pp 321–347, ISBN: 9781771888776.

Sharma, A. K.; Keservani, R. K.; Gautam, S. P. In *Herbal Product Development*; Apple Academic Press, CRC Press, Taylor & Francis Group, 2020; ISBN: 9781771888776.

Singh, P.; Kesharwani, R. K.; Keservani, R. K. Antioxidants and Vitamins: Roles in Cellular Function and Metabolism. In *Sustained Energy for Enhanced Human Functions and Activity*; Academic Press, 2017a; pp 385–407, ISBN: 9780128054130.

Singh, P.; Kesharwani, R. K.; Keservani, R. K. Protein, Carbohydrates, and Fats: Energy Metabolism. In *Sustained Energy for Enhanced Human Functions and Activity*; Academic Press, 2017b; pp 103–115, ISBN: 9780128054130.

Skerrett, P. J.; Willett, W. C. Essentials of Healthy Eating: A Guide. *J. Midwifery Womens Health* **2010,** *55* (6), 492–501.

Steiger, H.; Booij, L. Eating Disorders, Heredity and Environmental Activation: Getting Epigenetic Concepts into Practice. *J. Clin. Med.* **2020,** *9* (5), 1332.

Surana, K. R.; Ahire, E.D.; Sonawane, V. N.; Talele, S. G. Biomolecular and Molecular Docking: A Modern Tool in Drug Discovery and Virtual Screening of Natural Products. In *Applied Pharmaceutical Practice and Nutraceuticals*; Apple Academic Press, 2021; pp 209–223.

Tanofsky-Kraff, M.; Han, J. C.; Anandalingam, K.; Shomaker, L. B.; Columbo, K. M.; Wolkoff, L. E.; Kozlosky, M.; Elliott, C.; Ranzenhofer, L. M.; Roza, C. A.; Yanovski, S. Z. The FTO gene rs9939609 Obesity-Risk Allele and Loss of Control Over Eating. *Am. J. Clin. Nutr.* **2009,** *90* (6), 1483–1488.

Thunders, M.; Mangai, S.; Cooper, R. Nutrigenetics, Nutrigenomics, and the Future of Dietary Advice, 2013.

Tujague, J.; Bastaki, M.; Holland, N.; Balmes, J. R.; Tager, I. B. Antioxidant Intake, GSTM1 Polymorphism and Pulmonary Function in Healthy Young Adults. *Eur. Respir. J.* **2006,** *27* (2), 282–288.

Walker, W. A.; Blackburn, G. Symposium Introduction: Nutrition and Gene Regulation. *J. Nutr.* **2004,** *134* (9), 2434S–2436S.

Wu, X.; Yang, K.; Tang, X.; Sa, Y.; Zhou, R.; Liu, J.; Luo, Y.; Tang, W. Folate Metabolism Gene Polymorphisms MTHFR C677T and A1298C and Risk for Preeclampsia: A Meta-analysis. *J. Assist. Reprod. Genet.* **2015,** *32* (5), 797–805.

Yuan, L.; Ma, W.; Liu, J.; Meng, L.; Liu, J.; Li, S.; Han, J.; Liu, Q.; Feng, L.; Wang, C.; Xiao, R. Effects of GSTM1/GSTT1 Gene Polymorphism and Fruit & Vegetable Consumption on Antioxidant Biomarkers and Cognitive Function in the Elderly: A Community Based Cross-Sectional Study. *PloS One* **2014,** *9* (11), e113588.

Zhang, Y.; Tang, H. Q.; Peng, W. J.; Zhang, B. B.; Liu, M. Meta-analysis for the Association of Apolipoprotein E ε2/ε3/ε4 Polymorphism with Coronary Heart Disease. *Chinese Med. J.* **2015,** *128* (10), 1391–1398.

## CHAPTER 10

# THE IMPORTANCE OF NUTRIGENETICS AND NUTRIGENOMICS FOR FUNCTIONAL FOODS AND PERSONALIZED NUTRITION

DEEPAK S. BHAMBERE,[#] SAPANA P. AHIRRAO,[#]
UMESH D. LADDHA, JAYESH D. KADAM, and SURABHI B. PATIL

*MET's Institute of Pharmacy, Nasik, Maharashtra, India*

#Equally contributed.

## ABSTRACT

Dietary balance is among the major requirement to maintain the health of an individual. Many researchers have taken efforts to understand the impact of macro as well as micronutrients on the physiology of humans with the aim to avoid nutrients both in excess and shortage. However, conventional tools were not enough to maintain long-term health and well-being. Moreover, it also fails to consider that from person to person, the need of nutrition may vary which may also be genetic factor-dependent. There was a need of some novel tool that can focus on an individual need of nutrients. "Nutrigenetics and Nutrigenomics" is the recently developed tool that plays a key role in the study of how human genetic variation is responsible for different nutritional demands from every human being. This chapter provides overlook of the significance of Nutrigenetics and associated tools.

Nutrigenomics and Nutraceuticals. Raj K. Keservani, Eknath D. Ahire, Shubham J. Khairnar, Sanjay J. Kshirsagar, & Rajesh K. Kesharwani (Eds.)

## 10.1 INTRODUCTION

Conventional studies related to nutrition were focused on the impact of the excess or deficient amount of macronutrients, viz. lipids, proteins, carbohydrates, and micronutrients, such as vitamins and minerals on the physiology of humans. The foremost objective behind all studies related to nutrition was to avoid both deficiencies as well as the excess nutrients. However; later on, it was observed that the outputs of conventional studies were not adequate to maintain long-term health and well-being. As a result of which, copious methods were developed to define the optimal level of nutrients which results in a change in recommended daily amounts (RDAs) of essential nutrients and still it is an ongoing process (Ferguson and Fenech, 2012). Many studies identify the impact of a single nutrient at once, failing to pay attention to the role of nutrient interaction when a person takes whole food in a day-to-day life. Moreover, the majority of studies worked with the assumption that all individual requires the same amount and type of nutrition. However, many researches had divulged different needs for micro and macronutrients. Henceforth, it is meaningless to set one diet regimen which may be useful to the whole population. So many factors vary from one person to another, such as daily food intake, lack of or over-activation of enzymes, stress, diseased condition, and genetic variation (Heyman, 2006).

"Nutrigenetics" plays a key role in the study of how human genetic variation is responsible for different nutritional demands from every human being. Many studies have shown that inter-individual genetic variation can alter the effect of nutrients on metabolism. For example, single-nucleotide polymorphism (SNP) present in our body that encodes the enzyme MTHFR, that is, methylenetetrahydrofolate reductase is responsible for folate metabolism. Some individuals are homozygous for this SNP, which demands a higher quantity of dietary folic acid than normal individuals (Fenech, 2011). Albeit, a few genetic variants may be a promising part of personal genotyping, many of them are not. Genetic factors need to be explored more in context with the need of individual balanced food. Henceforth; while setting RDAs, there is a need to keep sufficient flexibility in context with population variability.

For the nutrients like folate, which is associated with a wide range of low and maximum doses, one can easily manage the level by alteration of food habits. Many of the nutrients at once are associated with very narrow safety windows, where above and below of which may lead to ruinous effects on health. Micronutrient "selenium (Se)" has come with a narrow dose window. Selenium plays a decisive role in oxidative stress management to cell and

DNA repair. Both low and excessive amount of "Se" leads to damaging effect. Furthermore, the window of "Se" changes as per the variants in several genes which make it more complicated to cover under nutrigenetics. An in-depth study of "nutrigenetics" will be useful in the future to understand how nutrients can change the expression of the huge number of genes in the human genome.

**Key benefits of nutrigenetics:**

1. Helps to understand the role of diet.
2. To stop chronic illnesses from arising.
3. Down regulate progress of existing chronic disease.
4. To maintain health and wellness of human being.

### *10.1.1 HUMAN GENETIC VARIATIONS*

All humans differ genetically. Twins developed from a single zygote also possess genetic and epigenetic differences to a certain extent. Single-nucleotide polymorphisms, that is, SNPs act as the major cause of genetic variations. On average, 10 million SNPs have been found in the human genome. When SNPs take place in a gene or regulatory region, they affect how the genes function. Many SNPs do not take a direct role in health and development. Only 3–5% of SNPs are functional and further responsible for phenotypic variation between individuals (Yamada, 2008). The knowledge of SNPs will be beneficial for various reasons in the future as summarized in Figure 10.1. Identification of SNPs responsible for the development of complex and life-threatening diseases, such as cancer, cardiovascular disease, and neurodegenerative diseases has attracted the attention of many researchers.

Recent studies have shown that non-SNP variations are responsible for more human genetic variations than SNPs. It includes CNV, that is, copy number variations (CNVs) develop from deletions, insertions, duplications, and inversions (Van Binsbergen, 2011). On average, CNVs vary between individuals by 0.5%. CNVs may be part of hereditary or may crop up at any time during life.

Epigenetics, the study of heritable changes in gene expression or cellular phenotype, is another important source of genetic diversity. Histone modification and DNA methylation are known examples of such amend. Additionally, one has to consider the impact of microbial genes, microbial food

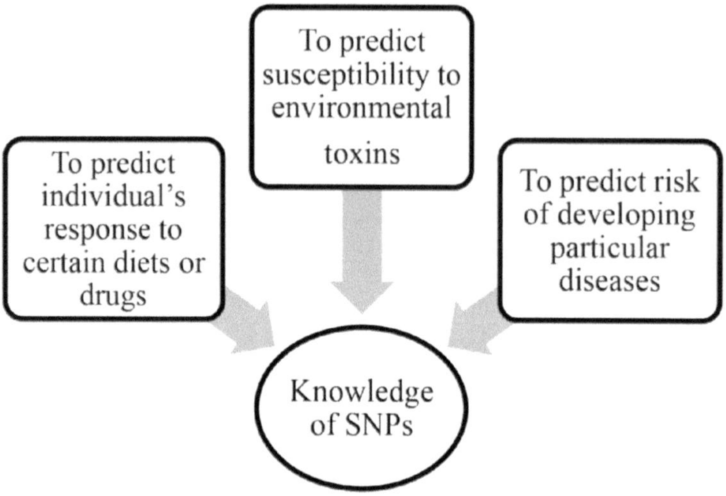


**FIGURE 10.1** Need of knowledge of SNPs.

safety, and ultimately the role of microbes in human fitness (Johnson and Tricker, 2010; Gentshew and Ferguson, 2012).

### *10.1.2 DESIRABLE HUMAN DIET*

Many developed countries have released a "dietary pyramid" for guiding people regarding a balanced diet. One such well-referred pyramid is developed by Walter Willett from the US Department of Agriculture which has been referred by many Western nations for food guidance to their people. However; the food pyramid was based on less scientific consideration which leads to people with poor health, weight gain, and early death. The dietary pyramid is shown in Figure 10.2.

The base of the pyramid consists of exercise and weight control which involves cognizant caloric restrictions with the help of limitations on fat intake. At the same time, Willett also considered the importance of certain fats in the diet and includes oils from the vegetable source at the low side of the pyramid. To enjoy the benefits of carbohydrates, the pyramid consists of whole grain food in preference to refined starches and simple sugars. Green vegetables and fruits should be consumed in good quantity on a daily basis for balanced health nuts, legumes, eggs, and poultry products should be consumed at moderate levels. Red meat should be consumed rarely. Dairy

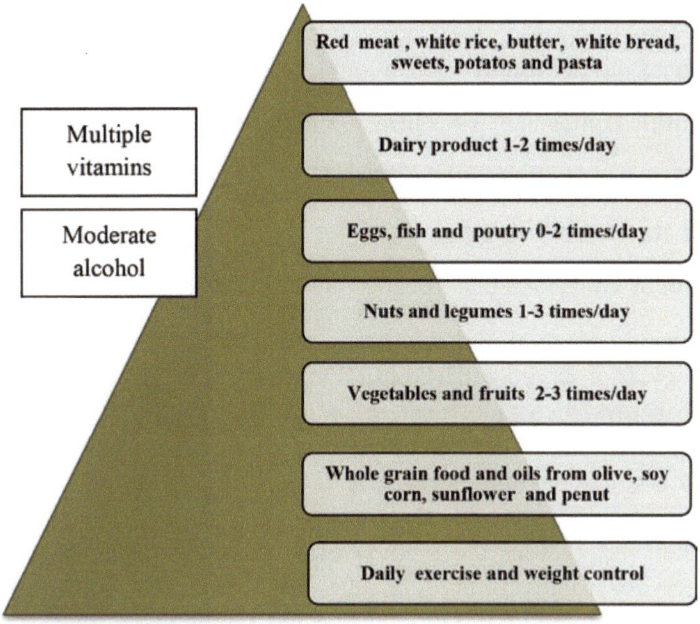


**FIGURE 10.2** Dietary pyramid.

products and calcium may be supplied as per the need with a regular intake of multivitamin tablets. A moderate amount of alcohol is also allowed in the healthy diet pyramid (Willett, 2006; Willett et al., 2001).

#### *10.1.2.1 EVIDENCE FOR A DESIRABLE HUMAN DIET*

Dietary pyramid have been prepared by using a database of Nurse's Health Study in the US, EPIC (European Prospective Investigation into Cancer and Nutrition) studies, and several other official clinical studies databases (Pan et al., 2012). Data obtained from these studies gave a strong backbone to develop a healthy eating pyramid. However, data failed to cover the problem of dietary recall. Also, the study failed to identify and establish the importance of a single or particular food or nutrient.

One such evidence-based clinical which indicated the importance of proper intake of food or supplement is "The Selenium and Vit. E Cancer Prevention Trials (SELECT). The study was carried out to check the significance of Vit. E and selenium individually as well as blend in control of prostate cancer.

Details of SELECT studies are given below.

**Number of participants:** 35, 533

**Geographical sites:** 427 sites from United States, Canadd, and Puerto Rico.

**Study Groups:** (1) Selenium, (2) Vit. E, (3) Selenium + Vit. E, (4) Placebo.

**Type of study:** Double-blind

**Duration of study:** August 22, 2001–June 24, 2004

**Outcomes:** In place of observing expected trends, a small increase in cases of cancer was observed with group 2. On the other hand, group 1 participants showed increased cases of diabetes. Furthermore; after 7 years (treatment time + off treatment) in comparison to placebo, 17% more cases of cancer were observed in the treatment group.

The above study indicates the significance of the involvement of biomarker endpoint in the study which helps to reduce the number of volunteers for a short duration of time. It also makes it more convenient to stop the trials at any time if any unexpected and opposite trend is observed. A recent study showed the use of serum cholesterol as a biomarker which recommended that a daily balanced diet should contain a fat intake of 30% of energy or less (Keys et al., 1984). Studies on biomarkers are yet going on which contribute to more accurate dietary feedback and refining the current healthy food information.

#### *10.1.2.2 DESIRABLE HUMAN DIET AND HUMAN GENETIC VARIATION*

The dietary pyramid lacks the assumption that a single diet plan is applicable to all. Genetic polymorphism plays a decisive role in the need of different nutrients in an individual. Key points of the pyramid are discussed below.

1. Regular exercise and weight control: The impact of exercise and weight control is affected by a number of genes. The influence of gene variation on food intake has not been considered in the pyramid (Corella et al., 2011).
2. Selection of lipid source: The nature of lipid and their plasma concentration acts as a leading factor in the development of many chronic diseases. PPAR family genes have the capacity to modify or alter the transportation of many lipids through the body. These

factors have not been focused on while selecting lipid sources in the pyramid (Alsaleh et al., 2011).

3. Selection of carbohydrate source in diet: Many researchers have shown the involvement of different genes in carbohydrate metabolism which varies among the population. These genes play an important role in the development of obesity. Henceforth, from individual to individual, there will be different needs for carbohydrate sources (Varma et al., 2010).
4. Red meat and dairy products: Occasional use of these products is not problematic as per the diet pyramid. However, both act as a source of undesired saturated fatty acids. The research done by Phillips et al. showed that any metabolic syndrome's emergence because of TCF7/L2 gene can further increase the complexity with the consumption of red meat (Phillips et al., 2012). Additionally, high red meat consumption results in a relatively high load of iron which is problematic for person suffering from genetic polymorphism of hemochromatosis (HFE) gene. It may end with liver cirrhosis and cardiomyopathy.
5. Nuts, legumes, fish, and poultry products: These are preferred sources of protein in comparison to red meat. These products possess better and desired lipid profiles. However, the daily needs of material vary from one to another individual as per genetic variation (Madden et al., 2011).
6. Low salt intake: Pyramid has suggested the involvement of less amount of salt in daily diet. It looks like a very general statement. Salt intake is more related to the variations in genes which are associated with rennin–angiotensin system (Poch et al., 2010).
7. Use of multivitamins supplement: Healthy diet pyramid has suggested the daily consumption of multivitamin supplements for all humans. The need of vitamins (Singh et al., 2017a), proteins (Singh et al., 2017b), and minerals definitely vary from one to another. There may not be a need of vitamin supplements to a person taking a balanced diet (van Ommen et al., 2010).
8. Alcohol: Moderate amount of alcohol is also allowed in the pyramid. However; certain study has shown an undesired impact of even a small quantity of alcohol. Furthermore, high concentration of alcohol may also accelerate the genetic polymorphism (Kim et al., 2012).

## 10.2 IMPACT OF INTESTINAL MICROBIOTA

One also needs to consider the impact of gut microbiota on the need of nutrients. For the case, US dietary guidelines have suggested that there is a very negligible chance to develop the deficiency of Vitamin biotin as it forms in excess of the daily need by gut microbes. For this reason, no RDA is also available for biotin. Two major limitations have not been considered in this case: (1) variation of the intestinal microbes from healthy to an unhealthy person suffering from inflammatory bowel disease; (2) impact of infections and use of antibiotics (Otten et al., 2006; Gentschew and Ferguson, 2012; Ahire et al., 2020).

## 10.3 NUTRIGENOMICS TOOLKIT

So far the development in genomics study has proved the relationship between the risk of development of disease and genetic variation. The mechanism responsible for gene variations' interaction with one another and with diet along with severity has not been fully understood. It is important to notice that there are possibly thousands of genetic polymorphisms which result in major or minor deviations in the nutritional needs. Tools used in the development of nutrigenomics include:

a) **Tools for measurement of transcriptome:** DNA microarrays, exon arrays, and tiling arrays.

b) **Tools for proteome:** mass spectrometry (MS), chromatography, and gel electrophoresis

c) **Tools for metabolome:** nuclear magnetic resonance (NMR) spectroscopy and MS, often in combination with gas chromatography (GC) and liquid chromatography (LC).

### *10.3.1 TRANSCRIPTOMICS*

The functional effect due to variations in the concentration of micronutrients in the diet is unpredictable unless it is below the level that leads to developing the diseased condition. It is difficult to determine the exact level of bioactive food components. Transcriptomics measures the degree of RNA expression in a given cell population which acts as an important tool to determine appropriate functional levels of micronutrients and foodstuff components that are bioactive as well as in risk assessment. The development of highly sophisticated methods like whole-genome transcriptome analysis is the most preferable method to obtain unbiased information.

### *10.3.2 PROTEOMICS*

Protein molecular proteins made up of building blocks are known as amino acids. Proteins are crucial for health, structural, metabolic, transport, immune system, signals, and regulatory functions. Australian Ph.D. candidate Marc Wilkins first used the term "proteome" in the 1994 series in Siena, Italy.

### *10.3.3 ORIGIN OF PROTEOMICS*

O'Farrell, Klose, and Scheele introduced the two-sided gel in 1975 and began mapping proteins from mice, guinea pigs, and *E. coli*, respectively (O'farrell, 1975; Klose, 1975; Scheele, 1975). In 1995, The term proteome was coined and is defined as the collection of all proteins found in cells, tissues, and living things. Proteomics is the study of how different proteins interact and what functions they play in an organism. Proteomics is an advanced technique that aims to produce a large amount of protein (proteome) in a single test, or single test. The concept of proteomics revolves around the principle that a large amount of protein can be detected in a single test (or a single proteomics test).

- **Proteomics:** the analysis of all the proteins that fill a cell, tissue, or organism in a certain predetermined set of circumstances – its present form also depends on decades of technological and metallurgical advances. These improvements have incorporated developments in mass spectrometry (MS) techniques, protein classification methods, and bioinformatics. One of the biggest challenges in proteomics is the uncertainty about its complexity.

  Proteomics relies on three existing technological stones: a method for classifying complex proteins or peptide compounds, MS to obtain the data needed to identify specific proteins and the use of bioinformatics to examine and combine data from MS (Yu, 2010).

- **Major proteomics testing steps:**

  Sample preparation, in which proteins are extracted from tissues/cells or biological fluids. Second, extracellular proteins are solved using gel electrophoresis. Third, the amount of peptide (precursor mass; MS) and/or quantity of precursor fragments (MS/MS) is measured by mass spectrometry. Fourth, the MS alone or MS/MS spectra are used to detect protein by comparing the target and theoretical spectra on a protein site using bioinformatics tools.

## 10.4 TYPES OF PROTEOMICS

Protein expression proteomics: A quantitative study of the structure of a protein between different samples is known as a proteomics expression in which the comparison of degree expression of proteins of the whole proteome or sub-proteomes between samples can be done. This method can detect novel proteins in signal transmission or target disease-specific proteins.

- **Structural Proteomics:** The study of proteomics aims to differentiate the structure of protein complexes or proteins present in a particular cell organelle known as a "cell map" or proteomics structure. Structural proteomics is used to identify all proteins within a complex protein organelle, determine their location, and demonstrate all protein and protein interactions. This type reflects the complete structure of cells and explains that the expression of a few proteins gives unique characteristics to the cell.
- **Functional proteomics:** Functional proteomics is the broad term for many specific, precise proteomics methods. In some cases, certain sub-proteomes are separated by affinity chromatography for further analysis. This may involve the separation of protein complexes or the utilization of protein ligands to separate certain protein types. This method makes it possible to identify and study a specific group of proteins, which can provide crucial details about protein expression, disease pathways, or drug interactions.
- **Metabolomics:** A novel method for examining the pathophysiology of human disease has been made possible by the development of genomics, metabolomics, proteomics, lipidomics, and transcriptomics. Modern advances in research and technology are directly linked to the use of metabolomics as an integral approach to understanding modern life sciences.

When Oliver et al., assessed the difference in the relative amounts of metabolites as a result of gene deletion or overexpression, the term "*metabolome*" first appeared in the literature in 1998. Thus, the term "*metabolome*" refers to the total set of metabolites that an organism expresses. By comprehensively and quantitatively analyzing every molecule of the biological system under investigation, Fiehn defined metabolomics in 2001. The term "*metabonomics*" was first used by Nicholson et al. The term "quantitative evaluation of the dynamic multiparametric metabolic response of biological systems to physiopathological stimuli or genetic alterations" was coined in 1999 to describe the "quantitative evaluation of the dynamic multiparametric metabolic response of biological systems to physiopathological stimuli or genetic alterations." The phrases "*metabolome*," "*metabolomics*," and "*metabonomics*" were proposed at the end of the 1990s, when several acronyms relating to *omics* tactics first arose. Living organisms are predetermined by their biological structures in how they could react to intrinsic or environmental challenges. While the genome contains a blueprint of an organism, the proteome and the metabolome are the functional agents. Proteomics and metabolomics, on the other hand, portray ongoing processes, whereas genomics can read out predictive information.

Analytical methodologies have improved in terms of sensitivity and reproducibility, metabolite coverage, and sample throughput as a result of metabolomics advancements. This in turn encouraged an increase in the number of applications in agriculture and food quality, biomedical research, and other fields. Research in metabolics is quite interdisciplinary. Analytical, bioinformatic, and study design components are all combined in metabolomics research.

One of the most powerful bioanalytical approaches is metabolomics which allows us to obtain an image of the metabolite of an organism during the biological process, considered a phenotyping tool. In short, the metabolic process involves identifying and determining a set of metabolites (or particular metabolites) in biological specimens (tissues, cells, fluids, or organisms) in the absence of disease compared with altered conditions promoted by disease, drug treatment, dietary interventions, or environmental change.

Although numerous classifications have been developed in the literature to classify metabolomics/metabonomics studies there is still no real agreement about the names. Targeted metabolic pathology is defined as quantitative analysis (focus determined) or semiquantitative analysis (affirmative affinity) of a few metabolites and/or metabolic reaction substrates that may be associated with normal chemical or linked chemical selections and metabolic pathways.

Unintentional metabolomics relies on qualitative or semiquantitative examination of the highest amount of metabolites from the various chemical and biological categories contained in a biological sample. This term includes both footprinting and fingerprinting metabolomics.

**Applications of metabolomics**

- Disease diagnosis
- Cancer research
- Metabolic mechanism
- Drug efficacy and screening
- Nutrition
- Agronomy and plant biology

As a crucial component of systems biology, metabolomics primarily analyses samples including blood, urine, and feces before studying the tiny molecules that make up the by-products of various metabolic pathways. Nuclear magnetic resonance, mass spectrometry, and chromatography are among the methods used in metabolomics. Clinical research, disease therapy, and MS-based metabolomics is used extensively in drug characterization, animal and plant research, agricultural research, and nutrition (Yang et al., 2019).

## 10.5 NUTRIGENOMICS AND THE MAINTENANCE OF HOMEOSTASIS

Both nutrigenetics and nutrigenomics are frequently used interchangeably in both professional, nutritional, and medical publications as well as in popular literature. The field of nutrigenetics, which has its roots in classical genetics, studies how a person's genotype composition affects how they respond phenotypically to dietary consumption. However, the Human Genome Project is where nutrigenomics first emerged. Its goal is to examine how diet and nutrition affect gene expression, namely by epigenomic (e.g., histone methylation), transcriptomic (e.g., RNA transcription), proteomic (e.g., protein synthesis), and metabolomic (e.g., exit tests) analyses.

Nutritional genomics (nutrigenomics), a link between health, nutrition, and genomics, can be considered a combination of molecular genomics and nutrition. Nutrigenomics will develop a greater comprehension of how metabolic processes and homeostatic regulation is affected by nutrition, how this regulation is disrupted in the early stages of disordered eating, and the degree to which sensitive genotypes contribute to such diseases.

The main focus of healthy eating research is to minimize chronic conditions including metabolic syndrome and cardiovascular disease, and cancer caused by chronic exposure to certain foods. Only long-term, time-consuming, and expensive interventions can determine the causal association between such bioactive food components and disease prevention or outcome, and the prevention or outcome of the disease can only be assessed through long-term, time-consuming, and expensive interventions. Metabolomics is the study of whole internal and external metabolites in a cell, organ, or body fluid and is a useful tool for producing individual metabolite profiles, such as plasma lipid (i.e., cholesterol, triglycerides) and vitamin profiles. This process is highly variable, and ensures accuracy and sensitivity, especially in the case of pre-disease.

Nutrigenomics involves the appearance of genetic products, their function in life, and their interactions. It emphasizes how nutrients affect the genome, proteome, and metabolome (Fig. 10.3) and explains the connection between these particular nutrients and the nutritional principles of human health.

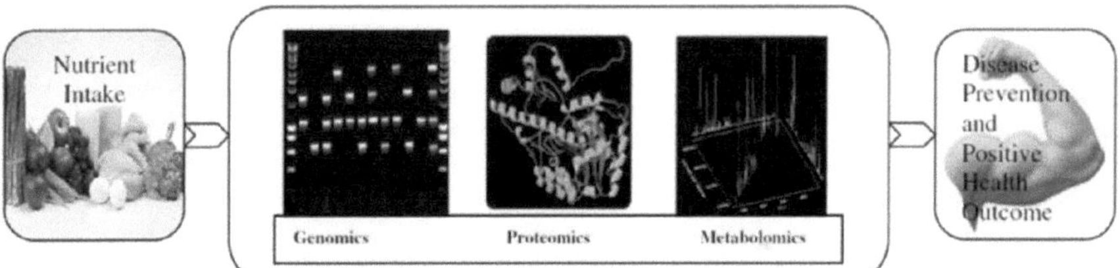


**FIGURE 10.3** Nutrigenomics approach in disease prevention.
*Source*: Adapted from Neeha and Kinth, 2012

The public's increased interest in preventative care has prompted scientific research into alternative medicines, with a focus on healthy nutrition and overall well-being. Nutrigenomics, which investigates the diet and disease interaction development based on the human genetic profile, will officially provide science to these approaches.

## 10.6 NUTRIGENOMICS AND PREVENTIVE HEALTH

Heart disease, metabolic syndrome, and cancer are among the most frequent chronic diseases caused by hereditary and environmental factors. Different genetic combinations can lead to the same phenotype disease, which further complicates imagery, and the effects of nutrition on genetic expression patterns are also difficult to detect. Important factors that can influence genetic expression in humans and animals for years are sex, healthy eating status, and other unknown pathogens (patho) – life limits. Very readily available information in human intervention trials is race and family history.

Transcriptomics, proteomics, and metabolomics are three genetic technologies that are complementary in terms of the sorts of information they create, but they are at distinct stages of development. Transcriptomics is a relatively mature technology compared with other "-omics" technologies. Plasma proteomics research is particularly fascinating for nutrition and nutrigenomics because it can yield information on essential proteins, such as cytokines or hormone levels, from a little plasma sample. Metabolomics is also in the early stages of growth. Metabolomics produces large amounts of data that require complex tools and software to allow researchers to extract valuable information from data (Lydia, 2006).

Nutrigenomics examines the diet and disease development interaction based on a person's genetic profile and will provide scientific suitability for such methods. SNPs and associated diseases are described in Figure 10.4.

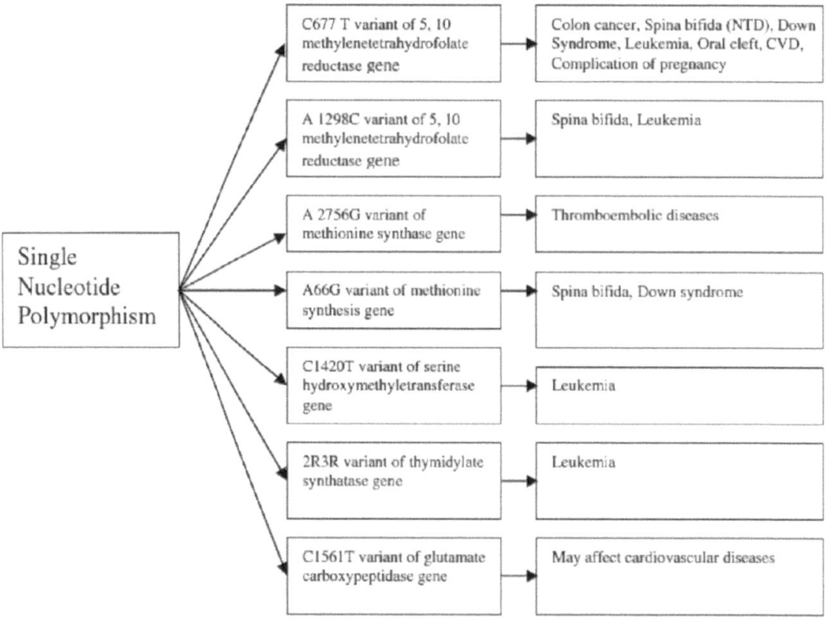


**FIGURE 10.4** Single-nucleotide polymorphism, its associated gene alteration, and diseases.
*Source*: Reprinted with permission from Neeha and Kinth, 2012. © 2012 Springer Nature.

## 10.7 NUTRIGENOMICS AND THE SLOWING OF DISEASE PROGRESSION

The main purpose of this emerging scientific field is to protect rather than cure. Nutrigenomics is a flexible way of looking at food, not just food, but as a medicine that can reverse disease. Component of the approach, nutrigenomics includes identifying the symptoms of the first stage of food-related diseases; this is the stage at which a nutritional intervention can restore a patient's health. A crucial objective of nutrigenomics research is studying the genome-wide effects of healthy eating, with a direct focus on the metabolic stress role in the genesis of metabolic syndrome, a set of phenotypes including inflammation, metabolic depression, insulin resistance, and diabetes. This goal is ambitious, but it is based on the premise that a healthy diet should be more focused on health and disease prevention and in line with treatment, which is aimed at the pathophysiological features of the disease. We looked into nutrigenomics research in India because of the growing number of food-related disorders in India (obesity, cardiovascular illness, diabetes, neural tube defects, and so on).

### *10.7.1 FUNCTIONAL FOODS*

In recent years, healthy eating has gained popularity within the health and fitness circles. Active food is very nutritious and is associated with a number of strong health advantages. For example, they can protect from disease, prevent malnutrition, and promote healthy growth and development (Keservani et al., 2010a; Keservani et al., 2010b; Keservani et al., 2017; Keservani et al., 2020).

### *10.7.2 WHAT ARE FUNCTIONAL FOODS?*

Functional foods are made up of components that offer more health advantages than your daily intake of nutritious foods. Some functional foods incorporate additives or additional ingredients intended to promote health. The idea originated in Japan in the 1980s when government institutions began to ratify proven nutritious foods in an effort to improve the lives of many. Other examples include foods rich in vitamins, minerals, probiotics, or fibers. Nutritional supplements, such as fruits, vegetables, nuts, seeds, and grains are considered to be useful foods. For instance, beta-glucan, a type of fiber found in oats, has been demonstrated to increase immunological response, lower inflammation, and enhance heart health. Similarly, fruits and vegetables are abundant in antioxidants, which are advantageous substances that work to fight against disease (Alasalvar and Taylor, 2002.).

### 10.7.2.1 EXAMPLES OF FUNCTIONAL FOODS

Active foods are usually divided into two categories: normal and modified. Normal foods are natural, nutritious whole foods rich in essential nutrients, such as vitamins, minerals, antioxidants, and healthy heart fat. Meanwhile, to boost the modified diet's health benefits, other substances have been added, such as fiber, probiotics, vitamins, and minerals (Borneo 2012).

Here are some examples of conventional functional foods:

- **Vegetables:** broccoli, cauliflower, kale, spinach, zucchini
- **Fruits:** pears, peaches, apples, oranges, bananas, berries, kiwi
- **Nuts:** almonds, cashews, macadamia nuts, Brazil nuts, pistachios
- **Legumes:** lentils, black beans, chickpeas, navy beans
- **Whole grains:** oats, barley, buckwheat, brown rice, couscous
- **Seafood:** salmon, sardines, anchovies, mackerel, cod
- **Seeds:** chia seeds, hemp seeds, pumpkin seeds, flax seeds
- **Fermented foods:** kombucha, kefir, sauerkraut, tempeh, kimchi
- **Herbs and spices:** cinnamon, ginger, cayenne pepper, turmeric (Sharma et al., 2020)
- **Beverages:** green tea, black tea, coffee

Here are some examples of modified functional foods:

- fortified juices
- fortified dairy products, such as milk and yogurt
- fortified milk alternatives, such as almond, rice, coconut, and cashew milk
- fortified grains, such as bread and pasta
- fortified cereal and granola
- fortified eggs

**Potential benefits:** Several potential health benefits of functional foods are as follows.

#### *10.7.2.1.1 Functional Foods may Prevent Nutritional Deficiencies.*

These foods are usually high in essential nutritional factors including healthy fats, fiber, minerals, and vitamins. Adding in the diet a number of active foods that include both regular and enriched can assist ensure that we obtain the nutrients we need and avoid nutritional shortages by eating a variety of ordinary and fortified foods. In fact, the prevalence of malnutrition has dropped dramatically worldwide. For example, after the introduction of cast iron flour in Jordan, iron deficiency in children is almost halved. Stabilization has also been used to minimize other conditions caused due to malnutrition, including rickets, goiter, and congenital malformations (Gormley, 2013).

### 10.7.3 MAY PROTECT AGAINST DISEASE

An active diet serves necessary nutrients that help against disease protection. Numerous are rich in antioxidants. These molecules help to reduce harmful substances called free radicals, which assist minimize cellular damage and a few chronic conditions, such as heart disease, cancer, and diabetes. Other active foods are also high in omega-3 FAs, a healthy fat that has been proven to reduce inflammation, strengthen brain functioning, and promote heart health. Some types are high in fiber, which can promote blood glucose regulation and protect against diseases such as diabetes, obesity, heart disease, and stroke. Fiber can also help in preventing digestive disorders, such as diverticulitis, stomach ulcers, hemorrhoids, and acid reflux (Guangchang, 2012).

### 10.7.4 MAY PROMOTE PROPER DEVELOPMENT AND GROWTH

Certain nutrients are necessary for the proper development of infants and children. Enjoying a variety of nutritious foods as part of a healthy diet can help in ensuring that nutritional necessities are met. Also, it can be advantageous to add foods rich in certain nutrients that are necessary for growth and development. Like, whole grains, cereals, and flour are often fortified with B-vitamins, such as folic acid, which are essential for a baby's health. Low folic acid levels can promote the danger of neural tube defects, which could affect the brain, spinal cord, or spinal cord. It is estimated that increasing folic acid consumption can reduce the spread of neural tube defects by 50–70%. Other nutrients found in active foods also have a key role in growth and development, including omega-3 FAs, iron, zinc, calcium, and vitamin B12 (Istva, 2008).

#### 10.7.4.1 USES

Various active foods should be abundant in nutrient-dense, nutritious foods, including nutritious foods such as fruits, vegetables, grains, and legumes. These foods not only give the body the necessary vitamins and minerals but also support your whole life. Organized, fortified foods can go into a nutritious diet. In fact, they can support your diet by filling in any voids to avoid nutrient deficits, as well as improve your health by increasing your intake of essential nutrients, such as vitamins, minerals, fiber, healthy heart fat, or probiotics (Kwak, 2001).

#### *10.7.4.2 SUMMARY*

An active diet is a diet component associated with a few dynamic health benefits. They can not only help in the prevention of malnutrition but also protect against diseases and promote normal growth and development. Also, to enjoy a variety of healthy foods, you can add solid foods to your diet to fill the gaps in nutrition and promote better health.

## 10.8 PERSONALIZED NUTRITION

It might appear that precision nutrition is a current phenomenon, but it is actually a trustworthy emerging community supported by the NIH under an umbrella of accurate medicine. Proper medication seeks improvement in the personal treatment of diseases, and proper nutrition is related to the food you eat. They both develop solutions to prevent or treat chronic diseases based on a person's unique traits, such as DNA, race, gender, health history, and lifestyle habits. Both focus to provide safe and effective ways of preventing and treating diseases by providing more accurate and targeted strategies. A healthy diet suggests that each person may have a different reaction to special foods and nutrients, so one person's best food may look very different from another. How many times have we been tempted to follow a fancy diet that someone recommended to help them lose 50 pounds [20 kg] or "cure" diabetes? The idea of a healthy diet can counteract this trend, as each of us may need a unique diet plan to succeed. An accurate and sensitive microbiome, billions of bacteria in our bodies play a vital role in a variety of daily internal functions. What types and how many viruses we have are different for each person. Our diet can determine what types of bacteria live in our digestive tract, and the exact opposite is true: the types of bacteria we ingest may determine how we break down certain foods and which foods benefit our bodies the most (Mattioli, 2016).

### *10.8.1 HOW DOES IT WORK?*

Precision nutrition, also called personal nutrition, is focused on the individual and not on the individual groups. A lot of research on nutrition and food types provides useful information to communities and the general public. High-nutrition diets have shown that for the average person, eating plenty of vegetables, whole grains, and lean protein (Jarouliya et al., 2019b) while eating high-sugar foods high in sugar and salt can help to reduce the risk of various diseases. However, when a study similar to the PREDICT 1 study focused on an individual's response to food, they found significant differences in their response to blood glucose and triglycerides even when people ate the same food. The human microbiome is proven to cause variations in blood triglyceride levels after meals. Non-dietary factors, such as sleep, exercise, and mealtime also play a role in causing alterations in blood glucose levels and triglycerides after meals. Therefore, one can see additional benefits if one follows the guidelines for personal nutrition beyond the usual health recommendations. Accurate nutrition examines human DNA, microbiome, and metabolic reactions to certain foods or food patterns to identify the best diet in the prevention or treatment of diseases (Shahidi, 1999).

### *10.8.2 APPLYING PRECISION NUTRITION TO DISEASE MANAGEMENT*

There are specific treatments based on genetic information that have been used for years: gluten-free foods to control celiac disease, lactose-free diets for those with lactose intolerance, and avoiding phenylalanine diet (amino acid protein foods and other synthetic additives) called genetic predisposition. -phenylketonuria (or PKU). However, the most accurate diet is in your early stages and very soon it can be started as a treatment for chronic diseases for most people. Research is being done on the proper diet for obesity (Jarouliya et al., 2019a), metabolic syndrome, certain cancers, and type 2 diabetes (T2DM). For example, traditional T2DM administration focuses on aspects of healthy living (exercise, diet, gaining a healthy weight), usually through medication use. Proper nutrition can improve diabetes management by focusing on the following areas:

- How human DNA can be linked to diet and the metabolism of certain nutrients, which can predict how a person reacts to certain foods.
- The ratio of metabolites (small molecules formed during digestion and digestion) reveals how a person eats for a long time, whether they eat fruits and vegetables regularly or eat a lot of saturated fat in their daily diet and butter diet. Accurate nutrition will see if there is a relation between a person's "signature" created by these dietary patterns and their danger of developing $T_2DM$.

- How the human body responds to particular diets can be established by metabolic testing used for nutritional accuracy. Finding the right types of gut bacteria that aid in better blood glucose regulation and changing one's diet to assist the development and maintenance of these particular gut bacteria will alter one's microbiome.
- Identifying types of intestinal bacteria that help improve blood sugar control, and using dietary patterns that alter the human microbiome to support the growth and maintenance of certain intestinal bacteria.

## 10.9 APPLYING PRECISION NUTRITION TO SPECIFIC FOODS

Proper nutrition examines how a person's DNA can be linked to his or her response to food. One example is coffee. Have you ever noticed that some people do not worry after drinking a cup after a cup of coffee during the day, when you may feel jittery and very depressed after just one cup? Coffee beans possess plant chemicals that have health benefits but not everybody can tolerate the side effects from caffeine content. The study found that genes are directly related to coffee consumption. These genes may determine how quickly or slowly the caffeine in coffee is digested. They are also associated with anti-inflammatory effects and a decreased risk of certain chronic diseases. A healthy diet uses genetic data to determine whether a person can benefit from drinking more or less coffee. Other studies include finding out if eating a high-fat diet can lead to obesity, or if drinking too much salt increases the risk of developing high blood pressure. However, the findings from these studies were not repeated for different individuals (Taylor, 2013).

## 10.10 CHALLENGES AND CONSIDERATIONS

The food industry is not yet ready for the main time due to a variety of challenges, the lack of well-designed clinical trials that show consistent results, and the expensive technology needed to collect and study human DNA, gut microbiome, and reactions to diet. Dietary interventions like these need high-quality proof of their effectiveness and compliance before it can be recommended for side effects or even replacement for common disease interventions. There may also be variations in clinical results from each metabolic response to a particular diet depending on the type of tests they use; this may result in variations in the personal nutrition recommendations provided. Physicians, primary care providers, registered nutritionists, and other providers who communicate directly with individuals need to be educated on nutritional accuracy, as it requires the concerted efforts of the entire health care team. Despite each difference in metabolic response to diet, personal dietary advice should be consistent with normal eating habits (e.g., eating more fruits and vegetables, eating more sugar from sugary drinks, reducing sodium, etc.). Although private companies have begun to offer genetic testing and microbiome to the public to make their food, further research needs to be done more efficiently and accurately. Ethical and legal aspects of proper nutrition should also be considered, including the protection of consumer privacy in the use of accurate nutrition and testing technologies. It is also important to note that personal approaches may be accessible to a small portion of the population, which may increase health disparities. We must not lose sight of the importance of developing a comprehensive food environment with practical policies, guidelines, and other human-based approaches that can help make healthy choices automatically (Xingliang, 2009).

## 10.11 CONCLUSIONS

Both micro as well as macronutrients play a pivotal role in the maintenance of health. Excess as well as deficient intake of nutrients may harm the person. While studying the need of nutrients one should always remember that individual to individual, the requirement of diet may get vary which can be more correlated with the genetic factor. Novel tools like "nutrigenetics" may help to address the limitations of conventional tools of calculation of diet. A balanced diet may help to avoid the development of chronic disease and downregulation of already existing diseased conditions. In conclusion, nutrigenetics is one of the most important tools to deal with personalized nutrition requirements.

## KEYWORDS

- **functional food**
- **nutrigenetics**
- **dietary pyramid**
- **transcriptomics**
- **proteomics**
- **nutrition**

## REFERENCES

Afman, L.; Müller, M. Nutrigenomics: From Molecular Nutrition to Prevention of Disease. *J. Am. Diet. Assoc.* **2006,** *106* (4), 569–576.

Ahire, E.; Talele, S. G.; Talele, G. S. Microbial Pigments: A Green Microbial Technology. In *Biologically Active Natural Products* Apple Academic Press, 2020; pp 225–244.

Alasalvar, C.; Taylor, Y. In *Seafoods – QualityTechnology and Nutraceutical Applications*; Springer: Verlag, Heidelberg, Germany, 2002; vol 508, p 2.

Ali, A.; Bhat, B. A.; Rather, G. A.; Malla, B. A.; Ganie, S. A. Proteomic Studies of Micronutrient Deficiency and Toxicity. In *Plant Micronutrients*; Springer, Cham, 2020; pp 257–284.

AlSaleh, A.; Frost, G. S.; Griffin, B. A.; Lovegrove, J. A.; Jebb, S. A.; Sanders, T. A.; O'Dell, S. D.; RISCK Study Investigators PPARγ2 gene Pro12Ala and PPARα Gene Leu162Val Single Nucleotide Polymorphisms Interact with Dietary Intake of Fat in Determination of Plasma Lipid Concentrations. *Lifestyle Genom.* **2011,** *4* (6), 354–366.

Borneo, R.; León, A. E. Whole Grain Cereals: Functional Components and Health Benefits. *Food Funct.* **2012,** *3* (2), 110-119. DOI: 10.1039/c1fo10165j.

Corella, D.; Arnett, D. K.; Tucker, K. L.; Kabagambe, E. K.; Tsai, M.; Parnell, L. D.; Lai, C. Q.; Lee, Y. C.; Warodomwichit, D.; Hopkins, P. N.; Ordovas, J. M. A High Intake of Saturated Fatty Acids Strengthens the Association between the Fat Mass and Obesity-Associated Gene and BMI. *J. Nutr.* **2011,** *141* (12), 2219–2225.

Fenech, M. Folate (vitamin B9) and Vitamin B12 and their Function in the Maintenance of Nuclear and Mitochondrial Genome Integrity. M*utat. Res./Fundam. Mol. Mech. Mutagen.* **2012,** *733* (1-2), 21–33.

Ferguson, L. R.; Fenech, M. F. Vitamin and Minerals that Influence Genome Integrity, and Exposure/Intake Levels Associated with DNA Damage Prevention. *Mutat. Res. Fundam. Mol. Mech. Mutagen.* **2012,** *733* (1-2), 1–3.

Gentschew, L.; Ferguson, L. R. Role of Nutrition and Microbiota in Susceptibility to Inflammatory Bowel Diseases. *Mol. Nutr. Food Res.* **2012,** *56* (4), 524–535.

Gormley, R. Fish as a Functional Food: Some Issues and Outcomes, 2013.Guangchang, P.; Xie, J.; Chena, Q.; Hu, Z. How functional Foods Play Critical Roles in Human Health. *Food Sci. Hum. Wellness* **2012,** *1* , 26–60.

Graves, P. R.; Haystead, T. A. Molecular Biologist's Guide to Proteomics. *Microbiol. Mol. Biol. Rev.* **2002,** *66* (1), 39–63.

Heyman, M. B. Committee on Nutrition. Lactose Intolerance in Infants, Children, and Adolescents. *Pediatrics* **2006,** *118*, 1279–1286.

Istva, N. S.; Polna, E. K.; Polna, B. K.;Lugasi, A. Functional Food. Product Development, Marketing Andconsumer Acceptance-A Review. *Appetite* **2008,** 51 (3), 456–467.

Jarouliya, U.; Keservani, R. K. Pathways Leading to Child Obesity, Chapter 12. In *Global Perspectives on Childhood Obesity: Current Status, Consequences and Prevention*, Second Ed.; Bagchi, D., Ed.; Academic Press, Elsevier, 2019a; pp 137–146, ISBN: 9780128128404.

Jarouliya, U.; Keservani, R. K. Protein Function as Cell Surface and Nuclear Receptor in Human Diseases, Chapter 1. In *Medicinal Chemistry with Pharmaceutical Product Development*; Bharti, S.; Mahapatra, D. kar, Ed.; Apple Academic Press, CRC Press, Taylor & Francis Group, 2019b; pp 1–32, ISBN: 9781771887106.

Johnson, L. J.; Tricker, P. J. Epigenomic Plasticity Within Populations: Its Evolutionary Significance and Potential. *Heredity* **2010,** *105* (1), 113–121.

Madden, J.; Williams, C. M.; Calder, P. C.; Lietz, G.; Miles, E. A.; Cordell, H.; Mathers, J. C.; Minihane, A. M. The Impact of Common Gene Variants on the Response of Biomarkers of Cardiovascular Disease (CVD) Risk to Increased Fish Oil Fatty Acids Intakes. *Ann. Rev. Nutr.* **2011,** *31*, 203–234.

Kaput, J.; Rodriguez, R. L. Eds.; In *Nutritional Genomics: Discovering the Path to Personalized Nutrition*; John Wiley & Sons, 2006.

Keservani, R. K.; Kesharwani, R. K.; Sharma, A. K.; Vyas, N.; Chadoker, A. Nutritional Supplements: An Overview. *Int. J. Curr. Pharm. Rev. Res.* **2010b,** *1* (1), 59–75.

Keservani, R. K.; Kesharwani, R. K.; Vyas, N.; Jain, S.; Raghuvanshi, R.; Sharma, A. K. Nutraceutical and Functional Food as Future Food: A Review. *Der Pharmacia Lettre* **2010a,** *2* (1), 106–116.

Keservani, R. K.; Sharma, A. K.; Kesharwani, R. K. An Overview and Therapeutic Applications of Nutraceutical and Functional Foods. In *Recent Advances in Drug Delivery Technology*, 2017; pp 160–201. ISBN:9781522507543.

Keservani, R. K.; Sharma, A. K.; Kesharwani, R. K. Eds.; *Nutraceuticals and Dietary Supplements: Applications in Health Improvement and Disease Management*; CRC Press, 2020; ISBN: 9781771888738.

Keys, A.; Menotti, A.; Aravanis, C.; Blackburn, H.; Djordevič, B. S.; Buzina, R.; Dontas, A. S.; Fidanza, F.; Karvonen, M. J.; Kimura, N.; Mohaček, I. The Seven Countries Study: 2,289 Deaths in 15 Years. *Prev. Med.* **1984,** *13* (2), 141–154.

Kim, J.; Cho, Y. A.; Kim, D. H.; Lee, B. H.; Hwang, D. Y.; Jeong, J.; Lee, H. J.; Matsuo, K.; Tajima, K.; Ahn, Y. O. Dietary Intake of Folate and Alcohol, MTHFR C677T Polymorphism, and Colorectal Cancer Risk in Korea. *Am. J. Clin. Nutr.* **2012,** *95* (2), 405–412.

Klose, J. Protein Mapping by Combined Isoelectric Focusing and Electrophoresis of Mouse Tissues. *Humangenetik* **1975,** *26* (3), 231–243.

Kwak, N. S.; Jukes, D. J. Functional Foods. Part 1: The Development of a Regulatory Concept. *Food Control* **2001,** *12* (2), 99–107.

Marcum, J. A. Nutrigenetics/Nutrigenomics, Personalized Nutrition, and Precision Healthcare. *Curr. Nutr. Rep.* **2020,** *9* (4), 338–345.

Mattioli, S. Bosco, A. D.; Martino, M.; Ruggeri, S.; Marconi, O.; Sileoni, V.; Falcinelli, B.; Castellini, C.; Benincasa, P. Alfalfa and Flax Sprouts Supplementation Enriches the Content of Bioactive Compounds and Lowers the Cholesterol in Hen Egg. *J. Funct. Foods* **2016,** *22* (9), 454–462.

Meyers, L. D.; Hellwig, J. P.; Otten, J. J. Eds.; Dietary Reference Intakes: The Essential Guide to Nutrient Requirements; National Academies Press, 2006.

Neeha, V. S.; Kinth, P. Nutrigenomics Research: A Review. *J. Food Sci. Technol.* **2012,** *50* (3), 415–428.

O'Farrell, P. H. High Resolution Two-dimensional Electrophoresis of Proteins. *J. Biol. Chem.* **1975,** *250* (10), 4007–4021.

Pan, A.; Sun, Q.; Bernstein, A. M.; Schulze, M. B.; Manson, J. E.; Stampfer, M. J.; Willett, W. C.; Hu, F. B. Red Meat Consumption and Mortality: Results from 2 Prospective Cohort Studies. *Arch. Intern. Med.* **2012,** *172* (7), 555–563.

Phillips, C. M.; Goumidi, L.; Bertrais, S.; Field, M. R.; McManus, R.; Hercberg, S.; Lairon, D.; Planells, R.; Roche, H. M. Dietary Saturated Fat, Gender and Genetic Variation at the TCF7L2 Locus Predict the Development of Metabolic Syndrome. *J. Nutr. Biochem.* **2012,** *23* (3), 239–244.

Poch, E.; González, D.; Giner, V.; Bragulat, E.; Coca, A.; de la Sierra, A. Molecular Basis of Salt Sensitivity in Human Hypertension: Evaluation of Renin-Angiotensin-Aldosterone System Gene Polymorphisms. *Hypertension* **2001,** *38* (5), 1204–1209.

Scheele, G. A. Two-dimensional Gel Analysis of Soluble Proteins. Charaterization of Guinea Pig Exocrine Pancreatic Proteins. *J. Biol. Chem.* **1975,** *250* (14), 5375–5385.

Shahidi, F.; Arachchi, J. K. V.; Jeon, Y. J. Food Applications of Chitin and Chitosans. *Trends Food Sci. Technol.* **1999,** *10*, 37–51.9.

Sharma, A. K.; Keservani, R. K.; Gautam, S. P. Herbal Product Development, Apple Academic Press, CRC Press, Taylor & Francis Group, 2020; ISBN: 9781771888776.

Singh, P.; Kesharwani, R. K.; Keservani, R. K. Antioxidants and Vitamins: Roles in Cellular Function and Metabolism. In *Sustained Energy for Enhanced Human Functions and Activity*; Academic Press, 2017a; pp 385–407, ISBN:9780128054130.

Singh, P.; Kesharwani, R. K.; Keservani, R. K. Protein, Carbohydrates, and Fats: Energy Metabolism. In *Sustained Energy for Enhanced Human Functions and Activity*; Academic Press, 2017b; pp 103–115, ISBN:9780128054130.

Taylor, V. J. K.; Siscovik, D. S.; Lemaitre, R. N. Circulating Omega-3 Polyunsaturated Fatty Acids Andsubclinical Brain Abnormalities on MRI in Older Adults: The Cardiovascular Health Study. *J. Am. Heart Assoc.* **2013**. DOI 10.1161/JAHA.113.000305.10.

Van Binsbergen, E. Origins and Breakpoint Analyses of Copy Number Variations: Up Close and Personal. Cytogenet. *Genome Res.* **2011,** *135* (3-4), 271–276.

van Ommen, B.; El-Sohemy, A.; Hesketh, J.; Kaput, J.; Fenech, M.; Evelo, C. T.; McArdle, H. J.; Bouwman, J.; Lietz, G.; Mathers, J. C.; Fairweather-Tait, S. The Micronutrient Genomics Project: A Community-Driven Knowledge Base for Micronutrient Research. *Genes Nutr.* **2010,** *5* (4), 285–296.

Varma, V.; Wise, C.; Kaput, J. Carbohydrate Metabolic Pathway Genes Associated with Quantitative Trait Loci (QTL) for Obesity and Type 2 Diabetes: Identification by Data Mining. *Biotechnol. J.* **2010,** *5* (9), 942–949.

Willett, W.; Skerrett, P. J.; Giovannucci, E. L. Eat, Drink, and be Healthy: the Harvard Medical School Guide to Healthy Eating; Simon and Schuster, 2017.

Xingliang, S.; Wang, J.; Zhu, J. Effect of Porogenic Solvent on Selective Performance of Molecularly Imprinted Polymer for Quercetin. *Mat. Res.* **2009,** *12* (3).

Yamada, R. Primer: SNP-Associated Studies and what they can Teach us. *Nat. Clin. Pract. Rheumatol.* **2008,** *4* (4), 210-217.

Yu, L. R.; Stewart, N. A.; Veenstra, T. D. Proteomics: The Deciphering of the Functional Genome. In *Essentials of Genomic and Personalized Medicine*; Academic Press, 2010; pp 89–96.

Yang, Q.; Zhang, A. H.; Miao, J. H.; Sun, H.; Han, Y.; Yan, G. L.; Wu, F. F.; Wang, X. J. Metabolomics Biotechnology, Applications, and Future Trends: A Systematic Review. *RSC Adv.* **2019,** *9* (64), 37245–37257.

# CHAPTER 11

# NUTRIENTS ARE AFFECTED BY GENETICS

VASIM T. PATHAN[1], SHUBHAM S. BAGADE[1], MANALI P. BORSE[1], SURAJ M. SARODE[2], and ANIL G. JADHAV[1]

[1]*Sandip Institute of Pharmaceutical Sciences, Nashik, Maharashtra, India*

[2]*Shree Sureshdada Jain Institute of Pharmaceutical Education and Research, Jamner, Maharashtra, India*

## ABSTRACT

In the scientific world, health and diet have taken center stage. Considerable progress has been observed in the notion of healthy food in recent years, and study on the subject has a significant impact on today's population. People have known that the food we consume has an effect on our health since the dawn of time. The human genome project, as well as advancements in science, have raised many questions about the relationship between what we consume and our genetic composition. The variations in genetics affect the food tolerances in various human population groups and may also impact dietary requirements, originating the new area of nutritional genomics and increasing the possibility of altering nutritional intake for the best possible prevention of diseases based on an individual's genome. Nutrigenomics deals with the interactions of genes and nutrients at molecular levels with the help of epigenomics and epigenetics. The common ground for pharmacogenomics and nutrigenomics is individual genetic composition. Each human being is genetically as well as phenotypically unique. The genetic composition and diet directly influence certain chronic diseases including

Nutrigenomics and Nutraceuticals. Raj K. Keservani, Eknath D. Ahire, Shubham J. Khairnar, Sanjay J. Kshirsagar, & Rajesh K. Kesharwani (Eds.)

cancer, obesity, hypertension, and diabetes. The relationship between food and genetics contributes to the array of disease risk factors in response to nutritional factors with the potential to boost or protect against disease. The understanding of these gene–nutrient interactions will benefit in the provision of personalized diets based on each person's genotype. As a result, it will be feasible to treat existing ailments or avoid future illnesses.

## 11.1 INTRODUCTION

Excellent nutrition is essential for good health, appropriate growth and development, and illness prevention. People have grown to recognize the food–health relationship through countless millennia, and many civilizations have integrated this notion into their approach to healing. We now know that nutrients as well as other dietary (Keservani et al., 2020) components received from a broad range of foods improve health, maintain metabolic homeostasis, and meet our energy needs; thanks to the advancement of nutritional sciences. Humans respond to nutrition in a variety of ways due to interindividual differences in genetic, epigenetic, and metabolic phenotypic status. Although genetics is a major factor, it only accounts for a percentage of the variance. The aggregate of metabolic activity operating under diverse conditions throughout the life cycle, including the complex relationships between genotype, metabolic phenotype, nutrition, lifestyle, and the environment, attain and maintain a person's overall phenotype, in addition to the health status. Metabolic control, from genes to metabolites, governs biochemical activities as well as an individual's nutritional and dietary requirements. The nutritional requirement varies according to different groups of the population and is not considered as a whole. As a result, genetic disposition and metabolic demands play a crucial role in selecting an individual's ideal diet. High-throughput methods in proteomics, metabolomics, and genomics may now detect and analyze DNA sequences, RNA transcripts, proteins, and nutrition metabolic fluxes in a single experiment in the post-genomic age (German and Young, 2004; Go and Wang, 2004).

The new availability of genomic data, as well as our improved knowledge about the relationships between genetic variants and food, allows for a quantitative evaluation of genetic variation's contribution to nutritional needs. One of the important elements that facilitate the research of gene–environment interactions is the human genome project. It gives a list of the approximately 30,000 genes in human DNA (Ventor and Adams, 2004), which is essentially a "parts list" for the proteins and chemicals that make up the processes and pathways that carry out all cellular operations, whether they be the interconversion of food into energy or the formation of ideas. With this parts list, we can put these networks together and learn how they operate, interact, and, most importantly, how they are governed. With this knowledge, molecular pathways may be altered for advantage by exposing them to exogenous stimuli, the most effective of which are food components and pharmacological drugs. The human genome project's upcoming phase focuses on cataloging and categorizing all of the variations found in the human genome. Each person is unique and phenotypically different, not just in appearance but also in physiology and reactivity to external stimuli. Single-nucleotide polymorphisms (SNPs) are a major component of human genetic diversity and serve as the molecular foundation for phenotypic variation. SNPs are changes in the DNA blueprint that might be single nucleotide base pair insertions, deletions, or base pair replacements. The human genome contains approximately 7 million SNPs (Hinds and Stuve, 2005), although only a tiny number of them have a functional impact.

Human nutrition assessments are incomplete unless the underlying genetic diversity is considered, which may manifest as variances in nutritional systems such as absorption, digestion, excretion, and receptor activation (Velazquez and Bourges, 1984). Inherited variances in enzyme and other functional protein activities lead to variations in dietary needs and the differential interaction of particular foods with genetically determined biochemical and metabolic variables. This inborn variation differs significantly from epigenetic variation during growth, pregnancy, and old age settings. Food preferences and dislikes, as well as nutrition, may be influenced by genetic diversity. The failure to taste the synthetic chemical phenylthiocarbamide, for example, is a widespread monogenic characteristic that causes a huge section of the population to be unable to taste this molecule, which others find highly unpleasant (Harris and Kalmus, 1941). Other instances, such as the variety in the taste of artificial sweeteners, have received less attention. The level of diversity for a specific gene product is an important nutritional topic. When providing dietary recommendations, little variations are often overlooked, but large variations cannot be ignored. Rare variations affecting just a few individuals may constitute a concern if their health is jeopardized under conditions that benefit the majority of the population's health. Nutritional considerations may have had an effect on human evolution by favoring specific genotypes (Neel, 1984). Thus, fasting may have benefited genotypes predisposed to hyperlipidemia and noninsulin-dependent diabetes by allowing for more rapid mobilization of lipids and glucose, resulting in a slightly improved probability of survival and reproduction. A same concept applies to obesity–predisposing genes. Such a theory would explain the unusually high frequency of these characteristics.

## 11.2 PHARMACOGENOMICS

The effects of pharmaceutical agents vary and show a wide array of effects. The effects range from being effective as well as beneficial in some subjects whereas have no effect in others. In certain cases, these agents can even have some side effects. The pharmacogenomics intends to elaborate the relationship between genetic sequencing and responses to a particular pharmaceutical product, with the main objective of better matching the drug to the individual to attain the intended outcome while avoiding the risk of an adverse consequence or side effect. The best example to understand what exactly pharmacogenomics is and its importance is well described by the relation between genetic polymorphism of the enzyme named thiopurine S-methyltransferase (TPMT) along with the toxic behavior exhibited when it is associated with 6-thiopurine. An immunosuppressant, 6-thiopurine, is utilized to cure acute lymphoblastic leukemia in children. When 6-thiopurine enters a cell, it is transformed into a nucleoside and competes with nucleic acid synthesis. 6-thiopurine is inactivated by TPMT, which is abundant in the liver and RBCs, and the normal dosage is large enough to account for this impact (Corominas and Baiget, 2004). TPMT activity, conversely, is subject to genetic variability, and about 1 in 300 persons acquire TPMT deficiency as an autosomal recessive characteristic (McLeod and Krynetski, 2000). TPMT-deficient individuals exhibit severe and in some instances fatal, myelosuppression when given a conventional dosage of 6-thiopurine; in these patients, a 10- to 15-fold lower dose is required for effective therapy (McLeod and Krynetski, 2000; Scheuermann and Keeler, 2004).

Pharmacogenomics is a pharmacological technique that takes the patient's genetics into consideration. It employs genomic approaches to investigate drug functioning and identify novel therapeutic targets, as well as genetically based high-throughput screens to identify new medications and large-scale genotyping to profile and assess research and treatment populations. Thus, pharmacogenomics employs genotyping to screen populations to discover which medications would be most effective in treating or avert disease in human subpopulations, as well as different genomic technologies (e.g., gene expression) to understand how the drug interacts with the genotype. The sole difference between pharmacogenetics and pharmacogenomics is the technology utilized and the range of the genes investigated (Corominas and Baiget, 2004).

## 11.3 RELATIONSHIP BETWEEN FOOD AND DRUGS

Humans developed in the existence of a broad diversity of foods, and human metabolism is well suited to the large variety of foods obtained from natural chemicals. In terms of human evolutionary history, medicines are a relatively recent innovation. However, one component of human physiology, notably the detoxification procedures mostly carried out by the liver, is well adapted to the metabolism of both dietary molecules and medications. These detoxification processes are primarily focused on plant-derived xenobiotics (Liska, 1998), but they also digest numerous medicines. This might be due to the increased number of medications produced from natural substances (Chin et al., 2006). Prescription medications may impair nutritional absorption, digestion, metabolism, use, or excretion. Similarly, nutritional status as well as food may influence medication action by affecting metabolism and function, and, more crucially, many dietary components can have pharmacological activity under particular conditions (Schmidt and Dalhoff, 2002; Lila and Raskin, 2005). Grapefruit is a well-known example of a drug/nutrient interaction since it may interact with a variety of medications including amiodarone, buspirone, carbamazepine, cyclosporine, and nifedipine. Monoamine oxidase blockers such as phenelcypromine, tranylcypromine, and isocarboxazid may also have harmful interactions with tyramine, a chemical present in several foods and drinks (Marcason, 2005). A negative interaction between the blood thinner warfarin as well as vitamin K has also been described (Camilo et al., 1998). Excessive usage of aluminium or magnesium hydroxide antacids may produce phosphate deficiency, which can lead to muscular weakness, anorexia, and even congestive heart failure. Diuretics such as thiazide and furosemide may deplete sodium, potassium, and magnesium, leading to anorexia and muscular fragility (Knauf et al., 2006). Methotrexate, a cancer-chemotherapeutic drug; triamterene, a diuretic; trimethoprim, an antibacterial agent; phenytoin, an anticonvulsant; and Sulphasalazine, an anti-inflammatory medication are all examples of folate antagonists (Nozaki et al., 2004). Phenytoin and Sulphasalazine are both folate antagonists and competitive inhibitors of folate transport. Weight loss and attenuation may result from a lack of folate. Penicillamine causes zinc deficiency, which may result in a loss of taste acuity and perhaps a reduction in food consumption (Floersheim et al., 1984). Alcoholism is also associated with deficits in thiamin, folate, vitamin B6, vitamin A, and zinc (Bowden et al., 1994). For their therapeutic impact, nutrients are occasionally employed in exceptionally high amounts. Niacin, for example, is a pharmacological agent used to decrease blood cholesterol levels (Morgan et al., 2004). Vitamin A retinoid compounds have been used effectively to treat severe acne and other disorders (Thielitz et al., 2006). There is no known genetic component in any of these cases, yet individuals born with gene deficits must generate crucial functioning enzymes and may need levels of specific nutrients well in excess of that necessary by most people. For virtually all vitamins, such inborn metabolic errors have been documented for enzymes required for absorption, metabolism, or storage. As the interconnections between foods and pharmaceuticals become more recognized, so will the connections among pharmacogenomics as well as nutrigenomics.

## 11.4 METABOLIC GENOTYPE–PHENOTYPE RELATIONSHIP

Throughout the previous century, the discovery of biochemical processes in human metabolism was a key accomplishment in biological science research. This biochemical–metabolism knowledge base has also had a significant influence on our food supply as well as the growth of the pharmaceutical and biotechnology sectors (German and Young, 2004). There has been a massive advance in technological innovation in genome sequencing, proteomics, and metabolomics, and also the bioinformatic handling of the massive data sets generated since Watson and Crick explored the structure of DNA and decoded the human genome in the latter half of the previous century, and the decoding of the human genome in the beginning of this century (Carpenter, 2003) (Fig. 11.1).

Functional investigation at the genome level (large-scale DNA sequencing that exposed inconsistency in cyphering segments of genes, ensuing polymorphisms), gene expansion (transcriptomics), protein translation (proteomics), and metabolite network and fluxes (metabolomics) is now completely established. These new “omic” tools enable researchers to investigate the many-sided molecular process from gene to phenotype in nutrition metabolism as a continuum, an interconnected complex metabolic network, or both (German and Young, 2004) (Fig. 11.2).

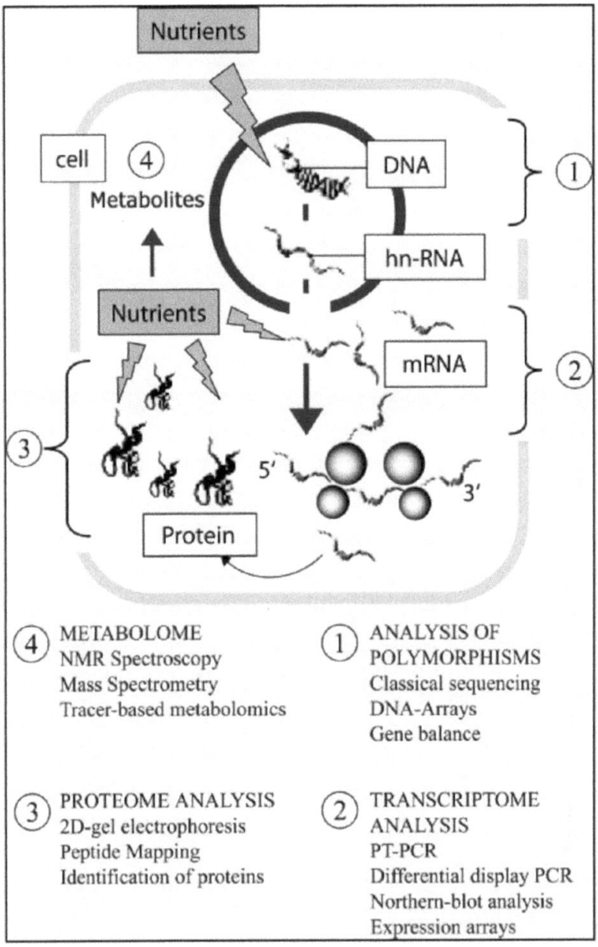


**FIGURE 11.1** A method used in metabolomics to investigate the biological response to dietary changes.

These genomic technologies have been employed in molecular nutrition research in two ways (Muller and Kersten, 2003). The classic hypothesis-driven method is used in the first strategy: particular nutrients impact the expression of specific genes and proteins, enabling the study of regulatory circuits, where data move from DNA to mRNA to protein. The other option is the systems biology approach, in which genes, proteins, metabolite nutrients, or dietary regimens are recorded and unified into a purposeful metabolism assessment tool to create indicators of early metabolic dysregulation and vulnerability impacted by diet. The study of the dietary genotype-phenotype

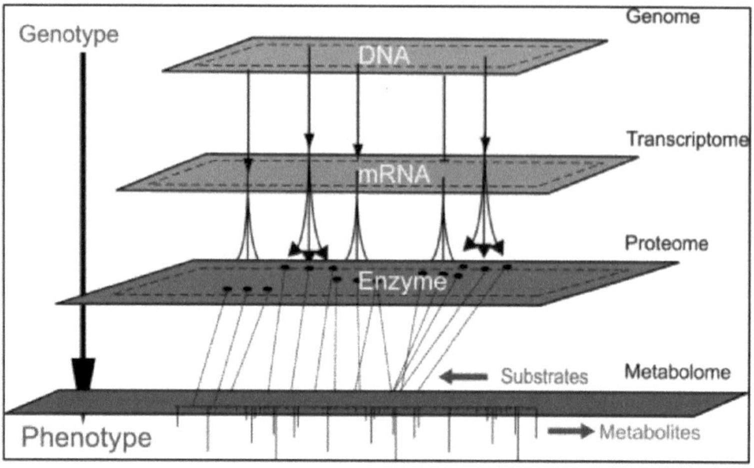


**FIGURE 11.2** The variety from genotype to phenotype.

association has been dubbed nutritional genomics, nutrigenetics, and nutrigenomics (DeBusk et al, 2005). The systems biology technique seeks to comprehend the genotype-phenotype link in biological systems connected by a metabolic network. All of the DNA sequences, mRNAs, proteins, and a huge multiple feedback loop from metabolites to proteins or transcripts involved in a certain physiological function are assessed using "omic" technology. This huge database is then merged and evaluated by powerful computer modules and bioinformatics approaches, which may result in the genotypic–phenotypic characterization of various cellular network systems as an answer to a given diet (Romero et al., 2004). However, functional analysis is inadequate unless metabolic fluxes are quantified using tracer-based metabolomics (Nelson, 2003; Lee and Go, 2005). Several databases have been constructed and employed in the progress of different biological networks to better understand the association between nutrition and cancer while also investigating the genotype-phenotype relationship (Chou et al., 2003; Martin et al., 2003; Eerligh et al., 2004). These techniques will result in a systems biology knowledge of the interactions between genetics, environment, and nutrition in health (Desiere, 2004). To achieve the goal of personal nutritional guideline recommendations throughout a person's life cycle, we hope to someday be able to examine personal variability in terms of specific diets and provide knowledge of how diet influences metabolic regulation in health and chronic disease control at the personal level (Fig. 11.3).

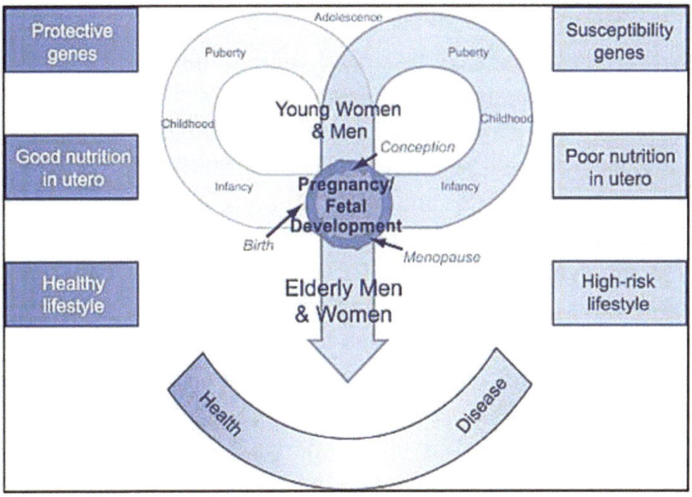


**FIGURE 11.3** Development of nutritional recommendations for health promotion and disease prevention during various periods of an individual's life cycle, including conception, infancy, adolescence, and aging, based on genotype and phenotype.

## 11.5 EFFECTS OF GENES ON NUTRITION

Nutrigenomics is an aspect of nutrition that employs molecular methods to search for, access, and comprehend the many responses produced by a particular diet applied to individuals or population groups (Cozzolino and Cominetti, 2013). This reaction will be determined by how genes modify their activity or gene expression (Fig. 11.4).

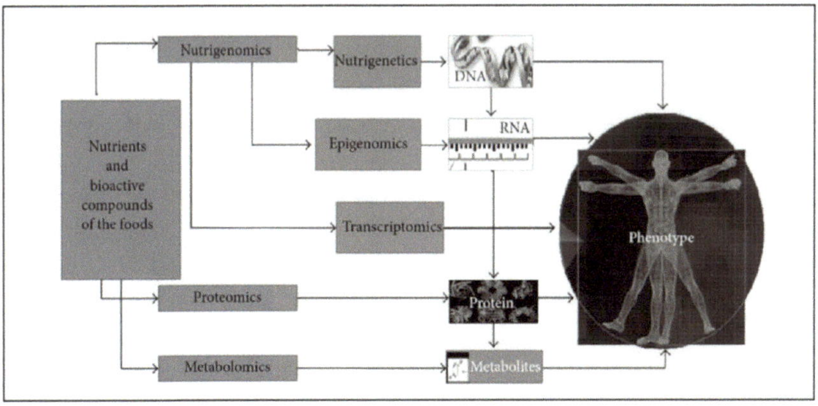


**FIGURE 11.4** "The disciplines of "omics" are utilized to better understand the relationship between diet, health, and illness.

*Source*: Modified from Anderson et al., 1985 and Bersaglieri et al., 2004.

## 1. From nutrition to nutrigenomics

Nutritional research may be traced back to 400 B.C., when Hippocrates proposed that a warm body temperature was natural. The so-called "analytical chemistry era" started about 1700 A.D. (Cruz et al., 2003). During this time, Lavoisier learned how the body digested food, producing water, energy, and carbon dioxide (Vasconcelos, 2010). Liebig discovered carbs, lipids, and other macronutrients that emit heat in the nineteenth century (Cruz et al., 2003; Vasconcelos, 2010). In the next era, known as the "analytical and chemical era of nutrition," which lasted from the 18th to the 20th centuries, Antonie Lavoisier made significant research on the metabolism of food and its relationship with the production of energy, including its implications for breathing and oxidation. The "postgenomic era" is now underway. This epoch is distinguished by the integration of three fields: biological, social, and environmental, which includes scientific advances in nutritional pathophysiology and metabolism (Cruz et al., 2003; Vasconcelos, 2010).

## 2. Nutrigenomics to other omics science

Following the HGP, talks, and initiatives started on a new biological age, the "postgenomic era, " in which bioinformatics evolution produced improvements in "omics" scientific study. In these fields, biotechnology is used to extract and analyze a greater number of biomolecules from the very same group, such as DNA, RNA, proteins, or metabolites. As a result, after genomics, additional omic sciences such as proteomics, metabolomics, and transcriptomics emerged as revolutionary tools (Dalmiel, 2012). As a result, the early definition of Nutrigenomics exclusively pertained to research on the impact of nutrients/bioactive foods on an individual's gene expression. Nowadays, this concept has broadened, and nutrigenomics also includes research on dietary aspects that safeguard the genome. As a result, this new research aims to comprehend the impact of dietary components on the genome, proteome, metabolome, and transcriptome (Ronteltap, 2009). Nutrigenomics may yield various answers in a single trial, therefore methods in this field must be unique (Cozzolino and Cominetti, 2013). Furthermore, nutrigenomics is capable of mining relevant biological data from the obtained information. The ultimate answer may only be obtained through a series of inquiries or surveys conducted among various groups and teams. As a result, there is a rising collaboration across countries/teams/research groups in the fields of nutrition, biology, medicine, genomics, and bioinformation (Ronteltap, 2009; Cozzolino and Cominetti, 2013; Vasconcelos, 2010).

## 3. Nutritional epigenomics

Epigenomics is the investigation of the whole collection of epigenetic alterations in a cell or tissue at a particular period. The epigenome is constituted of chemical substances that change or mark the genome in such a manner that it may signal what a cell can accomplish, as well as where and when it can do it. These are known as epigenetic markers (Sanhueza and Valenzuel, 2012). These epigenetic markers are transferred from cell toward cell as they divide and will therefore be handed down from generation to generation. These fingerprints are altered by genotype in the surrounding media (e.g., environment, nutrition, and medicines) and shape the phenotype (Sanhueza and Valenzuel, 2012; Dalmiel, 2012)

## 4. Epigenetics

The epigenetic study of DNA and protein modification that is connecting the DNA and histones, may produce changes in chromatin composition and not affect the nucleotide sequence (Moraes, 2009; 10]. Epigenetics is the transmission of information based on gene expression, and its modifications begin slowly but continue gradually and may be reversible (Moraes, 2009, Ordov′as and Carmena, 2012). For instance, in folate metabolism, folic acid, a vitamin connected to genetic integrity, guarantees a balanced number of deoxyribonucleotides for DNA replication (Ordov′as and Carmena, 2012). It acts as a cofactor for enzymes involved in the formation of nucleotides and thymidylate (a protein present in the RNA molecule), and finally as a universal donor in methyl and DNA methylation processes (Moraes, 2009; Ordov′as and Carmena, 2012). Changes in chromosomal shape may trigger epigenetic processes to alter gene expression (Cruz et al., 2003). histone acetylation and DNA methylation are two examples. Previous research has linked DNA methylation to chromatin remodeling, which is caused by the nutritional enzyme DNA methyltransferase (DNMT), which catalyzes the transfer of a $CH_3$ group from S-adenosylmethionine to particular locations on the DNA (Moraes, 2009; Cruz et al., 2003). Food molecules such as methionine, vitamins B6, B12, B2, folic acid, and choline are metabolized by S-adenosylmethionine (Cozzolino and Cominetti, 2013; Cruz et al., 2003). A lack of these may cause alterations in carbon metabolism and, as a result, affect DNA methylation, expanding the likelihood of developing NTCD. Once hypomethylation is related to malignancies such as hepatocellular carcinoma and prostate cancer, DNA hypermethylation decreases the gene accountable for transcription (Costa and Rosa, 2011; Cruz et al., 2003).

## 11.6 NUTRITIONAL GENOMICS

In much the same way as pharmacogeneticists seek to find polymorphisms that impact medication effectiveness and safety, the discipline of nutritional genomics seeks to identify the genetic variants that account for why some people respond in various ways to food components. For two reasons, the influence of genetic variation on dietary needs is always more subtle than that of pharmacological drugs. If the data used to generate the RDA and UL are drawn from a varied population, these standards should account for the majority of the genetic diversity that exists. If a genetic variant was discovered that necessitated a different RDA or UL, these criteria may be changed to meet the needs of these people (Brody et al., 2002).

The idea of the universal nutritional need is only jeopardized when the RDA of a nutrient for one person exceeds the UL of another. Second, the impact of diet lasts a lifetime, but the impact of pharmacological drugs lasts just a short time. As a result, any genetic variant that gives an unusual dietary need is nearly probably incompatible with life and early development, and so unviable. The SNPs that influence folate use, for example, in otherwise balanced populations, are potential causes for miscarriage and therefore not in Hardy–Weinberg equilibrium (Brody et al., 2002; Reyes-Engel et al., 2002; Nelen et al., 1997). In other words, a woman whose baby has two alleles that hinder adequate utilization of a certain nutrient is more likely to miscarry than a woman whose fetus has more frequent functional variations.

Nonetheless, specific genes and alleles have been shown that influence nutrition consumption. A well-studied mutation (Ala222Val) in the methylene tetrahydrofolate reductase (MTHFR) gene has been demonstrated to substantially affect folate metabolism (Reyes-Engel et al., 2002), elevating the risk of cardiovascular disease (CVD) as well as neural tube defects (NTDs) while decreasing the colon cancer risk (Smithells et al., 1983). Increased folate consumption may help to reduce biochemical disturbances and disease risk (Molly and Scott, 2001). This may be the finest example of a genetic variation that may alter an RDA to yet. It lends credence to the idea that genetic variation influences nutrient use and, possibly, dietary needs. Other polymorphisms have been discovered to affect homocysteine metabolism as well as folate absorption and transport (Brody et al., 2002). Polymorphisms in enzymes that use and metabolize vitamin $B_{12}$ have been linked to NTDs, the advancement of Down syndrome, and colon cancer, indicating that they may impact dietary needs (Wilson et at., 1999). Vitamin D receptor polymorphisms have been related to asthma in both children and adults (Raby et al., 2004). In Europeans, a frequent variation in the gene known as HEF (Cys282Tyr) is linked to the iron storage disorder hereditary hemochromatosis, and this may impact the UL for iron consumption (Beutler et al., 1996). Other gene polymorphisms influence lipid pathways, alcohol metabolism, and lactose metabolism. Many of these SNPs are connected with particular ethnic groups or geographic ancestral subpopulations, and they have positive selection genomic signatures, showing that, although they are deleterious now, they were likely advantageous in the ancestral environment where they originally emerged (Smithells et al., 1983; Raby et al., 2004).

Polymorphisms for the aldolase B enzyme, which metabolizes fructose, have been found. The gene encoding this specific enzyme is extensively variable, yet many of these variants were previously thought to be silent. Only when large amounts of fructose were introduced to the food supply as a sweetener did these polymorphisms appear as disease alleles. This is a paradigm of how a change in environment has pushed a typically quiet allele to the place where it starts to manifest as a disease allele (Cox, 2002).

In summary, the objectives of nutrigenomics are as follows:

1. Identifying transcription factors (as nutritional targets) and the genes they regulate.
2. Identification of cellular signaling pathways and characterization of the primary dietary signals.
3. Particular micronutrients and macronutrients that induce cell and organ-specific gene expression profiles.
4. Identification of linkages between nutrient-related regulatory mechanisms and proinflammatory stress processes to get a better knowledge of diet-related disorders.
5. Identifying genes that could play a role in the development of diet-related human illnesses (e.g., diabetes, hypertension, or atherosclerosis).
6. Identifying biomarkers for early sickness detection and susceptibility (stress signatures) that change in response to diet using dietary systems science.

## 11.7 CONSEQUENCE OF GENOMICS ON SIGNIFICANT DIETARY NECESSITIES

Those who are homozygous for such unique MTHFR T allele (T/T) require more folate than C allele carriers to reduce their risk of folate-related illnesses. The T/T genotype is prevalent in a wide range of populations around the world, despite the fact that there is no evidence that this polymorphism has been subjected to positive selection (Table 11.1).

**TABLE 11.1** The Existence of the MTHFR 667T/T Genetic Constitution in Several Cultural Sets.

| Populations | Prevalence (%) |
|---|---|
| African Americans (66) | 2 |
| Brazilians (68) | 7 |
| Chinese (68) | 16 |
| Dutch (64) | 5 |
| Europeans (68) | 17 |
| Irish (64) | 6 |
| Japanese (64) | 19 |
| Mexicans (64, 68) | 27–35 |
| Muslim Arab Israelis (65) | 16 |
| Pakistanis (68) | 4 |
| Sub-Saharan Africans (63, 68) | 0–1.5 |
| Toscanians (Italy) (67) | 30 |
| Yemenite Jews (65) | 2 |

This has the potential to alter public health policy since folate fortification may not be necessary in populations in which the T/T genotype is uncommon. However, the present RDA for folate covers both MTHFR genotypes, and there is no conclusive evidence that the existing RDA for folate should be adjusted for homozygous for the MTHFR 677T gene (Bailey and Gregory, 1999). The HFE gene Cys282Tyr polymorphism, contrary to this, has already had an impact on government policy, with two nations in Northern Europe suspending their iron fortification plans, in part due to possible harms to those at risk of hereditary hemochromatosis (Milman et al., 2003; Olsson et al., 1997). However, there is no conclusive evidence that the UL for those at risk of hereditary hemochromatosis exceeds the RDA. For the time being, it seems that there is no need to tailor iron consumption recommendations based on genotype (Moirand et al., 2002). Genetics is also influencing a rethinking of how we define nutritional deficiency. The term epigenetics mentions the inheritance of qualities that are not tied to DNA sequence but rather to DNA changes, one of which is DNA methylation, which impacts gene expression. Cooney et al. showed in 2002 how food intake may have a significant impact on epigenetics (Cooney et al., 2002). The scientists demonstrated that the folate consumption of the mother during pregnancy impacts the coat color of the progeny through an epigenetic impact on the expression of the agouti gene in inbred mice. Yellow-furred mice are observed when the agouti protein is expressed. However, methylation of the agouti gene promoter region during pregnancy inhibits agouti production, resulting in darker fur in the progeny (Cooney et al., 2002). The more methylation of the promoter area that occurs, the less yellow hair the mouse has. Methylation shows the results during gestation are metastable throughout the animal's lifetime; thus, by affecting methylation peaks during gestation, it is possible to manipulate the coat color of the offspring indefinitely. Folate metabolism is important for DNA methylation, and in the pups of moms that consumed a lot of folates while pregnant, the promoter region was methylated and the agouti gene was underexpressed. However, in pups whose moms were more folate-deficient during pregnancy, less DNA methylation happened during gestation, enabling the agouti to be expressed and the usual yellow coat color to emerge (Cooney et al., 2002; Waterland and Jirtle, 2003). This example shows how maternal diet may have a long-term impact on how genes are generated in the fetus and, eventually, the child, with potentially lifetime effects that influence a variety of health outcomes. Genomic stability is another kind of biomarker. According to twin studies, genetic variations account for 25% of the difference in life duration. As a result, environmental influences play a crucial role in influencing lifespan (Waterland and Jirtle, 2003). Mutations increase in our genomes as we age, and epigenetics has become dysregulated. Numerous B vitamins, along with several antioxidants, have an effect on the rate of mutation in mitochondrial and chromosomal DNA. Perhaps when deciding the RDA, we should take into account what concentration of a person's vitamins will reduce the mutation rate considerably to prevent disease (Fenech, 2002).

## 11.8 IDENTIFICATION OF GENETIC POLYMORPHISMS AFFECTING DIETARY REQUIREMENTS

Two techniques may be used to identify genetic variants that modify food requirements: the candidate gene methodology and evolutionary genomics.

The candidate gene strategy seeks connections between a specific pathologic disease and genetic variants that alter metabolic pathways linked with that disorder. This technique is effective in many circumstances; however, it is problematic in nutrition for three reasons:

Most nutrition-related disorders are complex diseases, which means they are multigenic and may be influenced by a variety of environmental variables. The sample size necessary for discovering the relationships between the allele of interest and the pathologic state is affected by the penetrance and prevalence of a polymorphism.

Many functional SNPs are not coding SNPs, but rather occur in gene promoter areas. In reality, it has been anticipated that the bulk of genetic variation is passed down via the generations that affect physiology in some manner is found in gene promoter regions (Rockman and Wray, 2004; Wray and Hahn, 2003). Two-thirds of SNPs found in human promoter areas have a 2-fold or higher effect on transcription rates (Rockman and Wray, 2004). Promoter regions may span thousands of base pairs and are more diverse than gene–coding regions (Rockman and Wray, 2004). The candidate gene technique is restricted by understanding the routes and networks that contribute to a certain pathologic condition: if all of the participants in a pathway or network are unknown, it will be impossible to discover all of the genes that impact it. The current accessibility of genome sequences from humans and nonhuman primates allows for the discovery of sequence "signatures" for positive selection (Clark et al., 2003)

The basic foundation for this technique is change in the DNA sequence, which occurs at a background rate of roughly $2.5 \times 10^{-8}$ mutations per nucleotide site in the truancy of selection pressure (Nachman and Crowell, 2000). However, sections of the sequence that are under selection pressure develop at a faster speed, which may be up to 400 times faster (Tishkoff and Verrelli, 2003). A fast-evolving gene is an adaptive sequence that allows an organism to accept the environmental difficulties, which might occur in different geographic locations. Nutrients and dietary components operate as environmental factors in nutritional genomics, forming today's human genome through millennia of selection pressure. This highly chosen genome may have an impact on dietary needs. When an adapted genome is placed in a new aspect, adaptive genes may become disease alleles, which must then be controlled or changed by dietary intervention (Schork et al., 1998).

The goal of evolutionary genomics is to uncover genes that evolve quickly within and across species. Once we have discovered these high-variable genes, we may infer selecting pressures based on their biological function. In fact, genes that quickly develop tend to cluster as immunity genes, metabolic genes, or reproduction genes (Clark et al., 2003). What are the indications that genes are being exposed to positive selection? Variants are predicted to congregate in certain geographic locations or ethnic groupings when there is a common selecting pressure and to be absent in other places or ethnic groups. We would witness an abundance of uncommon variations, substantial allele frequency disparities throughout the population, and a common haplotype that remained intact for many generations (Akey et al., 2004). Many genes found through evolutional genomics were formerly identified by the candidate gene technique, like the lactase gene polymorphism (Bersaglieri et at., 2004). This variation developed among northern Europeans and people living in desert parts of northern Africa, where being able to drink milk and eat dairy products as a source of nourishment may have provided a selection advantage. The *HFE Cys282Tyr* polymorphism, along with the alcohol metabolism gene, alcohol dehydrogenase (Osier et al., 2002), and the HbS allele in the "-globin" gene, exhibit indications of positive selection. The positive selection also exists for the $A^-$ and MED alleles of glucose-6-phosphate dehydrogenase because they give malaria resistance (Clark et al., 2003; Schork et al., 1998; Bersaglieri et at., 2004)

## 11.9 POLYMORPHISM IN MTHFR AND FOLATE NUTRITIONAL REQUIREMENTS

The MTHFR Ala222Val polymorphism, which impacts folate metabolism, is one example of how genetic variation may modify dietary needs. Folate is a B type of vitamin that transports and activates one-carbon units (Suh, 2001). The one-carbon units are passed by tetrahydrofolate's N-5 or N-10. The de novo production of the purine nucleotides as well as thymidylate, and the remethylation of homocysteine to methionine, need one-carbon metabolism. Methionine may be adenylated to generate S-adenosylmethionine, a cofactor for a variety of methylation processes, including those affecting gene regulation (Suh, 2001). MTHFR Ala222Val is a C-to-T transition in the coding area that causes the protein to change from an alanine to a valine. This has two impacts on one-carbon metabolism. First, it hinders homocysteine remethylation to methionine, which affects DNA methylation as well as gene expression (Oyama et al., 2004). Second, since the two routes compete, the conversion of deoxyuridine monophosphate (dUMP) to deoxythymidine monophosphate (dTMP) rises, resulting in greater folate-dependent thymidine production (Quinlivan et al., 2005). Folate deficiency raises the risk of various illnesses and abnormalities, including NTDs and CVD, and folate supplementation may reduce or eliminate the consequences of this polymorphism. As a result, the risk of NTDs is greatest in those who are low in folate. The same polymorphism is really protective against a number of malignancies, the most notable of which is colon cancer (Ma et al., 1997). The Physician's Health Study looked at the relationships of MTHFR mutation, plasma folate concentrations, and their interaction with colon cancer risk in 326 cancer-free controls and 202 colorectal cancer cases who were matched by smoking status and ag (Ma et al., 1997). The research found that among males with normal folate levels (serum folate 3 ng/mL) who were homozygous for the mutant allele, the incidence of colon cancer was one-third that of wild-type allele carriers (odds ratio (OR): 0.32; 95% CI: 0.15, 0.68). This protection, however, was lacking in individuals who were weak in folate. The same polymorphism and downstream impact on metabolism raise the risk of NTD while also being extremely protective against colon cancer (Ma et al., 1997).

A diet–gene interaction involving a variation in the gene metabolizing vitamin is the link between the gene encoding MTHFR and dietary folate. The existence of the variation seems to affect both cancer risk and folate needs. The autosomal recessive C677T mutation in the MTHFR gene results in a val/val genotype (Bailey and Gregory.1999; Frosst et al., 1996), which results in lower specific activity and increased thermolability of the enzyme. Homozygous people are reported to have 30% of normal enzyme activity, on the contrary, heterozygotes have 65% of normal enzyme activity (Frosst et al., 1996).

Because MTHFR aids in the conversion of 5, 10-methylenetetrahydrofolate, the vital intracellular folate pattern, to 5-methyltetrahydrofolate, the major form of folate in the circulation, the metabolic results of this genetic polymorphism can be seen in the dissemination of folate and related metabolites in body pools. Lower plasma folate and higher plasma homocysteine concentrations are linked to the MTHFR gene variation, indicating a change in the folate metabolic pathway (Bailey and Gregory, 1999; Frosst et al., 1996), differences in homocysteine and folate reactions to folic acid supplementation have been connected to this polymorphism (Bailey and Gregory.1999, Nelen et at., 1998). Due to the role of folate coenzyme activity in the transition of homocysteine to methionine, plasma homocysteine is a sensitive biomarker of folate status (Beresford and Boushey, 1997). A connection between the C677T MTHFR mutation and plasma folate and homocysteine levels has been discovered in many investigations. Those containing the homozygous C677T mutant genotype had complete plasma homocysteine concentrations that were 24% higher (P 0.05) than those with normal genotypes among those with low plasma folate values (15.4 nmol/L) (Jacques et al., 1996). In contrast, similar disparities across genotype subgroups are not detected in persons with higher folate levels (i.e., 15.4 nmol/L). According to studies including supplementary folic acid taken at amounts ranging from 0.5 to 5.7 mg/day, higher folate consumption may result in a 25% drop in homocysteine concentration (Malinow et al., 1999). As a result, it has been proposed that people with the homozygous C677T MTHFR genotype have a higher demand for folate to maintain optimum body pools.

The C677T polymorphism is found in varying degrees across racial and ethnic groupings. In recent research (Hessner et al., 1999), the PCR was used to analyze the incidence of numerous gene variants, including MTHFR C677T, in people from six racial/ethnic groups. MTHFR C677T allele frequencies were greater in Native Americans (29%), Koreans (47%), and Caucasians (27%), compared to African–Americans (12%) and Asian Indians (27%). In studies comparing African–Americans and Whites in the United States, homozygosity for the C677T genotype was shown to be very infrequent among African–American adults. For example, blood sample analysis in Ohio revealed that the C677T allele was present in three-tenth of Caucasians and one-tenth of African–Americans (McAndrew et al., 1996). Overall, the rates among Whites are 10–12% for homozygous and up to 50% for heterozygous for the MTHFR C677T mutation (Bailey and Gregory.1999), with large disparities amongst racial/ethnic minority groups in the United States.

Folate is vital for DNA production and methylation, both of which are physiologically important to carcinogenesis. Epidemiological studies have connected low folate levels or consumption with an increased risk of cervical cancer, colorectal cancer (Nelen et at., 1998), as well as lung cancer (Kluijtmans et al., 1996), and DNA methylation anomalies in colonic neoplasm is often reported (Makos et al., 1992).

In numerous case–control studies (Slattery et al., 1999), folate intake and dietary variables that impact folate status or methylation capacities (i.e., alcohol and dietary methionine) were observed to alter the pattern of menace for colorectal cancer that is normally linked with MTHFR polymorphisms. The C677T strain form of MTHFR, for example, was correlated with a low risk of colon cancer than any other types [RR 0.46 (95% CI: 0.25–0.84) in the Physicians' Health Study] (Ma J et al., 1996), but individuals with concentrated enzyme action appear to profit the most from a high folate and methionine diet and low alcohol (Chen et al., 1999). As the transformed folate metabolic route promotes normal DNA synthesis, people with the C677T strain of MTHFR may have a low risk of colorectal cancer, but they are also more likely to have trouble preserving normal DNA methylation due to a lower supply of folate or increased trauma on methylation capabilities caused by alcohol consumption (Chen et al., 1999; Jacques et al., 1996).

## 11.10 EXAMPLES OF DIET–GENE INTERACTION

### Metabolizing Enzymes

Xenobiotic metabolizing enzymes serve a substantial role in the activation and/or detoxification of foreign chemicals, including carcinogens. Interindividual heterogeneity in responsiveness to medicines and toxins has been documented a variety of dietary considerations. Both micronutrients and nonnutrient components of meals are well-known substrates or inducers of digesting enzyme systems, and numerous dietary items (containing both micronutrients and nonnutrient constituents of foods) operate as inducing agents via a variety of molecular pathways (Yang et al., 1992). In addition to the identified food inducers, other nutritional variables such as protein consumption, obesity, and fasting have been found in animal experiments to sway the activity of the oxidation enzymes (Guengerich, 1995). Polymorphisms occur in the genes which encode metabolizing enzymes, and the potential impact of these genetic changes on cancer risk due to diet–gene interactions have sparked a strong scientific interest.

The initial stage in the biotransformation of a foreign molecule or carcinogen is often the inclusion of more than one hydroxyl group to a relatively nonpolar hydrocarbon, which turns the substance into an electrophilic or more polar intermediate. These oxidation processes are done by activating enzymes (phase I), the cytochromes P450, which are encoded by CYP genes. The cytochrome enzymes, P450, also catalyze the oxidation of numerous endogenous substances, including steroid hormones and vitamin D metabolites. Phase II or conjugating enzymes accelerate conjugation reactions to substances such as glutathione, facilitating excretion. Phase II enzymes include glutathione S-transferases (GSTs), *N*-acetyltransferases (NATs), microsomal epoxide hydrolase, sulfotransferases, and uridine diphosphate (UDP)-glucuronosyltransferases (Lai and Sheilds, 1999). The exact enzymatic activity that is being promoted and the substrate involved determine whether a polymorphic version of these enzymes enhances or reduces cancer risk. Similarly, Increased metabolizing enzyme activity caused by dietary nutrients may result in the activation of a carcinogen or the purification of a reactive intermediate metabolite. Figure 11.5 depicts the relationships between the biotransformation enzyme systems.

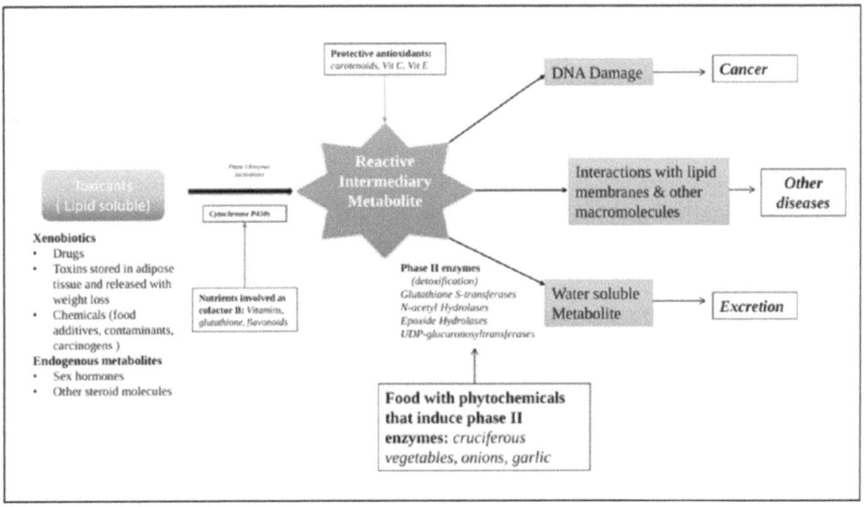


**FIGURE 11.5** Interrelations among the biotransformation enzyme organizations.

## P450 Cytochromes and CYP Genes

The cytochromes P450 (and related CYP genes) are a class of enzymes that are responsible for several endogenous chemical oxidation processes and the phase-I activation of a diverse variety of substrates (Guengerich, 1995; Yang et al., 1992). More than 30 enzymes of the same have been thoroughly studied, although just a limited number of P450s seem to be responsible for the majority of xenobiotic oxidation processes in humans (Guengerich, 1995). Polymorphic CYP genes cause variances in the capacity to oxidize substrates in the overall population. Chemical carcinogens, as well as certain foods and dietary ingredients, have been proven to induce P450s in people, and micronutrient supplements have been demonstrated to induce P450s in experimental animals (Andersonel al., 1985).

As previously noted, activating-enzyme reactions may result in the uptake of carcinogens to produce additional carcinogenic, activated intermediates. P4501A1, for example, stimulates polycyclic aromatic hydrocarbons (such as benzo[a]pyrene), while P4501A2 initiates heterocyclic amines, which are prevalent in high-temperature-cooked meats. Furthermore, these substrates and their suppliers (e.g., cigarette smoke, and well-done meat), as well as cruciferous vegetables (which include indole-3-carbinol), may stimulate P4501A2 activity. P450 activity may also be reduced by dietary elements such as naringenin, a flavonoid found in grapefruit juice (Yang et al., 1992).

Dietary influences on P450 endogenous substrates may also be important in the prevention of certain malignancies. Indole-3-carbinol, present in vegetables like cauliflower, cabbage, and Brussels sprouts, has been proven in animals and people to boost CYP1A1 and thereby increase estradiol 2-hydroxylase activity (Liu et al., 1994; Yang et al., 1992). In mice, feeding indole-3-carbinol at doses ranging from 34 to 700 mg/kg/day may boost estradiol 2-hydroxylase activity by a factor of five. In a small trial of 25 obese and nonobese women, oral dosages of 400 mg/day of pure indole-3-carbinol for 2 months raised the urine 2-hydroxyestrone to estriol ratio, an indirect indicator of 2-hydroxylation activity in the competitive estrogen metabolic pathway, by 82–93% (Michnovicz, 1998). Estradiol is converted to less powerful metabolites by 2-hydroxylation, and it is plausible that increasing this metabolic activity may lower the incidence of estrogen-related malignancies, however further research is certainly required to determine if the impact is clinically relevant.

The unexpected findings from two major clinical studies in which β-carotene and/or vitamin A supplementation was linked with an elevated risk of lung cancer in smokers (Omenn et al., 1996) may be attributable to an interaction between these micronutrients and cytochromes P450. P450 enzymes in the liver have been demonstrated to convert retinol and retinoic acid to polar metabolites (Ranerel al., 1996), and vitamin A, canthaxanthin (a non-provitamin A carotenoid), or a β-carotene metabolite (-apo-8′-carotenal) supplementation has been proven to trigger P450s in experimental animals (Michnovicz, 1998). In one recent research, large doses of β-carotene (500 mg/kg body weight) supplied to rats were demonstrated to stimulate CYP1A1, CYP1A2, and other comparable P450 genes, which might predispose a person to cancer risk from broadly bioactivated tobacco smoke procarcinogens (i.e., benzo[a]pyrene) (Paolini et al., 1999). At the moment, no human studies on the effect of carotenoids on P450 enzyme activity have been reported.

**S-Transferases for Glutathione**

GSTs are a significant cytosolic enzyme family that catalyzes the conjugation of reduced glutathione to a wide array of electrophilic chemicals produced by cytochrome P450 enzymes. GSTs serve an important function in shielding cells against the cytotoxic and mutagenic impacts of electrophiles because they may bind to DNA, creating adducts and possibly DNA mutations. So, it has been suggested that GST activation reduces overall cancer susceptibility (Bogaards et al., 1990), while poor detoxification by GST confers higher sensitivity to illness.

Based on their physicochemical and immunologic features, GSTs are classified into four primary classes: alpha (GSTA), pi (GSTP), mu (GSTM), and theta (GSTT). Within each class, many isozymes exist. GSTs have been detected in all human tissues investigated, however, the distribution of isozymes varies dramatically across tissues and organs. For example, of the total hepatic GST protein, 80% is GSTA, 10–20% is GSTM (in GSTM1-positive people), and 5% is GSTP (Rowe et al., 1997; Van Ommenel al., 1990). GSTP is the most usual class in the lung (85%), whereas GSTM and GSTA isozymes account for just 7–8% of the total (Van Ommenel al., 1990).

GST isozymes have overlapping substrate specificities; however, the GSTA family primarily conjugates organic hydroperoxides, some steroids, prostaglandins, and alkenals, whereas the GSTM family primarily conjugates epoxides and quinones, and polycyclic aromatic hydrocarbons, which are common carcinogens found in tobacco smoke, food, and combustion fumes. GSTT1 degrades epoxybutanes, monohalomethanes, and certain alkyl halides (Hayes and Pulford, 1995).

There are genetic polymorphisms in the GSTM, GSTT, and GSTP classes (27, 29, 62, 78). GSTM1 and GSTT1 have received the greatest attention to date since inherited homozygous deletions result in deficits of these two isozymes. The predominance of the GSTM1 null genotype varies greatly amongst ethnic groups. It varies from 22 to 35% among Africans and African–Americans to 38 to 67% among Caucasians and seems to be 100% among residents of the Republic of Kiribati (Oceania) (Rebbeck, 1997), which may explain for some of the worldwide disparities in cancer rates. The frequency of the GSTT1-null genotype in Africans and Asians is expected to be equivalent to that of the GSTM1-null genotype, but to be 20% in Caucasians (Rebbeck, 1997). GSTM1- and GSTT1-null genotypes, alone or in combination, increase the risk of numerous forms of cancer, especially among individuals exposed to carcinogen sources (e.g., cigarette smoke) (reviewed in Rebbeck, 1997). At the same time, a number of studies reveal that the GSTM1 genotype has no influence on risk, indicating that the GSTM1-null genotype is a moderately strong susceptibility factor (Brockmöller et al., 1998). However, when combined with CYP gene polymorphisms (e.g., CYP1A1 variations that result in enhanced P4501A1 activity), the GSTM1 null genotype may be a more significant risk factor. For example, among Japanese, GSTM1-null people with the CYP1A1 (MspI m2/m2) or m1/m2 genotype (OR: 5.2; 95% CI: 1.20–22.7), but not the wild type (m1/m1) (OR: 2.3; 95% CI: 0.39–12.6) genotype (OR: 2.3; 95% CI: 0.39–12.6) genotype (OR: 2.3; 95% CI (38). These gene-gene connections are weaker in non-Japanese groups, yet comparable associations have been documented (Hengstler et al., 1998).

GSTs have been shown to be modulated by diet in animal research and certain controlled feeding experiments in humans. In rats, different types of dietary fiber and fat have been demonstrated to influence hepatic GST activity (Reddy and Roa, 1993). Numerous phytochemicals, especially those found in cruciferous and allium vegetables, raise GST levels in rat liver and other tissues. In humans, 300 g of Brussels sprouts (rather than 300 g of glucosinolate-free veggies) elevated plasma GSTA concentrations by a factor of 1.4–1.5 in males, with no concurrent change in plasma GSTP concentrations (Nijhoff et al., 1995). Another feeding study discovered that drinking vegetable juice had also been closely linked with a twofold increase in peripheral lymphocyte GSTP expression in a subset of responders (Pool-Zobel et al., 1998); it has been proposed that the cancer-protective effect of vegetables is dependent on their ability to induce high levels of tissue GSTP. Cruciferous vegetables may have dual or conflicting effects in that their contents may both promote CYP1A2 (which may activate procarcinogens) and induce GSTM1 (which could potentially detoxify carcinogens). In fact, recent observational research found an interaction impact of GSTM1 genotype and cruciferous vegetable diet on CYP1A2 activity. GSTM1-null people had 21% greater CYP1A2 activity than GSTM1-positive persons who consumed equivalent quantities of cruciferous vegetables (P = 0.01) (Pool-Zobel et al., 1998), leading to increased excretion rates of the vegetable CYP1A2 inducers in the GSTM1-positive participants.

Recently, researchers have started to investigate the differential protective impact of food patterns based on GST genotypes. In one research that examined the relationship between colonic adenomas and cruciferous vegetables, only people who were GSTM1 null were protected against substantial levels of broccoli consumption ($p$ for trend, 0.001; $p$ for interaction, 0.01) (Roland et al., 1994). Grinberg–Funes et al. discovered an unfavorable connection between polycyclic aromatic hydrocarbon-DNA adducts and blood concentrations of vitamins E and C in GSTM1-null smokers ($r = 0.4$; $p = 0.06$). According to these findings, persons who lack the GSTM1 detoxifying mechanism and are exposed to high amounts of polycyclic aromatic hydrocarbons should consume more antioxidants and dietary elements that activate GSTs.

## Interactions of Diet with Other Metabolizing Enzymes

The NAT–gene interactions with food have been studied in various epidemiological studies, along with other metabolising enzymes that may be impacted by dietary variables. In humans, two NAT enzymes, NAT1 and NAT2, have been found, and both are demonstrated to be a polymorphic. These enzymes accelerate the metabolic stimulation of carcinogenic aromatic and heterocyclic amines. A NAT2 phenotype known as a slow acetylator is seen in 50–60% of Caucasians and 30–40% of African–Americans (Perera, 1997). Consumption of high-temperature-cooked meat, a source of heterocyclic amines and polycyclic aromatic hydrocarbons, and tobacco smoke exposure have been displayed to interact with the NAT2 genetic variant type, and thus may influence risk for colorectal and breast cancer, respectively (Zheng et al., 1999), though results from studies on the relationship between NAT2 and breast cancer risk have been somewhat inconsistent (Ambrosone et al., 1998). As NAT1 activity is detected in human mammary epithelial cells but not NAT2, it has been proposed that NAT1 is more important in the aetiology of breast cancer. Zheng et colleagues (Zheng et al., 1999) discovered a 30% greater risk (95% CI: 0.8–1.9) for breast cancer linked with the NAT1*10 allele and a nearly fourfold increased risk (95% CI: 1.5–10.5) linked with the NAT1*11 variant in a nested case–control analysis of postmenopausal Iowa women. The risen risk associated with the NAT1*11 allele was particularly noticeable among smokers (OR: 13.2; 95% CI: 1.5–116) and those who have taken the most red meat (OR: 6.1; 95% CI: 1.1–33.2). Similarly, in some (Chen et al., 1999) but not all epidemiological investigations, persons with the fast NAT2 acetylation type were shown to be at elevated risk of colorectal cancer in connection with greater amounts of meat intake.

Interactions between dietary variables and a variety of other metabolizing enzymes are also being investigated. For example, the existence of genetic variations of cytochromes P4503A3 and P4503A4 and the phase-II enzyme epoxide hydrolase may modify the risk of liver cancer associated with aflatoxin, fungal contamination of peanuts (Lai and Sheilds, 1999). As aflatoxin B1 is initiated and coupled through phases I and II enzymes, hereditary factors may contribute to the risk of cancer in reaction to aflatoxin exposure. The cytochrome P450 enzymes convert aflatoxin B1 to various compounds, at least one of which (8,9-exo-epoxide) seems to be mutagenic. The aflatoxin B1 epoxide metabolite is mutagenic when it combines with DNA, although it may also be detoxified by epoxide hydrolase. Although clinical relevance has yet to be shown, a link between mutant alleles of epoxide hydrolase, aflatoxin B1 adducts, and primary hepatocellular cancer (McGlynn et al., 1995) suggests that epoxide hydrolase polymorphisms may play a role in risk.

## 11.11 GENETIC FACTORS IN SOME CHRONIC DISEASES

### *11.11.1 CORONARY HEART DISEASE AND LIPIDS*

Hypercholesterolemia, high LDL concentrations, low HDL levels, and low apolipoprotein A1 levels have all been identified as risk factors for CHD, and they are all impacted by both hereditary and environmental factors. The presence of two distinct but mutually exclusive alleles at the apolipoprotein E locus has a significant impact on cholesterol and apolipoprotein B levels (Sing and Davignon, 1985). Mendelian inheritance is seen in familial hypercholesterolemia caused by multiple LDL receptor abnormalities (Goldstein and Brown, 1983), and genetic factors seem to play a role in the majority of hyperlipidemias. Familial mixed hyperlipidemia, a frequent disorder, seems to be passed as a monogenic autosomal dominant characteristic and is generally accompanied by an increase in apolipoprotein B (Brunzell and Motulsky, 1984). This condition's etiology seems to be diverse. Various apolipoprotein loci DNA markers have been linked to hyperlipidemia, CHD, or both, although the results are not totally consistent (Deeb et al., 1986). Considerable research is being conducted in these areas, which will likely assist in deciding the precise role of genetic variables in hyperlipidemias. The significance of such data in the provision of dietary guidance for the public is a critical problem. Some claim that, regardless of genetic diversity, reducing cholesterol levels in Western nations through dietary changes will significantly decrease the prevalence of CHD. The one with the lowest blood cholesterol levels had the lowest CHD mortality rate, which rises gradually as cholesterol levels rise (Martin et al., 1986). The majority of CHD fatalities and extra coronary events associated with increased cholesterol levels occur in the middle of the distribution curve, not at the top end. What are the consequences of reducing the population's cholesterol levels? For example, lowering an individual's cholesterol level from 226 to 210 mg/dL (through dietary adjustment) reduces the absolute risk of coronary death only marginally. A slight downfall in absolute risk for an individual, on the contrary, may have a significant impact on the community as a whole. In a hypothetical example, one may imagine that by making dietary changes, a person's chance of suffering a myocardial infarction during a specific time range can be reduced from 1 in 80 to 1 in 100. When applied to a population of 100,000 people, the estimated frequency of 1,250 myocardial infarcts (1/80) is decreased to 1,000 heart attacks (1/100). This 20% decrease (or the avoidance of 250 heart attacks) would have a substantial public health impact. The influence of nutrigenetic interactions on lipids is poorly understood. To address these issues, researchers must investigate the role of the LDL receptor, structural variation in apolipoproteins B and E, modulation of hepatic apolipoprotein B synthesis, and a variety of other factors. To determine the impact of nutrition in lowering cholesterol levels in people, further study in this vital field is necessary. However, it is evident that hereditary factors play a significant influence in determining cholesterol levels. The response to dietary restrictions on saturated fats, cholesterol, or lipid levels may vary, and not everyone will benefit similarly from dietary restraint. Some individuals will be susceptible, but others' lipid levels may be resistant to dietary modifications.

### *11.11.2 HYPERTENSION AND SALT*

Blood pressure levels are strongly influenced by genetics, as shown by research in families, adopted children, and twins (Burke and Motulsky, 1985). There is strong evidence that sodium has a role in the development and maintenance of high blood pressure. The prevalence of high blood pressure in communities is linked to average salt consumption. However, sodium loading does not raise blood pressure in everyone, while sodium restriction decreases blood pressure in many but not all hypertensives. As the so-called primitive populations have low hypertension, hypertension is an ecogenetic feature. Some individuals get high blood pressure when they relocate to a Western-style environment with a high salt consumption. African–Americans in the USA have higher mean blood pressure and a greater prevalence of hypertension than European-Americans. The absence of enhanced red-cell sodium/lithium countertransport in blacks, which is frequent in White hypertensives, is the most recently revealed discrepancy between Black and White hypertensive populations. This transport characteristic seems to be regulated by a single gene (Motulsky, 1987a). This, together with other research, implies that hypertension is a diverse entity with distinct genetic pathways. The salt restriction has been advised to minimize the frequency of high blood pressure, although it may not be equally beneficial for all people. Furthermore, clinical or laboratory criteria required to describe people's salt sensitivity are absent.

### 11.11.3 DIABETES MELLITUS

The majority of instances of diabetes mellitus in humans are classified as either noninsulin-dependent diabetes mellitus (NIDDM) or insulin-dependent diabetic mellitus (IDDM) (IDDM). Although twin and family studies clearly imply the existence of hereditary influences in both kinds, the precise processes remain unclear. Some of the individual genes implicated with IDDM (e.g., HLA-related genes) are better understood than those involved in the more frequent NIDDM variant. The relatively high correlation of identical twins for NIDDM shows that genes are particularly important in this condition (Barrett et al., 1981). Aside from the evident link of NIDDM to total calorie intake, nutrition as well as dietary components and not involved in the etiology of either type of the illness (Glinsmann et al., 1986).

Diabetes accounts for nearly 90% of all ailments worldwide (Cozzolino and Cominetti, 1993). Type-II diabetes is caused by a complex aetiology involving the interplay of hereditary and environmental variables (Phillips, 2013). According to genomic research, there are 65 SNPs related with the likelihood of acquiring type-II diabetes (Vorderstrasse et al., 2013). With improvements in genome sequencing and translating of the human genome, testing for detecting SNPs associated with type-II Diabetes was accessible to the general population. These checks may tell a patient whether they have a genetic tendency to develop the condition (Vorderstrasse et al., 2013). However, attention should be used while using this test in clinical practice, since studies have indicated that patients who got negative findings on this exam for the presence of type-II diabetes SNPs felt so confident in their results that they stopped taking care of their food diets. As a result, some people acquired diabetes later in life as a result of their food intake and insulin tolerance. Nonetheless, patients with a positive result for the existence of type-II Diabetes changed their lifestyle, particularly their food consumption, which subsequently reduced the disease's progression in this group (Vorderstrasse et al., 2013).

### 11.11.4 CANCER

Without question, those of us in cancer research were happy and pleased to witness a drop in the death rates of numerous main tumors 25 years following the signing of the National Cancer Act (Cole and Rodu 1996 ). Success, in contrast to failure, has a plethora of "fathers." Improved early exposure and treatment must take some of the credit. Simultaneously, we must investigate the influence that lifestyle modifications may have had. Cigarette-smoking cessation has clearly had a decisive role in the long-overdue lowering of lung cancer in men in the USA. Unfortunately, this is not the situation for females in the USA, where we are expected to see a continuous rise, in part because women tend to be at higher risk for lung cancer for a given quantity of tobacco exposure than males (Zand and Winder, 1996 ).

In terms of prostate, breast, and colon cancers, all three sites demonstrate a more than 6%-point decrease in mortality from 1989 to 1994. Could a comparable adjustment in lifestyle, such as changing food habits, have had an identical impact on all three sites? We have long believed that dietary fat, especially saturated as well as unsaturated fatty acids, has a substantial impact on the development and advancement of many malignancies. The evidence for this assertion is based primarily on topographical variations in cancer mortality that correlate with nutritional differences, as well as studies in laboratory animal models and mechanistic studies on the effect of specific fatty acid amounts and types on variables such as endocrine responses, immune functions, and gene expression. Beef and dairy products are two important sources of saturated fat in our diet. Beef and whole consumption of milk per capita have fallen significantly in recent decades (Stephen and Wald, 1990). Furthermore, several national surveys, including the United States Department of Agriculture (USDA) Nationwide Food Consumption Survey, the USDA ongoing Survey of Food Intakes of Individuals, and the National Health and Nutrition Examination Surveys, show that overall fat intake in the USA has steadily decayed over the last three decades, from more than 40% of total calories in the 1970s to approximately 36–37% in the 1980s and to 34% in the 1990s ( Stephen and Wald, 1990; Cole and Rodu 1996 ) Saturated fatty acids as of total daily calories fell from 15% in the 1970s to 13% in the 1980s and 12% in the 1990s. Cholesterol levels in plasma (mean mg/dL; corrected for ages 20–74 years) fell from 214 in the 1970s to 213 in the 1980s and 205 in the 1990s ( Johnson et al., 1993).

According to those who have bolstered the basis of behavioral medicine and have been monitoring lifestyle changes, the downward secular trend in dietary fat intake in the USA may be one of the "fathers" responsible for the recent decline in three major human cancers, a trend that may also be related to the marked decrease in mortality rates for myocardial infarction observed over the last three decades (Posner et al., 1995).

In 1900, the first year for which a reliable estimate is known, the CPM was 4%. It rose consistently after then, at least until 1990. The CPM is shown in Table 11.1 for 1970, 1980, and 1990 through 1995(Eerligh et al., 2004). The CPM has recently reached a plateau of about 23%. The CPM is not age-adjusted, and its long-term rise is mostly related to population aging. The rise mirrored decreases in fatalities from atherosclerotic heart problems and cerebrovascular disorders as well. Nonetheless, the proportional mortality of a group of diseases denotes the relative importance of the group to society over any short time interval. The practically constant CPM today suggests that the effect of cancer in the USA has stabilized. The mortality rate for all sources of death in Table 11.1 illustrates the well-established decrease in overall mortality and puts the decline in cancer rates into context. From 1990 to 1995, the overall mortality rate fell from 520 to 503 deaths per 100,000 person-years (py), a 0.7% decrease each year (Chou et al., 2003; Martin et al., 2003; Eerligh et al., 2004) The all-cancer mortality rate peaked at 135.0 in 1990 and then fell to 130.8 in 1995. The overall decrease in cancer mortality of 4.2 deaths per million people per year implies a 0.6% annual reduction. According to Table 11.1, a drop in lung carcinoma mortality accounted for 40% of the fall in cancer mortality rate or 1.6 deaths per million people per year. The remainder was attributed to decreases in many other types of cancer, with the mortality rate for "other cancer" falling by 2.6 deaths per 100,000 people each year. From 1950–1954 to 1986–1991 (36 years, mid-point to mid-point), the 5-year case casualty rate for all cancers fell by one-third, from 70% to 50%. A decrease in case fatality is difficult to interpret for several reasons, the most important of which is that cases diagnosed more recently tend to be diagnosed earlier in their course, resulting in both a true survival advantage and a lead-time bias when the endurance of recently treated patients is compared to that of patients diagnosed decades earlier. That is, newer patients actually live longer because they are detected sooner and get more sophisticated medical treatment. However, the increase in survival is overstated because early diagnosis introduces "lead-time bias, "which is the consequence of beginning the clock earlier when determining the length of survival. This bias was addressed by comparing the 5-year case fatality rate among a subset of 9 malignancies (or types of cancer) that are less likely to be identified earlier. Case fatality in these 9 disease entities decreased from 84% in 1950–1954 to 68% in 1986–1991. This equates to a 19% decrease in case fatality over 36 years, or around 0.5% every year. This pace of progress would have lowered the overall cancer mortality rate by around 0.6 fatalities per 100,000 people per year, or 3 deaths between 1990 and 1995 (Martin et al., 2003).

**TABLE 11.2** Cancer Proportional Mortality and Age-Adjusted Mortality Rates by Cause and Year: U.S. Total Population.

| Year | CMP (%) | All causes | All cancer | Lung cancer | Other cancer |
|---|---|---|---|---|---|
| 1970 | 17 | 714 | 129.9 | 28.4 | 101.5 |
| 1980 | 21 | 586 | 132.8 | 36.4 | 96.4 |
| 1990 | 24 | 520 | 135.0 | 41.4 | 93.6 |
| 1991 | 24 | 514 | 134.5 | 41.1 | 93.4 |
| 1992 | 24 | 505 | 133.1 | 40.8 | 92.3 |
| 1993 | 23 | 513 | 132.6 | 40.8 | 91.8 |
| 1994 | 24 | 508 | 132.1 | 40.1 | 92.0 |
| 1995 | 23 | 503 | 30.8 | 39.8 | 91.0 |
| **1990–1995 Reduction** | | | | | |
| total | – | 17 | 4.2 | 1.6 | 2.6 |
| annual | – | 3.4 | 0.8 | 0.3 | 0.5 |
| Annual (%) | – | 0.7 | 0.6 | 0.7 | 0.5 |

### *11.11.5 OBESITY*

Obesity (Jarouliya et al., 2019) is the most common chronic illness in the world, leading to the advancement of cardiovascular disease, diabetes mellitus type 2, metabolic syndrome, and other ailments. Slow body weight growth, which leads to obesity, is a long-term aftereffect of a long-term positive energy balance, which develops as a consequence of decreased physical activity and increased calorie consumption. The trend of decreasing physical activity and increasing the caloric content of food consumption is most likely the primary cause of the increase in patients with obesity, but it is essential to note that this tendency arises due to genetic variance in the population. The amount of scientific material pertinent to the issue of genetic predilection testing for obesity is rapidly expanding. This article presents a summary of contemporary research on the genetics of obesity, and also the importance of genetic testing of patient gene polymorphisms and genes related to carbohydrate and lipid metabolism problems (FTO, ADRB2, ADRB3, PPARG, and a number of others).

### *11.11.6 EPIDEMIOLOGIC RESEARCH*

Obesity has been recognized for many decades to be a genetic feature. Most obese individuals have usually one obese parent; nevertheless, members of a family or relatives share several nutritional and environmental factors. Other characteristics such as age, gender, and socioeconomic class, for example, were significant predictors of obesity in a study of the 10-State Nutrition Survey (Garn and Clark, 1976). Garn (1985) has observed that most offspring of fat parents become overweight, while those of slim parents are slimmer. Monozygotic and dizygotic twin studies provide the strongest evidence for genetic influences (Borjeson, 1976; Bouchard et al., 1985; Fabsitz et al., 1980; Feinleib et al., 1977; Medlund et al., 1977; Stunkard et al., 1986a, b). Borjeson (1976) analyzed 40 monozygotic twin pairs and 61 same-sex dizygotic twin pairs and found obesity heritability to be 88% (heritability may vary from 0%, showing no genetic influences, to 100%, indicating that a characteristic is fully influenced by genetics). Brook et al. came to a similar result (1975). A much-bigger twin research based on the vast Swedish Twin Registry found that monozygotic twins had a high concordance of body fatness but not indizygotic twins (Fabsitz et al., 1980). A new eating and activity challenge research in mono and dizygotic twins suggests a strong relationship between genetic predisposition and environment (Poehlman et al., 1986). The relative outcomes of environment & heredity have also been investigated by comparing obesity in adoptees to obesity in their adoptive and biological progenitors. There were no evident tendencies discovered in numerous early adoption studies (Annest et al., 1983; Withers, 1964). Obesity was discovered in one research among children with fat adoptive parents (Annest et al., 1983). The majority of these research lacked information on biological parents. Children spent a significant amount of time with a biological parent prior to adoption, and adoptive placement may have been based on physical similarities between adoptive parents and the adopted kid, which was a typical practice many decades ago. However, two recent investigations, one grounded on a Danish adoption registry (Stunkard et al., 1986a) and the other on Iowa fatness data (Price et al., 1987), containing information on both the patents, that is, biological as well as adoptive. The association of BMI in the Iowa research was independent of height and extended to the whole spectrum of obesity. The best relationships were predicted by a prototype that took both nonfamily and genetic environmental variables into account in stepwise multiple regression analyses. Nonetheless, most of the variation remained unexplained. Additional research on biological parents and siblings has shown a varying genetic and genetic-environmental influence (Bouchard, 1988). Nonetheless, the findings of these research studies imply that multiple polygenic variables have a role in human obesity. This formulation does not really rule out the potential of previously unknown essential genes being involved. Ethnic differences provide credence to but do not prove, the role of genes in obesity. The increased predominance of obesity among black women compared to White women may be due, at least in part, to hereditary factors (VanItallie, 1985). This might also apply to Mexican-American children, who are heavier than White or Black youngsters (Mueller, 1988). Excess calories alone cannot explain the high incidence of obesity among some Native American tribes and Pacific Island people. A twin study and (Bouchard, 1988) other ethnic group studies (Mueller, 1988) show that patterns of body fat distribution are inherited. This is particularly noticeable in children. Mexican-American children, for example, have a strong upper body fat distribution that is occasionally autonomous of adiposity per se (Mueller, 1988), while children of European heritage seem to have a more peripheral fat distribution (Mueller, 1988). The nature of the precise genes responsible for such events remains unclear.

### 11.11.7 ANIMAL RESEARCH

Obesity may be produced by a variety of single-gene mutations in mice or rats (Bray and York, 1971). Some pig and rat strains grow fat under certain dietary conditions but control animals do not grow fat. The precise biochemical or metabolic abnormalities responsible for this have yet to be identified. Intensive research studies are conducted to clone relevant potential genes. As the genetic process of human fat does not seem monogenic, it is doubtful that any of the animal models of monogenic obesity would adequately explain the genetic components operating in human obesity. However, it is probable that some of the genes inculpated in monogenic animal obesity have a role in human obesity. The discovery of the nature of the genes implicated in animal obesity will allow for direct assessment of the participation of comparable genes in humans utilising molecular genetics methods.

## 11.12 CONCLUSIONS

The nutrient intake in our day-to-day life affects our overall health. Each human being has a unique set of genes and hence responds to the nutrients differently. The human genome projects have given us a major insight on the effect of genes on the nutrients and how a proper diet would help avoid the serious health issues including hypertension, diabetes Mellitus, Cancer as well as obesity. The understanding of the gene–nutrient interaction will play a vital role among the treatment of these serious ailments. Pharmacogenomics takes into consideration the interaction of genes with a pharmaceutical product and hence plays and important role in determining how nutrients are affected by genes. Similarly, nutrigenomics is specifically the study of nutrients and genes and how the nutrient intake is affected by the phenotype of an organism. Pharmacogenomics is an important aspect when taking into mind the concepts of nutrigenomics. The genetic polymorphism can lead to modification in the reaction to the dietary changes by altering the ADME process. Epigenetics may cause alteration in the DNA methylation pattern and can influence the complete gene expression in response to the nutrient uptake. With the advancement in the knowledge about genes and their effects on the nutrient intake, we can get a clearer picture about the genetic concept at the fundamental level.

## KEYWORDS

- **nutrients**
- **genetics**
- **health**
- **human genome project**
- **epigenetics**
- **proteomics**
- **nutrigenomics**

## REFERENCES


Moraes, A.C.; Pereira, F. E. L. Nutrigenomics of the Soy. In *Parenteral Nutrition in Clinical Practice*; Waitzberg, D., Enteral, L., Eds., 4th ed.; Atheneu: S˜ao Paulo, Brazil, 4th edition, 2009; pp 2039–2047.

Akey, J. M.; Eberle, M. A.; Rieder, M. J. et al. Population History and Natural Selection Shape Patterns of Genetic Variation in 132 Genes. *PLoS Biol* [Serial Online] **2004,** *2*, e286.

Ambrosone, C. B.; Fredenheim, J. L.; Sinha, R.; Graham, S.; Marshall, J. R. et al. Breast Cancer Risk, Meat Consumption and *N*-acetyletransferase (NAT2) Genetic Polymorphisms. *Int. J. Cancer* **1998,** *75*, 825–830.

Anderson, K. E.; Pantuck, E. J.; Conney, A. H.; Kappas, A. Nutrient Regulation of Chemical Metabolism in Humans. *J. FASEB* **1985,** *44*, 130–133.

Cruz, B. M.; Taufer, M.; Schwanke, C. H. A. Genomics in the Era of Aging and Its Potential Application in Gerontology and Geriatrics. In *Institute of Geriatrics and Gerontology*

*PUCRS: The Cradle of Academic Geriatrics in Brazil*; Souza, A. C. A., Ed., 1st ed.; Pontif'ıcia Universidade Cat'olica do Rio Grande do Sul (EDIPUCRS), Porto Alegre: Brazil, 2003; pp 83–84.

Bailey, L. B.; Gregory, J. F., 3rd. Polymorphisms of Methylenetetrahydrofolate Reductase and Other Enzymes: Metabolic Significance, Risks and Impact on Folate Requirement. *J. Nutr.* **1999,** *129*, 919–922.

Barrett, A. H., Eff, C.; Leslie, R. D. G.; Pyke, D. A. Diabetes in Identical Twins: A Study of 200 Pairs. *Diabetologia* **1981,** *20*, 87–93. [PubMed: 7193616]

Beresford, S. A. A.; Boushey, C. J. Homocysteine, Folic Acid, and Cardiovascular Disease Risk. In *Preventive Nutrition: The Comprehensive Guide for Health Professionals*; Bendich, A.; Deckelbaum, R. J.; Humana: Totowa, 1997; pp. 193–224.

Bersaglieri, T.; Sabeti, P. C.; Patterson, N. et al. Genetic Signatures of Strong Recent Positive Selection at the Lactase Gene. *Am. J. Hum. Genet.* **2004,** *74*, 1111–1120.

Beutler, E.; Gelbart, T.; West, C. et al. Mutation Analysis in Hereditary Hemochromatosis. *Blood Cells Mol Dis* **1996,** *22*, 187–194; discussion 194a–b.

Bogaards, J. J. P.; van Ommen, B.; Falke, H. E.; Willems, M. I.; van Bladeren, P. J. Glutathione S-Transferase Subunit Induction Patterns of Brussels Sprouts, Allyl Isothiocyanate and Goitrin in Rat Liver and Small Intestinal Mucosa: A New Approach for the Identification of Inducing Xenobiotics. *Food Chem. Toxicol.* **1990,** *28*, 81–88.

Bowden, S.; Bardenhagen, F.; Ambrose, M.; Whelan G. Alcohol, Thiamin Deficiency, and Neuropsychological Disorders. *Alcohol Alcohol Suppl* **1994,** *2*, 267–272.

Brockmöller, J.; Cacorbil, I.; Kerb, R.; Sachse, C.; Roots, I. Polymorphisms in Xenobiotic Conjugation and Disease Predisposition. *Toxicol. Lett.* **1998,** *102*, 173–183.

Brody, L. C.; Conley, M.; Cox, C. et al. A Polymorphism, R653Q, in the Trifunctional Enzyme Methylenetetrahydrofolate Dehydrogenase/Methenyltetrahydrofolate Cyclohydrolase/ Formyltetrahydrofolate Synthetase Is a Maternal Genetic Risk Factor for Neural Tube Defects: Report of the Birth Defects Research Group. *Am. J. Hum. Genet.* **2002,** *71*, 1207–1215.

Brunzell, J. D.; Motulsky, A. G. Status of "Familial Combined Hyperlipidemia." In *Genetic Epidemiology of Coronary Heart Disease: Past, Present, and Future*; Rao, D. C., Elston, R. C., Kuller, L. H., Feinleib, M., Carter, C., Havlik, R., Eds.; Alan R. Liss: New York, 1984; pp 541–548.

Burke, W.; Motulsky, A. G. Hypertension—Some Unanswered Questions. *J. Am. Med. Assoc.* **1985,** *253*, 2260–2261. [PubMed: 3974121]

Phillips, C. M. Nutrigenetics and Metabolic Disease: Current Status and Implications for Personalised Nutrition. Switzerland. *J. Nutr.* **2013,** *5* (1), 32–57.

Camilo, M. E.; Paiva, S. A.; O'Brien, M. E.; Booth, S. L.; Davidson, K. W.; Sokoll, L. J. et al. The Interaction Between Vitamin K Nutriture and Warfarin Administration in Patients with Bacterial Overgrowth Due to Atrophic Gastritis. *J Nutr Health Aging* **1998,** *2*, 73–78.

Carpenter, K. J. A Short History of Nutritional Science: Part 3 (1912–1944). *J. Nutr.* **2003,** *133* (10), 3023–3032.

Chen, J.; Giovannucci, E. L.; Hunter, D. J. MTHFR Polymorphism, Methyl-Replete Diets and the Risk of Colorectal Carcinoma and Adenoma Among, U.S. Men and Women: An Example of Gene–Environment Interactions in Colorectal Tumorigenesis. *J. Nutr.* **1999,** *129*, 560–564S.

Chin, Y. W.; Balunas, M. J.; Chai, H. B.; Kinghorn, A. D. Drug Discovery from Natural Sources. *AAPS J.* **2006,** *8*, E239–E253.

Chou, W. H.; Yan, F. X.; Robbins-Weilert, D. K.; Ryder, T. B.; Liu, W. W.; Perbost, C.; Fairchild, M.; de Leon, J.; Koch, W. H.; Wedlund, P. J. Comparison of Two CYP2D6 Genotyping Methods and Assessment of Genotype-Phenotype Relationships. *Clin. Chem.* **2003,** *49* (4), 542–551.

Clark, A. G.; Glanowski, S.; Nielsen, R. et al. Inferring Nonneutral Evolution from Human-Chimp-Mouse Orthologous Gene Trios. *Science* **2003,** *302*, 1960–1963.

Cole, P.; Rodu, B. Declining Cancer Mortality in the United States. *Cancer* **1996,** *78*, 2045–2048.

Cooney, C. A.; Dave, A. A.; Wolff, G. L. Maternal Methyl Supplements in Mice Affect Epigenetic Variation and DNA Methylation of Offspring. *J. Nutr.* **2002,** *132*, 2393S–2400S.

Corominas, H.; Baiget, M. Clinical Utility of Thiopurine S-Methyltransferase Genotyping. *Am. J. Pharmacogenom.* **2004,** *4*, 1–8.

Cox, T. M. The Genetic Consequences of Our Sweet Tooth. *Nat. Rev. Genet.* **2002,** *3*, 481–487.

DeBusk, R. M.; Fogarty, C. P.; Ordovas, J. M.; Kornman, K. S. Nutritional Genomics in Practice: Where Do We Begin? *J. Am. Diet. Assoc.* **2005,** *105* (4), 589–598.

Deeb, S.; Failor, A.; Brown, B. G.; Brunzell, J. C.; Albers, J. J.; Motulsky, A. G. Molecular Genetics of Apolipoproteins and Coronary Heart Disease. In *Cold Spring Harbor Symposia on Quantitative Biology, Vol. LI. Molecular Biology of Homo Sapiens*; Cold Spring Harbor: New York, 1986; pp. 403–409. [PubMed: 2884066]

Desiere, F. Towards a Systems Biology Understanding of Human Health: Interplay Between Genotype, Environment and Nutrition. *Biotechnol. Annu. Rev*. **2004,** *10*, 51–84.

Eerligh, P.; Koeleman, B. P.; Dudbridge, F.; Jan Bruining, G.; Roep, B. O.; Giphart, M. J. Functional Genetic Polymorphisms in Cytokines and Metabolic Genes as Additional Genetic Markers for Susceptibility to Develop Type 1 Diabetes. *Genes Immun.* **2004,** *5* (1), 36–40.

Vasconcelos, F. A. G. The Science of Nutrition in Transit: From Nutrition and Dietetics to Nutrigenomics. *Revista de Nutricao* **2010,** *23* (6), 935–945.

Fenech, M. Micronutrients and Genomic Stability: A New Paradigm for Recommended Dietary Allowances (RDAs). *Food Chem. Toxicol.* **2002,** *40*, 1113–1117.

Floersheim, G. L.; Bianchi, L.; Probst, A.; Chiodetti, N.; Honegger, C. G. Influence of Zinc, D-Penicillamine and Oxygen on Poisoning with *Amanita phalloides*: Zinc Accelerates Liver Regeneration and Prevents the Depletion of Brain Noradrenaline Caused by the Mushroom. *Agents Actions* **1984,** *14*, 24–30.

Frosst, P.; Blom, H. J.; Milos, R.; Goyette, P.; Sheppard, C. A. et al. A Candidate Genetic Risk Factor for Vascular Disease: A Common Mutation in Methylenetetrahydrofolate Reductase. *Nat. Genet.* **1995,** *10*, 111–113.

Garn, S. M.; Clark, D. C. Trends in Fatness and the Origins of Obesity: Ad Hoc Committee to Review the Ten-State Nutrition Survey. *Pediatrics* **1976,** *57*, 443–456. [PubMed: 1264536]

German, B.; Young, V. R. Nutrition and Genomics. *Nestle Nutr. Workshop Ser. Clin. Perform Programme* **2004,** *9*:243–261; discussion 261–263.

Glinsmann, W. H.; Irausquin, H.; Park, Y. K. Evaluation of Health Aspects of Sugars Contained in Carbohydrate Sweeteners: Report of Sugars Task Force. *J. Nutr.* **1986,** *116*, SI–S216. [PubMed: 3543257]

Go, V. L.; Wang, Y.; Yang, H.; Lee, W. N. Neuro-Hormonal Integration of Metabolism: Challenges and Opportunities in the Postgenomic Era. *Nestle Nutr. Workshop Ser. Clin. Perform Programme* **2004,** *9*, 227–238; discussion 238–242.

Goldstein, J. L.; Brown, M. S. In Stanbury, J. B., Wyngaarden, J. B., Fredrickson, D. S., Goldstein, J. L., Brown, M. S., Eds.; *The Metabolic Basis of Inherited Disease*, 5th ed.; McGraw-Hill: New York, 1983; pp 672–712.

Guengerich, F. P. Influence of Nutrients and Other Dietary Materials on Cytochrome P-450 Enzymes. *Am. J. Clin. Nutr.* **1995,** *61*, 651S–658S.

Harris, H.; Kalmus, H. The Measurement of Taste Sensitivity to Phenylthiocarbamide. *Ann. Eugen.* **1941,** *15*, 24–31.

Hayes, J. D.; Pulford, D. J. The Glutathione S-Transferase Supergene Family: Regulation of GST and the Contribution of the Isoenzymes to Cancer Chemoprotection and Drug Resistance. *Crit. Rev. Biochem. Mol. Biol.* **1995,** *30*, 445–600.

Hengstler, J. G.; Arand, M.; Herrero, M. E.; Oesch, F. Polymorphisms of *N*-Acetyl Transferases, Glutathione S-Transferases, Microsomal Epoxide Hydrolase and Sulfotransferases: Influence on Cancer Susceptibility. *Recent Results Cancer Res.* **1998.**

Hessner, M. J.; Luhm, R. A.; Pearson, S. L.; Endean, D. J.; Friedman, K. D.; Mongomery, R. R. Prevalence of Prothrombin G20210A, Factor V G1691A (Leiden), and Methylenetetrahydrofolate Reductase (MTHFR) C677Tin Seven Different Populations Determined by Multiplex Allele-Specific PCR. *Thromb. Haemost.* **1999,** *81*, 733–738.

Hinds, D. A.; Stuve, L. L.; Nilsen, G. B. et al. Whole-Genome Patterns of Common DNA Variation in Three Human Populations. *Science* **2005,** *307*,1072–1079.

Ordov´as, J. M.; Carmena, R. Nutrigenetics. 2012. http://www.fundacionmhm.org/pdf/Mono9/Articulos/articulo1.pdf

Sanhueza, J.; Valenzuela, A. Nutrigenomics: Revealing the Molecular Aspects of a Personalized Nutrition. *Chilean J. Nutr.* **2012,** *39* (1), 1–12.

Jacques, P. F.; Bostom, A. G.; Williams, R. R.; Ellison, R. C.; Eckfeldt, J. H. et al. Relation Between Folate Status, a Common Mutation in Methylenetetrahydrofolate Reductase, and Plasma Homocysteine Concentrations. *Circulation* **1996,** *93*, 7–9.

Jarouliya, U., Keservani, Raj, K. Pathways Leading to Child Obesity. In *Global Perspectives on Childhood Obesity: Current Status, Consequences and Prevention*; Bagchi, D., Ed., 2nd ed.; Academic Press, Elsevier, 2019; Chapter 12, pp 137–146. ISBN: 9780128128404.

Johnson, C. L.; Rifkind, B. M.; Sempos, C. T.; Carroll, M. D.; Bachorik, P. S.; Briefel, R. R. et al. Declining Serum Total Cholesterol Levels Among US Adults: The National Health and Nutrition Examination Surveys. *JAMA* **1993,** *269*, 3002–3008.

Kawajiri, K.; Eguchi, H.; Nakachi, K.; Sekiya, T.; Yamamoto, M. Association of CYP1A1germ Line Polymorphisms with Mutations of the p53 Gene in Lung Cancer. *Cancer Res.* **1996,** *56*, 72–76.

Keservani, R. K.; Sharma, A. K.; Kesharwani, R. K., Eds. *Nutraceuticals and Dietary Supplements: Applications in Health Improvement and Disease Management*; CRC Press, 2020. ISBN: 9781771888738.

Kluijtmans, L. A.; van den Heuvel, L. P. W. J.; Boers, G. H. J.; Frosst, P.; Stevens, E. M. B. et al. Molecular Genetic Analysis in Mild Hyperhomocysteinemia: A Common Mutation in the Methylenetetrahydrofolate Reductase Gene Is a Genetic Risk Factor for Cardiovascular Disease. *Am. J. Hum. Genet.* **1996,** *58*, 35–41.

Knauf, H.; Bailey, M. A.; Hasenfuss, G.; Mutschler E. The Influence of Cardiovascular and Anti-Inflammatory Drugs on Thiazide-Induced Hemodynamic and Saluretic Effects. *Eur. J. Clin. Pharmacol.* **2006.** 10.1007/s00228-006-0190-3

Dalmiel, L.; Vargas, T.; Molina, A. R. Nutritional Genomics for the Characterization of the Effect of Bioactive Molecules in Lipid Metabolism and Related Pathways. *Electrophoresis J.* **2012,** *33* (15), 2266–2289.
Lai, C.; Shields, P. G. The Role of Interindividual Variation in Human Carcinogenesis. *J. Nutr.* **1999,** *129*, 552S–555S.
Le Marchand, L.; Sivaraman, L.; Pierce, L.; Seifried, A.; Lum, A. et al. Associations of CYP1A1, GSTM1, and CYP2E1 Polymorphisms with Lung Cancer Suggest Cell Type Specificities to Tobacco Carcinogens. *Cancer Res.* **1998,** *53*, 4858–4863.
Lee, W. N.; Go, V. L. Nutrient Gene Interaction: Tracer-Based Metabolomics. *J. Nutr.* **2005,** *135*, 3027–3032.
Lila, M. A.; Raskin, I. Health-Related Interactions of Phytochemicals. *J. Food Sci.* **2005,** *70*, R20–R27.
Liska, D. J. The Detoxification Enzyme Systems. *Altern. Med. Rev.* **1998,** *3*, 187–198.
Liu; Qian, S. B. Translational Regulation in Nutrigenomics. *Am. Soc. Nutr.* **2011,** *2*, 511–519.
Liu, H.; Wormke, M.; Safe, S. H.; Bjeldanes, L. F. Indole[3, 2-b]Carbazole: A Dietary-Derived Factor That Inhibits Both Antiestrogenic and Estrogenic Activity. *J. Natl. Cancer Inst.* **1994,** *86*, 1758–1765.
Dauncey, M. J. Recent Advances in Nutrition, Genes and Brain Health. *Proc. Nutr. Soc.* **2012,** *71* (4), 581–591.
Ma, J.; Stampfer, M. J.; Giovannucci, E. et al. Methylenetetrahydrofolate Reductase Polymorphism, Dietary Interactions, and Risk of Colorectal Cancer. *Cancer Res.* **1997,** *57*, 1098–1102.
Makos, M.; Nelkin, B. D.; Lerman, M. I.; Latif, F.; Abar, B.; Baylin, S. B. Distinct Hypermethylation Patterns Occur at Altered Chromosome Loci in Human Lung and Colon Cancer. *Proc. Natl. Acad. Sci. USA* **1992,** *89*, 1929–1933.
Malinow, M. R.; Bostom, A. G.; Krauss, R. M. Homocysteine, Diet, and Cardiovascular Diseases. *Circulation* **1999,** *99*, 178–182.
Marcason, W. What Is the Bottom Line for Dietary Guidelines When Taking Monoamine Oxidase Inhibitors? *J. Am. Diet Assoc.* **2005,** *105*, 163.
Martin, L. J.; North, K. E.; Dyer, T.; Blangero, J.; Comuzzie, A. G.; Williams, J. Phenotypic, Genetic, and Genome-Wide Structure in the Metabolic Syndrome. *BMC Genet.* **2003,** *4* (Suppl 1), S95.
Martin, M. J.; Hulley, S. B.; Browner, W. S.; Kuller, L. H.; Wentworth, D. Serum Cholesterol, Blood Pressure, and Mortality: Implications for a Cohort of 361, 662 Men. *Lancet* **1986,** *2*, 934–936. [PubMed: 2877128]
McAndrew, P. E.; Brandt, J. T.; Pearl, D. K.; Prior, T. W. The Incidence of the Gene for Thermolabile Methylenetetrahydrofolate Reductase in African Americans. *Thrombosis Res.* **1996,** *83*, 195–198.
McGlynn, K. A.; Rosvold EAT, Lustbader, E. D.; Hu, Y.; Clapper, M. L. et al. Susceptibility to Hepatocellular Carcinoma Is Associated with Genetic Variation in the Enzymatic Detoxification of Aflatoxin B1. *Proc. Natl. Acad. Sci. USA* **1995,** *92*, 2384–2387.
McLeod, H. L.; Krynetski, E. Y.; Relling, M. V.; Evans, W. E. Genetic Polymorphism of Thiopurine Methyltransferase and Its Clinical Relevance for Childhood Acute Lymphoblastic Leukemia. *Leukemia* **2000,** *14*, 567–572.
Michnovicz, J. J. Increased Estrogen 2-Hydroxylation in Obese Women Using Oral Indole-3-Carbinol. *Int. J. Obes.* **1998,** *22*, 227–229.

Milman, N.; Byg, K. E.; Ovesen, L.; Kirchhoff, M.; Jurgensen, K. S. Iron Status in Danish women, 1984–1994: A Cohort Comparison of Changes in Iron Stores and the Prevalence of Iron Deficiency and Iron Overload. *Eur. J. Haematol.* **2003,** *71*, 51–61.

Moirand, R.; Guyader, D.; Mendler, M. H. et al. HFE Based Re-evaluation of Heterozygous Hemochromatosis. *Am. J. Med. Genet.* **2002,** *111*, 356–361.

Molloy, A. M.; Scott, J. M. Folates and Prevention of Disease. *Public Health Nutr.* **2001,** *4*, 601–609.

Morgan, J. M.; Carey, C. M.; Lincoff, A.; Capuzzi, D. M. The Effects of Niacin on Lipoprotein Subclass Distribution. *Prev. Cardiol.* **2004,** *7*, 182–187.

Muller, M.; Kersten, S. Nutrigenomics: Goals and Strategies. *Nat. Rev. Genet.* **2003,** *4* (4), 315–322.

Munoz-Moran, E.; Dieguez-Lucena, J. L.; Fernandez-Arcas, N.; Peran-Mesa, S.; Reyes-Engel, A. Genetic Selection and Folate Intake During Pregnancy. *Lancet* **1998,** *352*, 1120–1121.

Nachman, M. W.; Crowell, S. L. Estimate of the Mutation Rate Per Nucleotide in Humans. *Genetics* **2000,** *156*, 297–304.

Neel, J.V. Genetics and Nutrition: An Evolutionary Perspective. In *Genetics Factors in Nutrition*; Velazquez, A.; Bourges, H., Eds.; Academic Press: Orlando, 1984; pp 3–16.

Nelen, W. L.; Blom, H. J.; Thomas, C. M.; Steegers, E. A.; Boers, G. H.; Eskes, T. K. Methylenetetrahydrofolate Reductase Polymorphism Affects the Change in Homocysteine and Folate Concentrations Resulting from Low Dose Folic Acid Supplementation in Women with Unexplained Recurrent Miscarriages. *J. Nutr.* **1998,** *128*, 1336–1341.

Nelen, W. L.; Steegers, E. A.; Eskes, T. K.; Blom, H. J. Genetic Risk Factor for Unexplained Recurrent Early Pregnancy Loss. *Lancet* **1997,** *350*, 861.

Nelen, W. L. D. M.; Blom, H. J.; Thomas, C. M. G.; Steegers, E. A. P.; Boers, G. H. J.; Eskes, T. K. A. B. Methylenetetrahydrofolate Reductase Polymorphism Affects the Change in Homocysteine and Folate Concentrations Resulting from Low Dose Folic Acid Supplementation in Women with Unexplained Recurrent Miscarriages. *J. Nutr.* **1998,** *128*, 1336–1341.

Nielsen, J. It Is all About Metabolic Fluxes. *J. Bacteriol. V*, *185* (24), 7031–7035.

Nijhoff, W. A.; Mulder, T. P. J.; Verhagen, H.; Van Poppel, G.; Peters, W. H. M. Effects of Consumption of Brussels Sprouts on Plasma and Urinary Glutathione S-Transferase Class-$\alpha$ and -$\pi$ in Humans. *Carcinogenesis* **1995,** *16*, 955–957.

Nozaki, Y.; Kusuhara, H.; Endou, H.; Sugiyama, Y. Quantitative Evaluation of the Drug-Drug Interactions Between Methotrexate and Nonsteroidal Anti-Inflammatory Drugs in the Renal Uptake Process Based on the Contribution of Organic Anion Transporters and Reduced Folate Carrier. *J. Pharmacol. Exp. Ther.* **2004,** *309*, 226–234.

Olsson, K. S.; Vaisanen, M.; Konar, J.; Bruce, A. The Effect of Withdrawal of Food Iron Fortification in Sweden as Studied with Phlebotomy in Subjects with Genetic Hemochromatosis. *Eur. J. Clin. Nutr.* **1997,** *51*, 782–786.

Omenn, G. S.; Goodman, G. E.; Thornquist, M. D.; Balmes, J.; Cullen, M. R. et al. Effects of a Combination of Beta Carotene and Vitamin A on Lung Cancer and Cardiovascular Disease. *N. Engl. J. Med.* **1996,** *334*, 1150–1155.

Osier, M. V.; Pakstis, A. J.; Soodyall, H.; et al. A Global Perspective on Genetic Variation at the ADH Genes Reveals Unusual Patterns of Linkage Disequilibrium and Diversity. *Am. J. Hum. Genet.* **2002,** *71*, 84–99.

Oyama, K.; Kawakami, K.; Maeda, K.; Ishiguro, K.; Watanabe, G. The Association Between Methylenetetrahydrofolate Reductase Polymorphism and Promoter Methylation in Proximal Colon Cancer. *Anticancer Res.* **2004,** *24*, 649–654.

Paolini, M.; Cantelli-Forti, G.; Perocco, P.; Pedulli, G. F.; Abdel-Rahman, S. Z.; Legator, M. S. Co-Carcinogenic Effect of β-Carotene. *Nature* **1999,** *398*, 760–761.

Perera, F. P. Environment and Cancer: Who Are Susceptible? *Science* 1997, *278*, 1068–1073.

Pool-Zobel, B.; Bub, A.; Leigibel, U. M.; Treptow-Van Lishaut, S.; Rechkemmer, G. Mechanisms by Which Vegetable Consumption Reduces Genetic Damage in Humans. *Cancer Epidemiol. Biomarkers Prev.* **1998,** *7*, 891–899.

Posner, B. M.; Franz, M. M.; Quatromoni, P. A.; Gagnon, D. R.; Sytkowski, P. A.; D'Agostino, R. B. et al. Secular Trends in Diet and Risk Factors for Cardiovascular Disease: The Framingham Study. *J. Am. Diet Assoc.* **1995,** *95*, 171–179.

Quinlivan, E. P.; Davis, S. R.; Shelnutt, K. P. et al. Methylenetetrahydrofolate Reductase 677C- T Polymorphism and Folate Status Affect One-Carbon Incorporation into Human DNA Deoxynucleosides. *J. Nutr.* **2005,** *135*, 389–396.

Raby, B. A.; Lazarus, R.; Silverman, E. K. et al. Association of Vitamin D Receptor Gene Polymorphisms with Childhood and Adult Asthma. *Am. J. Respir. Crit. Care Med.* **2004,** *170*, 1057–1065.

Raner, G. M.; Vaz, A. D.; Coon, M. J. Metabolism of all-trans, 9-cis, and 13-cis Isomers of Retinal by Purified Isozymes of Microsomal Cytochrome P450 and Mechanism-Based Inhibition of Retinoid Oxidation by Citral. *Mol. Pharmacol.* **1996,** *49*, 515–522.

Rebbeck, T. R. Molecular Epidemiology of the Human Glutathione S-Transferase Genotypes GSTM1 and GSTT1in Cancer Susceptibility. *Cancer Epidemiol. Biomarkers Prev.* **1997,** *6*, 733–743.

Reddy, B. S.; Rao, C. V.; Rivenson, A.; Kelloff, G. Chemoprevention of Colon Carcinogenesis by Organosulfur Compounds. *Cancer Res.* **1993,** *53*, 3493–3498.

Reference:

Reyes-Engel, A.; Munoz, E.; Gaitan, M. J. et al. Implications on Human Fertility of the 677C3T and 1298A3C Polymorphisms of the MTHFR Gene: Consequences of a Possible Genetic Selection. *Mol. Hum. Reprod.* **2002,** *8*, 952–957.

Rockman, M. V.; Wray, G. A. Abundant Raw Material for Cis-Regulatory Evolution in Humans. *Mol. Biol. Evol.* **2002,** *19*, 1991–2004.

Roland, N.; Nugon-Baudon, L.; Flinois J-P, Beaune, P. Hepatic and Intestinal Cytochrome P-450, Glutathione S-Transferase and UDP-Glucuronosyl Transferase are Affected by Six Types of Dietary Fiber in Rats Inoculated with Human Whole Fecal Flora. *J. Nutr.* **1994,** *124*, 1581–1587.

Romero, P.; Wagg, J.; Green, M. L.; Kaiser, D.; Krummenacker, M.; Karp, P. D. Computational Prediction of Human Metabolic Pathways from the Complete Human Genome. *Genome Biol.* **2005,** *6* (1), R2. Epub 2004 Dec 22.

Ronteltap, J. C. M. van Trijp; Renes, R. J. Consumer Acceptance of Nutrigenomics-Based Personalised Nutrition. *Br. J. Nutr.* **2009,** *101* (1), 132–144.

Rowe, J. D.; Nieves, E.; Listowsky, I. Subunit Diversity and Tissue Distribution of Human Glutathione S-Transferases: Interpretations Based on Electrospray Ionization-MS and Peptide Sequence-Specific Antisera. *Biochem. J.* **1997,** *325*, 481–486.

Cozzolino, S. M.F.; Cominetti, C. *Biochemical and Physiological Bases of Nutrition in Different Stages of Life in Health and Disease*, 1st ed.; Monole: S˜ao Paulo, Brazil, 2013.

Scheuermann, T. H.; Keeler, C.; Hodsdon, M. E. Consequences of Binding an S-Adenosylmethionine Analogue on the Structure and Dynamics of the Thiopurine Methyltransferase Protein Backbone. *Biochemistry* **2004,** *43*, 12198–12209.

Schmidt, L. E.; Dalhoff, K. Food–Drug Interactions. *Drugs* **2002,** *62*, 1481–1502.

Schork, N.; Cardon, L. R.; XuX. The Future of Genetic Epidemiology. *Trends Genet* **1998,** *14*, 266–272.

Sing, C.F.; Davignon, J. Role of the Apolipoprotein E Polymorphism in Determining Normal Plasmalipid and Lipoprotein Variation. *Am. J. Hum. Genet.* **1985,** *37*, 268–285. [PMC free article: PMC1684560] [PubMed:3985008]

Slattery, M. L.; Potter, J. D.; Samowitz, W.; Schaffer, D.; Leppert, M. Methylenetetrahydrofolate Reductase, Diet, and Risk of Colon Cancer. *Cancer Epidemiol. Biomarkers Prev*. **1999,** *8*, 513–518.

Smithells, R. W.; Nevin, N. C.; Seller, M. J. et al. Further Experience of Vitamin Supplementation for Prevention of Neural Tube Defect Recurrences. *Lancet* **1983,** *1*, 1027–1031.

Stephen, A. M.; Wald, N. J. Trends in Individual Consumption of Dietary Fat in the United States, 1920–1984. *Am. J. Clin. Nutr.* **1990,** *52*, 457–469.

Suh, J. R.; Herbig, A. K.; Stover, P. J. New Perspectives on Folate Catabolism. *Annu. Rev. Nutr.* **2001,** *21*, 255–282.

Surana, K. R.; Ahire, E. D.; Sonawane, V. N.; Talele, S. G. Biomolecular and Molecular Docking: A Modern Tool in Drug Discovery and Virtual Screening of Natural Products. In *Applied Pharmaceutical Practice and Nutraceuticals*; Apple Academic Press, 2021; pp 209–223.

Thielitz, A.; Krautheim, A.; Gollnick, H. Update in Retinoid Therapy of Acne. *Dermatol. Ther.* **2006,** *19*, 272–279.

Tishkoff, S. A.; Verrelli, B. C. Patterns of Human Genetic Diversity: Implications for Human Evolutionary History and Disease. *Annu. Rev. Genomics Hum. Genet.* **2003,** *4*, 293–340.

Van Ommen, B.; Bogaards, J. J. P.; Peters, W. H. M.; Blaauboer, B.; van Bladeren, P. J. Quantification of Human Hepatic Glutathione S-Transferases. *Biochem. J.* **1990, 269,** 609–613.

Velazquez, A.; Bourges, H. , Eds. *Genetic Factors in Nutrition*; Academic Press: Orlando, 1984; 434 pp.

Venter, J. C.; Adams, M. D.; Myers, E. W. et al. The Sequence of the Human Genome. *Science* **2001,** *291*, 1304–1351.

Vorderstrasse, A. C.; Voils, C. E.; Orlando, L. A.; Ginsburg, G. S. Personalized Medicine Clinical Utility of Genetic Risk Testing in Primary Care the Example of Type 2 Diabetes. *Personalized Med.* **2013,** *10*, *6*, 549–563.

Wang, X. D.; Liu, C.; Bronson, R. T.; Smith, D. E.; Krinsky, N. I.; Russell, R. M. Retinoid Signalling and Activator Protein-1 Expression in Ferrets Given β-Carotene Supplements and Exposed to Tobacco Smoke. *J. Natl. Cancer Inst.* **1999,** *91*, 60–66.

Watanabe, M. Polymorphic CYP Genes and Disease Predisposition—What Have the Studies Shown So Far? *Toxicol. Lett.* **1998,** *102–103*, 167–171.

Waterland, R.; Jirtle, R. L. Transposable Elements: Targets for Early Nutritional Effects on Epigenetic Gene Regulation. *Mol. Cell Biol.* **2003,** *23*, 5293–5300.

Wilson, A.; Platt, R.; Wu, Q. et al. A Common Variant in Methionine Synthase Reductase Combined with Low Cobalamin (Vitamin B12) **1999,** *67*, 317–323.

Wray, G. A.; Hahn, M. W.; Abouheif, E. et al. The Evolution of Transcriptional Regulation in Eukaryotes. *Mol. Biol. Evol.* **2003,** *20*, 1377–1419.

Yang, C. S.; Brady, J. F.; Hong, J. Y. Dietary Effects on Cytochromes P450, Xenobiotic Metabolism, and Toxicity. *J. FASEB.* **1992,** *6*, 737–744.

Zang, E. A.; Wynder, E. L. Differences in Lung Cancer Risk Between Men and Women: Examination of the Evidence. *J. Natl. Cancer Inst.* **1996,** *88*, 183–192.

Zheng, W.; Deitz, A. C.; Campbell, D. R.; Wen, W. Q.; Cerhan, J. R. et al. N-Acetytransferase 1 Genetic Polymorphism, Cigarette Smoking, Well-Done Meat Intake, and Breast Cancer Risk. Cancer Epidemiol. *Biomarkers Prev.* **1999,** *8*, 233–239.

# CHAPTER 12

# NUTRIGENETICS DIFFERENCES AND THEIR ORIGINS

MUSAB TANZEEL, GAYATRI N. KANADE, PAVAN B. UDAVANT, MASOOD SIDDIQUI, and SHUBHAM J. KHAIRNAR

*Department of Pharmacology, MET's Institute of Pharmacy, Bhujbal Knowledge city, Nashik, India*

## ABSTRACT

The revelation of gene–nutrient interactions has allured researchers' interest over the last few years. In 2017, diets high in salt and low in whole grains and fruits were accountable for about half of all diet-related deaths, with 10 million deaths from cardiovascular disease. Nutrigenetics is a branch of personalized nutrition that studies how genetic differences, particularly those related to single-nucleotide polymorphisms (SNPs), influence people's food responses. Epigenetic changes can be caused by food solely or in combination with other environmental parameters. These variations in the genes have a big impact on cellular functions, such as hormonal levels, cell signaling, carcinogen metabolism, apoptosis, cell cycle control, energy levels, and angiogenesis. Nutrigenomics and pharmacogenomics will be used in the treatment and management of a range of diseases, including personalized prescriptions and lifestyle modifications. Despite decades of research, current instrumentation, and approaches are still insufficient in this sector. A shift away from candidate genes and toward genome-wide screening will be beneficial. Gaining a clearer picture may lead to the prescription of personalized diets for each individual. The most difficult task is preventing these diseases in populations that do not lack food but require adequate micronutrient consumption. The implementation and headway of nutrigenetics, and

Nutrigenomics and Nutraceuticals. Raj K. Keservani, Eknath D. Ahire, Shubham J. Khairnar, Sanjay J. Kshirsagar, & Rajesh K. Kesharwani (Eds.)

customized nutrition in general, is not only reliant on the quality of the data but also on social acceptability. Individuals known to possess a higher illness threat through genetic testing may be more encouraged to adhere to a dietary intervention.

## 12.1 INTRODUCTION

We prioritize a diverse blend of meals throughout our lives that may have a beneficial or detrimental influence on our bodies. Complex metabolic activities generate the energy and other essential components that allow us to develop and function, and many previously inconsequential chemicals have since been identified as advantageous. The issue, for both researchers and users, is that the benefits do not seem to be uniform (Astley, 2007). Although, since the completion of the Human Genome Project (HGP), additional hypotheses concerning the impact of nutrients on individuals 's diets were proposed, including: (i) will gene expression in response to metabolic processes at the cellular level affect an individual's health? (ii) Is the interplay between genotype and environment/nutrient the cause of gene regulation and metabolic reaction? (iii) Gaining a better picture of how genes and nutrients interact could result in the prescribing of customized diets for each person. As a consequence, Nutrigenomics was implemented to respond to the questionnaire (Dauncey, 2012; Sales et al., 2014). Dietary reference values (DRVs) serve as a reference for deciding the proper nutrient intake to avoid illness induced by nutritional deficiencies (e.g., scurvy in the case of vitamin C deficiency) or surplus (e.g., iron-overload disease, which can be deleterious when there is too much iron in the cell system). It is critical to identify these intense situations connected with nutritional disorders these days, and the toughest problem is preventing these types of developmental and degenerative diseases in populations that do not lack food, fortified food, or supplements, but require therapy through proper intake of micronutrients individually or in combination (nutriomes) to optimize cellular and organism efficiency on both a personal and genetic scale. The suppression of destruction to the nuclear and mitochondrial genomes is crucial for cellular function optimization (Fenech, 2005a, 2010a; Mondal, 2020).

## 12.2 HISTORY

The provenance of the nutrigenetics/nutrigenomics notions is missed in time, and even though one of the first written references pointing to phenotypic variation in response to dietary factors and its effects on human health is the well-known Greek adage "What is food for one, is bitter poison for others" (Lucretius). In recent times, one of the earliest publications establishing a link between genes and nutrition was published in 1945. The researchers found that a diet rich in whole grains and whole dry milk increased the survivability of W-Swiss mice infected with *S. enteritidis* when compared with a "synthetic" diet (Schneider and Webster, n.d.). The potential of diet to develop innate immunity, nevertheless, was discovered to be dependent on the biological characteristics of the mice examined. The phrase nutrigenetics arise afterward, in the by-line of a 1975 book by Brennan and Mulligan, which was one of the first to use it. Although, it was not until the 1980s that we developed the technical capability to probe specific genes and uncover variations linked to food response variability. From then, the field has incorporated new technology breakthroughs, and we are now seeing the utilization of genome-wide association studies (GWAS) linked with dietary data to determine the genetic link between diet and lifestyle pattern differences (Daniele et al., 2017).

## 12.3 EQUATION OF NUTRITION AND HEALTH

Diet has long been known to have an impact on one's health. "Leave your pharmaceuticals in the chemist's pot if you can heal your patient with food," Hippocrates counseled health care providers in 400 BC. Additionally, it is generally recognized that humans have different nutritional requirements (Elsamanoudy et al., 2016). The health benefits of nutrients and food bioactive components are influenced by a number of physiological mechanisms, such as absorption, transport, biotransformation, uptake, binding, storage, and excretion, along with cellular underlying mechanisms, such as nuclear receptor binding or transcription factor regulation. Each of these mechanisms can include a number of genes, each of which may have common variants that affect their activity, and as a result, the physiological response to a food item. Diet–gene correlations investigation has also looked into how genetics affect dietary requirements by influencing sensory, reward, and energy homeostatic systems (Garcia-Bailo et al., 2009). Diet and nutrients are thought to be the most prominent environmental triggers. However, the origins of nutrition may be traced back years, contemporary science has revealed that nutrients are not just vital, but that the precise amount of each nutrient is also required for maximum health. The advancement of molecular tools and nutritional studies has aided our knowledge of diet and its impact on health. The 20th century aided in the study of macro and micronutrients on a fundamental level. Though one perspective was typified, and strategies like My Plate and the Food Guide Pyramid were highly beneficial in eliminating malnourishment and poor nutritional status disorders. Numerous gene variants have been linked to distinct dietary features in GWAS, and they are responsible for many diet–gene interactions and the development of many human disorders. The metabolism of numerous nutrients is affected by several factors in the diet, as well as the influence of various genotypes, which might lead to negative diet–gene interactions (Chilton et al., 2017). Nutritional genetics is the study of the link between genes and diet that has resulted from nutritional studies. Nutrigenomics and nutrigenetics are two independent areas of research that have been combined. Though they are inextricably linked, the approaches to assessing the interaction among genes and nutrition are very distinct. The purpose of nutritional research is to find out who benefits from nutritional changes and what options are available for others who do not. This is beneficial for both the prevention and treatment of chronic diseases by tailoring dietary habits to the individual's genetic structure (Bahinipati et al., 2021).

Nutrition can either actively or passively play a significant role in disease development or manifestation. Nutrients and meals often interfere with genes in a beneficial way; however, this interaction can sometimes be fatal. Humans are impacted by the interaction between genetic and environmental variables, and both must be taken into account in order to maintain a normal health status. Previous research has primarily focused on either the influence of environmental factors alone or the effect of genes alone, rather than both. Recently, an investigation has been conducted to investigate the relationship between genes and nutrition (Bull and Fenech, 2008a; Elsamanoudy et al., 2016). Nutrigenomics looks into how nutrients help to keep the body's equilibrium in check at the cellular, tissue, and organ levels. The mechanism of nutrient-dependent interactions must be understood at the genetic molecular, protein synthesis, and metabolic profile levels. As a corollary, nutrition research has progressed beyond epidemiological and physiological elements to molecular biology, genetics, and nutritional genomics (Arab, 2004a; Elsamanoudy et al., 2016). The interaction of genetics and the environment is responsible for the etiology of any disease. Genetics determines an individual's vulnerability to disease, but environmental circumstances ascertain who among those is vulnerable to the ailment (Bahinipati et al., 2021). A poor diet is responsible for one out of every five fatalities worldwide. Diets rich in salt and poor in whole grains and fruits were responsible for about half of all diet-related deaths worldwide in 2017, with 10 million deaths from cardiovascular disease (CVD), 913000 deaths from cancer, and 339000 deaths from Type 2 Diabetes Mellitus (T2DM; 1). Chronic illnesses are becoming more prevalent over the world at an alarming rate. It was responsible for 60% of the world's 56.5 million total recorded fatalities in 2001, accounting for almost 46% of the global illness burden. By 2020, this risk is predicted to rise to 57%. Not only are chronic diseases such as CVD, obesity, and T2DM on the rise, but they're also showing up considerably earlier in life (World Health Organization., 2003).

## 12.4 NUTRIGENETICS

### *12.4.1 GENETIC VARIATION AND DIETARY RESPONSE*

The finding of gene–nutrient interconnections has piqued researchers' curiosity throughout the last several years. Considering technological obstacles at the time, the early investigations focused on one or a few genes, as they did in other branches of biology. But subsequently, technological hurdles have been broken down, making it feasible to analyze a huge number of genes, and in some cases, the entire genome in a single experiment. The ideas of nutrigenetics and nutrigenomics have emerged in this setting (Bouchard and Ordovas, 2012). The terms have been more extensively employed in a consistent way by the community of active scientists in recent times, following the founding of the International Society of Nutrigenetics/Nutrigenomics and the introduction of the Journal of Nutrigenetics and Nutrigenomics in 2008. While not all theoretical concerns have been cleared up, enough progress has been made to allow both terms to be used in order to improve the overall communication clearly. Specifically described, nutrigenetics is the science of how Genomic sequences' variability affects dietary responses, whereas nutrigenomics is the study of how nutrients affect the expression of genes (Bouchard and Ordovas, 2012; Ordovas & Corella, 2004). Dr. R.O Brennan coined the term "nutrigenetics" in his book "Nutrigenetics" in 1975 (Simopoulos, 2010). Nutrigenetics is a discipline of individualized diet that looks at how genetic differences, primarily those related to single-nucleotide polymorphisms (SNPs), affect people's responses to food (Mustapa et al., 2020). Nutrigenetics (NGx) is a branch of pharmacogenomics that studies the links between genetic variations, nutrition, and health consequences. NGx offers the ability to integrate genetic data effectively that assist patients to attain positive health outcomes by further individualizing nutrition recommendations. NGx testing as part of clinical therapy seems effective, although it is not done consistently in the clinical context (Dong, 2021). It has now been effectively used to raise animal growth rates and production efficiency, and could even serve a foundation for breeding improved nutrient efficiency and identifying genotype-dependent supplementation that improves health (Ricigliano et al., 2021). Functional foods, defined as foods that contain biochemically active ingredients that offer a health benefit in relation to the basic nutritional function, can then interfere with the genes (Henry, 2010). These components are known as bioactive chemicals, and they have the ability to lessen the risk of chronic diseases even though existing at modest levels. It is preferable to acquire these chemicals in their native state. Polyphenols, such as resveratrol, quercetin, curcumin, and genistein are examples of bioactive components (Karasawa and Mohan, 2018). For nutrigenetics to flourish, two major challenges must be solved, and as this volume demonstrates, both issues are being outlined in stages.

One is the use of high-throughput genotyping and sequencing techniques to conduct genome-wide investigations of DNA variant–nutrient relationships. The required sample numbers are well exceeding the scope of any single investigation. A switch away from candidate genes toward the genome-wide tests will result in newer randomized targets with the potential to affect our basic picture of the underpinning biology and devise novel frameworks and processes of interest. Whereas, much of this optimism is based on the idea that we will be able to gather accurate and consistent dietary data. Amidst decades of study, current instrumentation and methodologies in this field are still insufficient. The second concern is significantly more difficult to solve. Not just observational studies, but also controlled trials and experimental approaches, require improvement. Cross-sectional investigations are beneficial for identifying relatively intriguing benchmarks, but they can be hampered by covariates, which restrict their utility, especially when addressing DNA identifiers and gene–nutrient interactions, which have tiny effect sizes (Kaput and Rodriguez, 2004; Kussmann and Fay, 2008; Ordovas and Corella, 2004).

### *12.4.2 GENETIC VARIATION AND INDIVIDUALIZED DIETS*

Individualized nutrition allows us to customize our diets for a one's genotype, making it easier to provide nutritional counsel that is precisely suited to their needs and thus more helpful in suppressing chronic conditions and enhancing wellness.

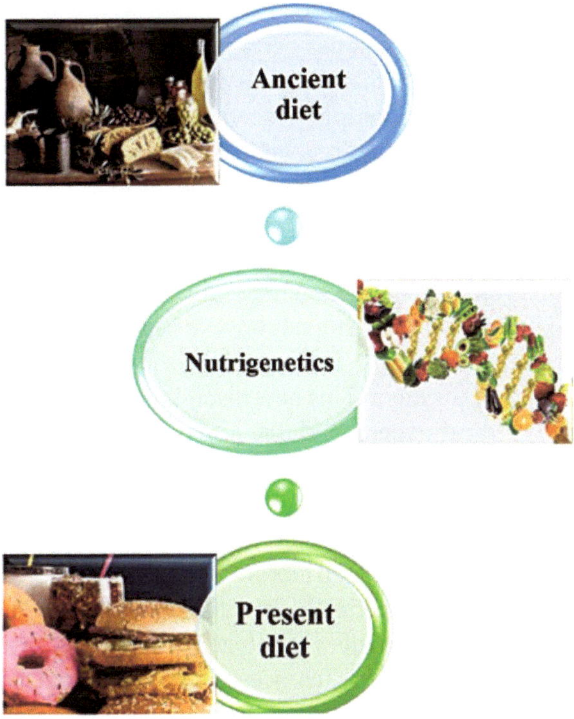


**FIGURE 12.1** DNA variant–nutrient relationship.

Nutrigenetics is the assessment of how an individual's genetic makeup influences their nutrition. The study of gene–nutrient interactions is still in its infancy. The concept of a negative diet/genome interaction causing disease is not unfamiliar, and an improper diet for any genotype could be a contributing factor for monogenetic and polygenetic disease (Farhud and Zarif Yeganeh, 2010). With the sequence analysis of a eukaryotic genome, we have made considerable progress in our view of the human genome. Single-gene disorders, such as galactosemia and phenylketonuria account for 97% of human diseases among the genes discovered thus far. A few of these monogenic disorders can be prevented by changing how certain dietary elements are consumed. Galactosemia is caused by a recessive trait in galactose-1-phosphate uridyl transferase, which causes galactose to accumulate in the blood. A galactose-free diet is the best way to nutritionally treat them (Arab, 2004). Afterward, an utmost pressing question is whether nutrigenetics can lead to better outcomes than population-based suggestions. The evidence for nutrigenetics has now been claimed to be too underdeveloped to be employed in practice (Camp and Trujillo, 2014). Nevertheless, it has been suggested that nutrigenetics is held to a higher standard of proof than general nutritional advice, perhaps ignoring high-quality evidence for multiple gene–diet relationships. Considering the intricacy of gene–nutrient relations and the reported variation in nutrition-related health outcomes, irrespective of genetics, we need to think about whether we've set this bar quite greater. Nutrition is complicated, and results are hard to predict, whether or not there is a hereditary aspect (Beckett et al., 2017a; Görman et al., 2013). The interplay of our genes with the meals we eat is investigated. It precisely investigates how varied diets influence individuals with multiple genetic makeups. This technique seeks to match people to diets that are most suitable for them. This new discipline pulls together researchers, legislators, and medical experts with the aim of 1 day applying personalized nutrition guidance and developing functional foods that maximize health according to individual needs. It combines how we eat, what our parents ate, and how it affects our cells and biological systems' growth paths (Fenech et al., 2011; Murgia and Adamski, 2017).

The significance of a person's genetic composition on their bodily responses to consumed nutrients is analyzed. Individuals are often thought to be gene variants. Subtle differences in people's genetic composition lead to differing reactions to specific nutrients. Several interactions, based on nutrient consumption, can lead to the onset of various illness problems (Pavlidis et al., 2015a). Genetics is the study of genes, which are passed down through generations as hereditary molecules. Genes are responsible for producing proteins in our bodies and determining individual variances (*Finishing the Euchromatic Sequence of the Human Genome International Human Genome Sequencing Consortium**, 2004). Additionally, it is similar to how diet will direct our anatomical development in different ways depending on our genetic makeup. Nutrigenomic profiling will aid in identifying the processes that underpin differences between individuals in dietary requirements and responsiveness to food-related therapies (Uthpala et al., n.d.). Nutrigenetics' underlying intent is to offer dietary advice for people based on their genetic makeup (Murgia and Adamski, 2017). When it comes to dietary genomics, there are a few things to keep in mind. The first is to acknowledge that for some people, nutrition is a substantial risk factor for certain diseases. 2 Furthermore, common dietary nutrients can alter gene expression or structure in the genetic material, potentially actively or passively (Mariman, 2006; Pavlidis et al., 2015). Phenylketonuria (PKU) was the first "inborn error of metabolism" induced by a single-gene flaw that could be treated with a low-phenylalanine diet. There are about 6000 single-gene diseases, with 2000 of them actually been recognized. CHD, high blood pressure, diabetes, cancer, and obesity are multigenetic and multifactorial illnesses. Numerous heterogeneous genes may underlie the same symptom. The interaction of genetic and environmental variables is accountable for producing the phenotype in polygenic disorders. Single-gene disorders are scarce with a prevalence of fewer than 1 in 1000 births or people; they are often under negative selection pressure ( Marian and Roberts, 2001).

## 12.5 NUTRIGENOMICS AND NUTRIGENETICS: WHAT'S THE CORRELATION?

The influence of genetic variations on dietary response, as well as the involvement of nutrients and bio–active chemicals in gene regulation, are characterized as nutritional sciences (Simopoulos, 2010).

Breakthroughs in the disciplines of food supplements, nutritional supplements, and their implications on human wellness have been integrated into current molecular nutrition, entitled "Nutrigenomics or Nutrigenetics." Nutrigenomics is a nascent discipline of genomics that studies the impact of dietary consumption on the entire genome (entire genetic make-up, including epigenetics), the proteome (the total number of proteins), and the metabolome (the sum of all metabolites). Galactosemia, for example, is a disease caused by a hereditary genetic deficit in one of the three enzymes involved in galactose breakdown (Ferland-Mccollough et al., 2012). Nutrigenetics is the study of the impact of genotypes on the prevalence of nutritional diseases. It begins with the notion that several genetic alterations that alter nutritional deficit Genetic variants are uncovered and examined in the field of nutrigenetics to investigate whether they alter the correlations between dietary consumption and risks. Nutrigenomics, which emphasizes on the implication of nutrition or food components on transcriptional activity, proteomics, and metabolic activity. In this case, the intent is that diet has an impact on disease via mechanisms regulating expression of genes. Nutrigenomics is more challenging than nutrigenetics to use in nutritional studies (Arab, 2004). Nutrigenomic interactions differ from nutrigenetic interactions in that the environment regulates the expression of genes rather than the other way around. Nutrigenomics is dealing with the impact of bioactive components interacting with a gene-on-gene expression, either upregulating or downregulating (turning it up or down), stimulating or repressing it (on or off). Direct or indirect nutrigenomic interactions are possible. The molecules in direct contact are of low molecular weight, carrier-mediated, and lipid-soluble. The molecules in indirect contact are bigger and more hydrophilic. At the cellular membrane, these molecules interact (Joffe and Houghton, 2016).

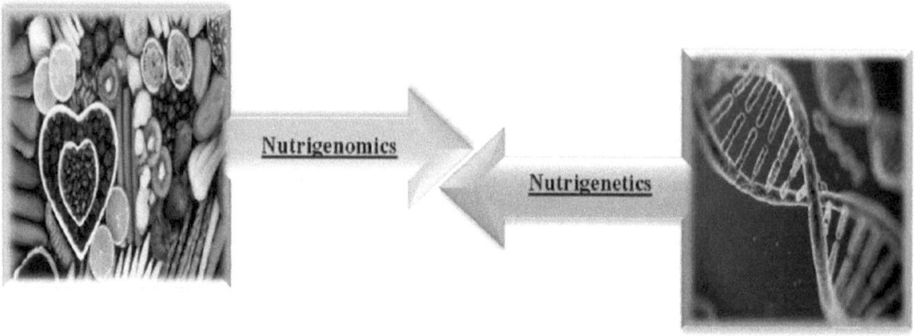


**FIGURE 12.2** Nutrigenetics and nutrigenomics.
*Source*: Adapted from Mutch et al., (2005).

Diet and lifestyle repercussions on gene regulation (DNA damage at the molecular and chromosome level), epigenome modifications (DNA methylation), RNA and micro-RNA expression (transcriptomics), protein expression (proteomics), and metabolite changes (metabolomics) are all explored separately or in combination to assess the overall health and/or disease flight path in the field of nutrigenomics. Only DNA damage, among all of these benchmarks, is an unambiguous indicator of an underlying illness that can be alleviated by promoting cell death in genetically abnormal cells or limiting the cadence of DNA damage accumulation. Shifts in the epigenome, transcriptome, proteomics, and metabolome may merely signify adjustable homeostatic reactions to increase dietary intake, and might not be enough to demonstrate definitive irreparable pathological alteration at the genome level on their own. DNA damage can be detected in a variety of methods, including the following: destruction of single bases (e.g., DNA adducts related to oxidative stress, such as the conjunction of a hydroxyl radical to guanine); abasic sites in the DNA sequence (quantifiable through the use of the aldehyde-reactive probe); DNA strand breaks (typically identified using the Comet assay); telomere narrowing (measured by end restriction fragment length analysis, quantitative PCR, or flow cytometry); chromosome breakage (usually measured as deletions or base damage in the circular mitochondrial DNA sequence) (Bull and Fenech, 2008; Fenech, 2005; Fenech, 2010).

## 12.6 PERSPECTIVES ON THE CURRENT STATUS OF NUTRIGENETICS AND NUTRIGENOMICS

On October 4, 2009, a 1-day conference entitled "Fundamentals of Nutrigenomics and Its Applications" was conducted in Bangkok in conjunction with the 19th International Congress of Nutrition. The International Life Sciences Institute (South East Asia region) collaborated with the Commonwealth Scientific and Industrial Research Organization (Australia) to assist healthcare providers attain better insight and consideration of the scientific knowledge of nutrigenetics and nutrigenomics, as well as promising effects for boosting health outputs in genetic segments and individuals.

Hereunder are the basic speculations that endorse the research of nutrigenetics and nutrigenomics:

- Nutrition can affect the quality of care directly by switching gene expression in vital biochemical processes and/or obliquely by impacting the extent of genetic mutation at the base pair or chromosomal level, which leads to changes in gene dosage and expression.
- The health impacts of nutrients and nutriomes (nutrient combinations) are influenced by hereditary gene mutations that affect nutrient absorption and processing, as well as the molecular linkages of enzymes with their nutrient cofactors, and thereby biochemical reaction activity.
- Improved health results can be attained if feed intake is tailored to each individual relying on life stage, dietary changes, and health condition, taking into account both hereditary and developed genetic traits (Fenech et al., 2011).

## 12.7 POPULARITY AMONG THE POPULATION

The implementation and headway of nutrigenetics, and customized nutrition in general, is not only reliant on the quality of the data, the elucidation of mechanisms, and the determination of the necessary degree of evidence required for implementation, but also on social acceptability. Amongst the most difficult aspects of any health and nutrition strategy is persuading people to amend their eating habits. There should be fluidity in use and access, as well as a certainty of worth. Generally speaking, genetic screening has its own set of hurdles to adoption, and most of these persist in the perspective of nutrigenetics, despite results (Ronteltap et al., 2009). There have been myriad investigations on the adoption of genetic screening in general. While screening is getting more ubiquitous in the presence of severe single polymorphisms, such as the BRCA1 gene in risk of breast cancer, the correlations in nutrigenetics are extensively weaker and also less obvious. Nonetheless, views vary considerably on demographic changes, and the findings have been inconsistent. Men have been found to be more inclined to undergo genetic testing than women (Cherkas et al., 2010), but the contrary has also been seen. It has also been claimed that age influences receptivity of genetic testing; however, several research has suggested that elderly individuals are more inclined to participate (Stewart-Knox et al., 2009), although some have discovered that younger adults are the most likely (Morin, 2009; Weir et al., 2010). Given that nutrition is a considerably controllable enabler of disease and health has been highlighted as a key stepping stone for undertaking genetic testing (Cherkas et al., 2010; Stewart-Knox et al., 2009) which may expedite the adoption of nutrigenetics. Adoption may also be closely associated with present health; for instance, individuals with high cholesterol levels or central adiposity were more likely to admit they were willing to embrace genetic testing in general, and notably for proper nutrition, than those who did not. There are, although, a number of alternative possibilities. Individuals known to possess a higher illness threat through genetic testing may be more encouraged to adhere to a dietary intervention; yet, awareness of a genetic predisposition may result in a pessimistic mindset and poorer compliance (Joost et al., 2007). Additionally, an undesirable outcome could contribute to a decrease in incentives as people are alerted that they will not develop sickness. To guarantee that responses to information are not damaging to outcomes, it will be critical to communicate the plausible nature of risk and risk adjustment through nutrigenetics. Individuals with the at-risk genotypes were less prone to trust that adopting a reduced fat diet would lessen their cholesterol levels, and more probably to believe that medicine would. Affirmation of genetic testing and tailored diets may be based on the consumers' perception of the implications for their own health rather than on the science itself (Beckett et al., 2017; David 2008, n.d.).

## 12.8 THE SIGNIFICANCE OF NUTRIGENETICS AND NUTRIGENOMICS IN BOOSTING HEALTH AND QUALITY OF LIFE

Even though nutrionists perceive those children and adolescents have different nutritional demands, and also that males and females have different nutritional necessities, nutrition research has primarily centered on the presumption that all persons have identical dietary requirements.

In most nations, dietary recommendations are based on the need to combat deficiency disorders. These are sometimes referred to as RDAs, because they specify how much of a nutrient is required per day for the majority of people to remain healthy. The RDA is modified on a regular basis to ensure new research, although chronic disorders linked to nutrition have been on the rise in the last several years. The introduction of the online dietary tool 'Mypyramid' marked a significant breakthrough in nutritional comprehension and tools that can be employed to create specific nutritional needs. This allows for only a limited recognition of individual distinctions. Thus, an individual dietary plan can be drawn by inputting age, gender, height, bodyweight, and exercise habits (*Mypyramid*, n.d.). Whereas experimental models and bioengineering methodologies can be utilized to comprehend the effects of these components on gene pathways, extrapolation to humans and clinical application are not currently common. Indeed, utilizing nutrigenetic techniques, it is now feasible to bring an insight into the genetic foundation of health and disease to the realistic approach of customized diet. Nutritional specialists are learning to analyze the data on genes and genetic variants, diet, lifestyle, and environment in order to design nutritional recommendations based on genetic composition, often in the form of SNPs, but not primarily (Adeyemo and Rotimi, 2009; Fenech, 2010, Surana et al., 2021).

## 12.9 THE FUTURE OF NUTRIGENETICS

Epigenetic changes can be caused by food solely or in combination with other environmental parameters (Fenech et al., 2011), and these variations in the genes have a big impact on cellular functions such as hormonal levels, cell signaling, carcinogen metabolism, apoptosis, cell cycle control, energy levels, and angiogenesis. Furthermore, embryonic development and long-term health might have an impact on children. As a result, it's crucial to know your health state and how it relates to your nutritional requirements (Felisbino et al., 2021). Nutrigenetic exploratory study is an immense strategy for discovering the full capability of your body. Nutritionists can use nutrigenetics to determine which nutrients your body requires to function at its best. Human beings may thrive on tailored nutrition plans from childhood if this method of science-based nutrition was used. This might result in people living longer, healthier lives, fewer chronic illnesses, and healthier lifestyles all throughout the world. Since they are unable to acquire a thorough diagnosis, many people suffer for years, if not decades, from avoidable and controllable illnesses. The study of nutrigenomics can help identify many of these diseases, deficits, and intolerances. Embracing nutrigenetic and nutrigenomic research in routine life and integrating the view that our living standard is reliant on the food we consume in our body is the prospect of nutrigenetic and nutrigenomic research (*What Is The Difference between Nutrigenomics and Nutrigenetics?* 2021). Nutrigenetics is just getting started in terms of realizing its promise. One might see beverages and foods being developed as illness prevention or treatment agents for individuals, families, or subgroups who are prone to a certain condition. The adoption of ketogenic diets in the therapy of children with persistent epilepsy has a precedent in pediatrics. Patients with celiac disease, PKU, and other inborn metabolic abnormalities are commonly prescribed specific foods and regimens. Patients with chronic inflammatory disorders, such as osteoarthritis, respiratory problems, peptic ulcers, lupus, and atherosclerotic CHD should consume a diet rich in essential fatty acids (omega-6 and omega-3) (Shane Broughton et al., 1997; Simopoulos, 2011).

We can forecast the mentioned milestones in nutrigenetics and nutrigenomics in 15–25 years in the 21st century:

- Most diseases will have gene/nutrient correlations revealed.
- Nutrigenomics and pharmacogenomics will be combined and used in the treatment and management of a range of diseases, including personalized prescriptions and lifestyle modifications.
- To anticipate the reliability of suitable nutrition and therapeutic interventions, people will be classified into categories depending on genotypes and dietary patterns.
- For coronary artery disease, diabetes, high blood pressure, arthritis, asthma, mental diseases, and other conditions, gene-based designer meals will be produced.
- Nutrigenomic analyses calorie restriction, defined carbohydrate and fat characteristics, and exercise interventions have identified mechanisms that lead to activation and downregulation of gene expression to promote good health, particularly in individuals with metabolic syndrome, obesity, and type 2 diabetes. Increased physical activity and omega-3/omega-6-balanced diets can serve as cornerstones for primary care in a variety of chronic diseases.

There are hereditary predispositions to all diseases. Genetic variations that influence the risk of complicated diseases are being discovered as a result of GWAS research. Nutrient data are lacking from these studies, which is crucial for establishing dietary advice for disease prevention and

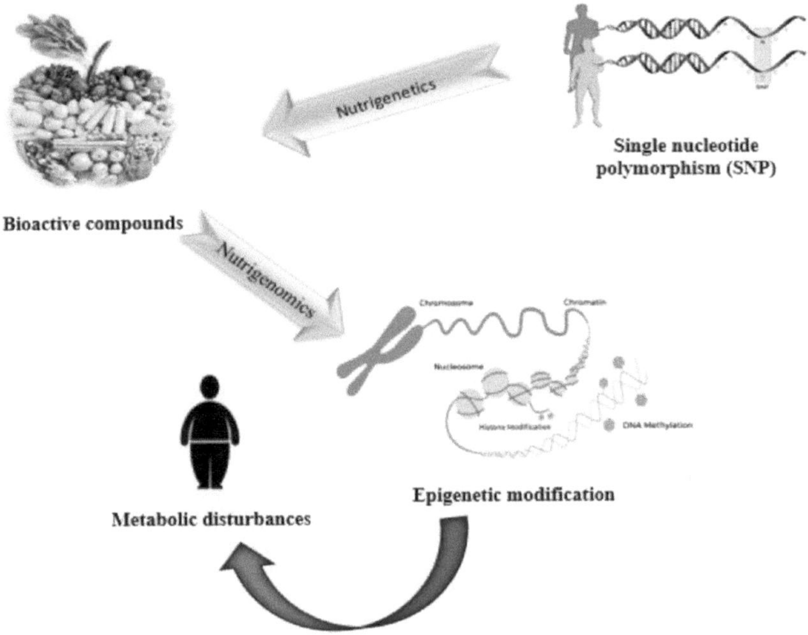


**FIGURE 12.3** Genes and diet play a role in halting chronic diseases for better future.

*Source*: Adapted from Franzago et al., 2020

treatment, as well as individualized nutrition. Age, sex, health condition, obesity, region, and other factors influence nutritional reactions in addition to genetic variants (Simopoulos, 2010).

## 12.10 CONCLUSIONS

The sequence analysis of human genetics and recent advancements in molecular biology techniques are generating a breakthrough in a diverse array of sciences, notably nutrition. Nutrigenomics is a branch of study concerned with detecting and comprehending molecular level connections between food and the genome. Better eating regimens or newer treatments for chronic conditions may be developed with the help of nutrigenomics research. It employs methods, such as SNPs, microarrays and genomics, proteomics, and bioinformatics to boost the population nutritional status. Nutrigenetics examines how an individual's genetic makeup interacts with specific dietary components. Diet–gene correlation studies are already divulging information on which to establish gene-specific lifestyle intervention trials to corroborate conclusions, attributable to nutritional genomics techniques. Nutrigenetics studies can enlighten us about genetic predispositions to undernutrition and toxicities, and they can also help with diagnosis. Despite the fact that nutrigenomics and nutrigenetics have a promising future, users should approach them cautiously because the subject is seemingly novel.

## KEYWORDS

- **nutrigenetics**
- **nutrigenomics**
- **diets**
- **nutrition**
- **health**
- **genetic variation**

## REFERENCES

Adeyemo, A.; Rotimi, C. Genetic Variants Associated with Complex Human Diseases Show Wide Variation Across Multiple Populations. *Public Health Genom.* **2009,** *13* (2), 72–79. https://doi.org/10.1159/000218711

Arab, L. Individualized Nutritional Recommendations: Do We Have the Measurements Needed to Assess Risk and Make Dietary Recommendations? *Proc. Nutr. Soc.* **2004a,** *63* (1), 167–172. https://doi.org/10.1079/pns2003325

Astley, S. B. An Introduction to Nutrigenomics Developments and Trends. *Genes Nutr.* **2007,** *2* (1), 11–13. https://doi.org/10.1007/s12263-007-0011-z

Bahinipati, J.; Sarangi, R.; Mishra, S.; Mahapatra, S. Nutrigenetics and Nutrigenomics: A Brief Review with Future Prospects. *Biomedicine* **2021,** *41* (4), 714–719. https://doi.org/10.51248/.v41i4.445

Beckett, E. L.; Jones, P. R.; Veysey, M.; Lucock, M. Nutrigenetics—Personalized Nutrition in the Genetic Age. *Exploratory Res. Hypothesis Med.* **2017,** *2* (4), 1–8. https://doi.org/10.14218/erhm.2017.00027

Bouchard, C.; Ordovas, J. M. Fundamentals of Nutrigenetics and Nutrigenomics. In *Progress in Molecular Biology and Translational Science*, Vol. 108; Elsevier B.V., 2012, pp. 1–15. https://doi.org/10.1016/B978-0-12-398397-8.00001-0

Bull, C.; Fenech, M. Genome-Health Nutrigenomics and Nutrigenetics: Nutritional Requirements or "nutriomes for Chromosomal Stability and Telomere Maintenance at the Individual Level. *Proc. Nutr. Soc.* **2008,** *67* (2), 146–156. https://doi.org/10.1017/S0029665108006988

Camp, K. M.; Trujillo, E. Position of the Academy of Nutrition and Dietetics: Nutritional Genomics. *J. Acad. Nutr. Diet.* **2014,** *114* (2), 299–312. https://doi.org/10.1016/j.jand.2013.12.001

Cherkas, L. F.; Harris, J. M.; Levinson, E.; Spector, T. D.; Prainsack, B. A Survey of UK Public Interest in Internet-Based Personal Genome Testing. *PLoS ONE* **2010,** *5* (10). https://doi.org/10.1371/journal.pone.0013473

Chilton, F. H.; Dutta, R.; Reynolds, L. M.; Sergeant, S.; Mathias, R. A.; Seeds, M. C. Precision Nutrition and Omega-3 Polyunsaturated Fatty Acids: A Case for Personalized Supplementation Approaches for the Prevention and Management of Human Diseases. *Nutrients* **2017,** *9* (11). https://doi.org/10.3390/nu9111165

Daniele, N. di, Noce, A.; Vidiri, M. F.; Moriconi, E.; Marrone, G.; Annicchiarico-Petruzzelli, M.; D'urso, G.; Tesauro, M.; Rovella, V.; de Lorenzo, A. Impact of Mediterranean Diet on Metabolic Syndrome, Cancer and Longevity. *Oncotarget* **2017,** *8* (5). www.impactjournals.com/oncotarget/

Dauncey, M. J. Recent Advances in Nutrition, Genes and Brain Health. *Proc. Nutr. Soc.* **2012,** *71* (4), 581–591. https://doi.org/10.1017/S0029665112000237

Dong, O. M. Using the Diffusion of Innovation Theory to Understand the Challenges and Opportunities to Advancing Use of Nutrigenetics in Clinical Practice. *Lifestyle Genomics* **2021,** *14* (4), 124–128). https://doi.org/10.1159/000517760

Elsamanoudy, A.; Mohamed Neamat-Allah, M.; Hisham Mohammad, F.; Hassanien, M.; Nada, H. The Role of Nutrition Related Genes and Nutrigenetics in Understanding the Pathogenesis of Cancer. *J. Microsc. Ultrastruct.* **2016,** *4* (3), 115. https://doi.org/10.1016/j.jmau.2016.02.002

Farhud, D.; Zarif Yeganeh, M. Nutrigenomics and Nutrigenetics. *Iranian J. Publ. Health* **2010,** *39* (4).

Felisbino, K.; Granzotti, J. G.; Bello-Santos, L.; Guiloski, I. C. Nutrigenomics in Regulating the Expression of Genes Related to Type 2 Diabetes Mellitus. *Front. Physiol.* **2021,** *12.* https://doi.org/10.3389/fphys.2021.699220

Fenech, M. F. Dietary Reference Values of Individual Micronutrients and Nutriomes for Genome Damage Prevention: Current Status and a Road Map to the Future. *Am. J. Clin. Nutr.* **2010,** *91* (5). https://doi.org/10.3945/ajcn.2010.28674D

Fenech, M. The Genome Health Clinic and Genome Health Nutrigenomics Concepts: Diagnosis and Nutritional Treatment of Genome and Epigenome Damage on an Individual Basis. *Mutagenesis* **2005a**, *20* (4), 255–269. https://doi.org/10.1093/mutage/gei040

Fenech, M.; El-Sohemy, A.; Cahill, L.; Ferguson, L. R.; French, T. A. C.; Tai, E. S.; Milner, J.; Koh, W. P.; Xie, L.; Zucker, M.; Buckley, M.; Cosgrove, L.; Lockett, T.; Fung, K. Y. C.; Head, R. Nutrigenetics and Nutrigenomics: Viewpoints on the Current Status and Applications in Nutrition Research and Practice. *J. Nutrigenet. Nutrigenom.* **2011,** *4* (2), 69–89. https://doi.org/10.1159/000327772

Ferland-Mccollough, D.; Fernandez-Twinn, D. S.; Cannell, I. G.; David, H.; Warner, M.; Vaag, A. A.; Bork-Jensen, J.; Brøns, C.; Gant, T. W.; Willis, A. E.; Siddle, K.; Bushell, M.; Ozanne, S. E. Programming of Adipose Tissue miR-483-3p and GDF-3 Expression by Maternal Diet in Type 2 Diabetes. *Cell Death Differ.* **2012,** *19* (6), 1003–1012. https://doi.org/10.1038/cdd.2011.183

*Finishing the Euchromatic Sequence of the Human Genome International Human Genome Sequencing Consortium**. 2004. http://www.genome.gov/10000923.

Garcia-Bailo, B.; Toguri, C.; Eny, K. M.; El-Sohemy, A. Genetic Variation in Taste and Its Influence on Food Selection. In *OMICS J. Integr. Biol.* **2009,** *13* (1), 69–80. https://doi.org/10.1089/omi.2008.0031
Görman, U.; Mathers, J. C.; Grimaldi, K. A.; Ahlgren, J.; Nordström, K. Do We Know Enough? A Scientific and Ethical Analysis of the Basis for Genetic-Based Personalized Nutrition. *Genes Nutr.* **2013,** *8* (4), 373–381. https://doi.org/10.1007/s12263-013-0338-6
Henry, C. J. Functional Foods. *Eur. J. Clin. Nutr.* **2010,** *64* (7), 657–659. https://doi.org/10.1038/ejcn.2010.101
Joffe, Y. T.; Houghton, C. A. A Novel Approach to the Nutrigenetics and Nutrigenomics of Obesity and Weight Management. *Curr. Oncol. Rep.* **2016,** *18* (7), 1. https://doi.org/10.1007/s11912-016-0529-6
Joost, H. G.; Gibney, M. J.; Cashman, K. D.; Görman, U.; Hesketh, J. E.; Mueller, M.; van Ommen, B.; Williams, C. M.; Mathers, J. C. Personalised Nutrition: Status and Perspectives. *Br. J. Nutr.* **2007,** *98* (1), 26–31. https://doi.org/10.1017/S0007114507685195
Kaput, J.; Rodriguez, R. L. Nutritional Genomics: The Next Frontier in the Postgenomic Era. *Physiol. Genom.* **2004,** *16,* 166–177. https://doi.org/10.1152/physiolgenomics.00107.2003.-The
Karasawa, M. M. G.; Mohan, C. Fruits as Prospective Reserves of bioactive Compounds: A Review. *Nat. Prod. Bioprospect.* **2018,** *8* (5), 335–346. https://doi.org/10.1007/s13659-018-0186-6
Kussmann, M.; Fay, L. B. Nutrigenomics and Personalized Nutrition: Science and Concept. *Personalized Med.* **2008,** *5* (5), 447–455. https://doi.org/10.2217/17410541.5.5.447
Marian, A. J.; Roberts, R. The Molecular Genetic Basis for Hypertrophic Cardiomyopathy. *J. Mol. Cell. Cardiol.* **2001,** *33* (4), 655–670. https://doi.org/10.1006/jmcc.2001.1340
Mariman, E. C. M. Nutrigenomics and Nutrigenetics: The 'Omics' Revolution in Nutritional Science. *Biotechnol. Appl. Biochem.* **2006,** *44* (3), 119. https://doi.org/10.1042/ba20050112
Mondal, S. P. D. Nutrigenomics: An Interface of Gene-Diet-Disease Interaction. In *Mineral Deficiencies—Electrolyte Disturbances, Genes, Diet and Disease Interface*; Mózsik, G., Díaz-Soto, G., Eds.; IntechOpen: London, 2020.
Morin, K. Knowledge and Attitudes of Canadian Consumers and Health Care Professionals Regarding Nutritional Genomics. *OMICS J. Integr. Biol.* **2009,** *13* (1), 37–41. https://doi.org/10.1089/omi.2008.0047
Murgia, C.; Adamski, M. M. Translation of Nutritional Genomics Into Nutrition Practice: The Next Step. *Nutrients* **2017,** *9* (4). https://doi.org/10.3390/nu9040366
Mutch, D. M., Wahli, W., & Williamson, G. (2005). Nutrigenomics and nutrigenetics: the emerging faces of nutrition. The FASEB Journal, 19(12), 1602–1616. https://doi.org/10.1096/fj.05-3911rev
Mustapa, M. A. C.; Amin, L.; Frewer, L. J. Predictors of Stakeholders' Intention to Adopt Nutrigenomics. *Genes Nutr.* **2020,** *15* (1). https://doi.org/10.1186/s12263-020-00676-y
Ordovas, J. M.; Corella, D. Nutritional Genomics. *Annu. Rev. Genom. Hum. Genet.* **2014,** *5,* 71–118. https://doi.org/10.1146/annurev.genom.5.061903.180008
Pavlidis, C.; Patrinos, G. P.; Katsila, T. Nutrigenomics: A Controversy. *Appl. Transl. Genom.* **2015a,** *4,* 50–53. https://doi.org/10.1016/j.atg.2015.02.003
Ricigliano, V. A.; Ihle, K. E.; Williams, S. T. Nutrigenetic Comparison of Two Varroa-Resistant Honey Bee Stocks Fed Pollen and Spirulina Microalgae. *Apidologie* **2021,** *52* (4), 873–886. https://doi.org/10.1007/s13592-021-00877-3

Ronteltap, A.; van Trijp, J. C. M.; Renes, R. J. Consumer Acceptance of Nutrigenomics-Based Personalised Nutrition. *Br. J. Nutr.* **2009,** *101* (1), 132–144. https://doi.org/10.1017/S0007114508992552

Sales, N. M. R.; Pelegrini, P. B.; Goersch, M. C. Nutrigenomics: Definitions and Advances of This New Science. *J. Nutr. Metabol.***2014,** *2014*. https://doi.org/10.1155/2014/202759

Schneider, D. A.; Webster, L. T. Nutrition of the Host and Natural Resistance to Infection I. The Effect of DRM.t on the Response of Several Genotypes of Musculus to Salmonella Enteritidis Infection, n.d.

Shane Broughton, K.; Johnson, C. S.; Pace, B. K.; Liebman, M.; Kleppinger, K. M. *Reduced Asthma Symptoms with n-3 Fatty Acid Ingestion Are Related to 5-Series Leukotriene Production13* **1997,** *65*. https://academic.oup.com/ajcn/article/65/4/1011/4655589

Simopoulos, A. P. Genetic Screening: Programs, Principles, and Research—Thirty Years Later. Reviewing the Recommendations of the Committee for the Study of Inborn Errors of Metabolism (SIEM). *Public Health Genom.* **2009,** *12* (2), 105–111. https://doi.org/10.1159/000156114

Stewart-Knox, B. J.; Bunting, B. P.; Gilpin, S.; Parr, H. J.; Pinhão, S.; Strain, J. J.; de Almeida, M. D. V.; Gibney, M. Attitudes Toward Genetic Testing and Personalised Nutrition in a Representative Sample of European Consumers. *Br. J. Nutr.* **2009,** *101* (7), 982–989. https://doi.org/10.1017/S0007114508055657

Surana, K. R.; Ahire, E. D.; Sonawane, V. N.; Talele, S. G. Biomolecular and Molecular Docking: A Modern Tool in Drug Discovery and Virtual Screening of Natural Products. In *Applied Pharmaceutical Practice and Nutraceuticals*; Apple Academic Press, 2021; pp 209–223.

Uthpala, T.; Fernando, H.; Thibbotuwawa, A.; Jayasinghe, M. *Importance of Nutrigenomics and Nutrigenetics in Food Science*, n.d. https://doi.org/10.15406/mojfpt.2020.08.00250

Weir, M.; Morin, K.; Ries, N.; Castle, D. Canadian Health Care Professionals' Knowledge, Attitudes and Perceptions of Nutritional Genomics. *Br. J. Nutr.* **2010,** *104* (8), 1112–1119. https://doi.org/10.1017/S0007114510002035

*What Is The Difference between Nutrigenomics and Nutrigenetics?* 2021.

World Health Organization. *Diet, Nutrition, and the Prevention of Chronic Diseases: Report of a Joint WHO/FAO Expert Consultation*; World Health Organization, 2003.

CHAPTER 13

# HUMAN GENOME ADAPTATION TO DIETARY CHANGES

NEELAM L. DASHPUTRE, RAHUL R. SABLE, SHUBHAM J. KHAIRNAR, SURABHI B. PATIL, and JAYESH D. KADAM

*Department of Pharmacology, MET's Institute of Pharmacy, Bhujbal Knowledge City, Adgaon, Nashik, India*

## ABSTRACT

Human evolution has advanced in recent decades, thanks to the large-scale cataloging of gene mutations among current and extinct populations. It appears that the relatives of chimpanzees and hominins diverged 7–9 million years ago (Mya), with some earlier hominins emigrating out of Africa; *Homo sapiens* gradually arose when climate change resulted in warmer, less wooded African environments. African population increased and evolved in a range of climates, with modest variation and selection levels in the human DNA but much higher mutation in mitochondrial DNA. Human populations, both in and out of Africa, adapted to various elements of their local environment, such as altitude, weather, UV exposure, nutrition, and diseases, leaving definite indications of gene variation in some instances. With some crossbreeding with archaic early humans and the alcohol sensitivity development as humans traveled throughout Eurasia, further substantial mitochondrial mutations occurred. The Neolithic period witnessed the most crucial nutritional alterations in human history, with new means of subsistence emerging and new nutrients being introduced to the diet. Several key nutritional modifications in the human evolution course took place due to the cultural development, ecosystem, and habitat changes, including meat

Nutrigenomics and Nutraceuticals. Raj K. Keservani, Eknath D. Ahire, Shubham J. Khairnar, Sanjay J. Kshirsagar, & Rajesh K. Kesharwani (Eds.)

eating, cooking food, and those linked to the animals' and plants' domestication. The discovery of adaptation signals in the genomes of extant primates (including humans) may offer light not only on our species' evolutionary past but also on the processes that support major metabolic problems in present human populations.

## 13.1 INTRODUCTION

Important dietary changes took place during the human evolution process, including the introduction of meat consumption, cooking, and modifications related to the animal and plants domestication. Decades of anthropological study have gone into understanding this food history, in part because these changes were likely linked to substantial physical and societal changes (e.g., the modern civilization's development through agriculture and the growth in brain size comparatively). This reconstruction, on the contrary, is essential for comprehending the evolutionary context of our Western diets and the diseases that are frequently connected with them (Luca et al., 2010). Several recent investigations have focused on natural selection signatures in nonhuman and human species—both at the genomic sequence level and in data of expression of genes. Also, a lot of research has been effective in detecting unique signals of adaptation to dietetic modifications in the genomes of modern humans. This research has shown that particular coding, regulatory, and copy number adaptive modifications in the human DNA can be identified concerning dietary changes. However, identifying the wider gene sets or processes that have been under selection because of modifications in the human diet, in comparison with other primates, particularly big apes, is the next difficulty. Gene set enrich analyses, which look at broader categories of genes with positive selection signatures, demonstrate a consistent pattern across research, with numerous selected differences linked to nutrition and metabolism. Selection acted to optimize differing metabolic requirements in nonhumans and human species, according to this consistent result. Now, the problem is how to find and understand the specific genes or processes that underpin these metabolic adjustments biologically. To solve this issue, it is proposed to combine two types of information: Anthropological and comparative genomic evaluations. In the field of anthropology, there are a lot of popular hypotheses about the importance of diet in human evolution. According to the comparisons made with extant great apes, along with the archaeological and fossil record, the most noticeable change in the diet during human evolution was increased consumption of animal foodstuffs (meat and fat) as well as starchy plants goods (Valente et al., 2015). Diet and foraging habits are believed to play a significant impact on human phenotype, with nutritional differences being blamed for the major morphological alterations seen in contemporary humans. These dietary variations may help us to know better about the differences in sequence and gene expression reported in metabolic genes between primates. In this study, we plan to combine the data of genomic investigations into this well-developed anthropological context. The recent evidence for nutrition adaptation is then addressed, both at the sequencing and gene expression levels. Finally, we comment on how genetic technologies can aid in the identification of specific adaptations resulting from metabolic variations among humans and nonhuman primates (Babbitt et al., 2010).

## 13.2 HOMININ EVOLUTION IN DIETARY SHIFTS

Anthropologists have used a variety of strategies and techniques for archaeological and fossil data to conclude about ancient diets. The shapes of molar cusps, for example, may mention shifting to a leaf- and/or fruit-based diet, or a more general diet. The meals' fracture–resistance qualities consumed in the period of a few weeks before death are reflected at a microlevel in tooth surfaces' wear patterns. The stability of isotope ratios (e.g., 18O/16O) within bone and teeth includes a broad indication of what was eaten at the time those tissues were developed. Following an approximate historical ordering, we present a summary and critical appraisal of the current evidence for the most important nutritional shifts during hominin evolution (Luca et al., 2010).

### *13.2.1 HOMININS FROM THE BEGINNING*

It's difficult to recreate the food of the earliest hominins due to the limited amount of early hominin fossils and the almost complete lack of chimpanzee lineage. Although the closest existing evolutionary cousins to humans—chimps and bonobos—have largely frugivorous diets, using them as precise models for our common ancestor is possibly an oversimplification. Recent descriptions of the fossils of *Ardipithecus ramidus*, the earliest (4.4-million-years old) moderately complete human fossil recovered to date, give new insight into prehistoric diets.

The oldest hominins were believed to have migrated from a forested habitat to a savanna environment, accompanied by a transition to the harder and rougher food items that were more common in the new place. A detailed simulation of the habitats of completely bipedal *A. ramidus*, on the contrary, reveals that this hominid lived in a woodland patchy forest rather than a savanna. The habitat of the hominid *Orrorin tugenensis*, who lived about 6 Mya, was said to be identical. As a result, different biped and hominin origins scenarios should have been considered and analyzed. Leading hominins moved into more open settings and encountered more technically difficult diets. *Australopithecines*' massive, thick-enameled teeth (4.1–1.4 Mya) indicate a hard food-rich diet. However, there is a significant dispute as to which foods such as dental morphology would have been used. *Australopithecines* may have utilized their premolars for the opening of the hard shells (endocarps) of rather big seeds, as per current craniofacial strain biomechanical modeling technique studies. Other scholars, on the contrary, see underground storage organs (USOs) rich in starch-like corms and bulbs as a substantial factor in the australopithecine diet. These two hypotheses (huge seeds vs USOs as significant food sources) for explaining australopithecine craniodental morphology are not completely exclusive, as no hominin species ate only one type of food. Certain facts and certain studies of craniodental morphology in *Australopithecines* reveal important interindividual dietary heterogeneity. When foods of choice were unobtainable, it was ordered to examine the idea that even relatively infrequently consumed items were critical for survival. Being able to use such "fallback food" sources during human evolution could have been a major selective pressure (Luca et al., 2010).

### *13.2.2 SCAVENGING AND HUNTING SEEKING MEAT*

Chimpanzees sometimes hunt small primates and nonprimate species, and meat is a small source of food. Meat, together with plant foods, is an important dietary resource for modern human hunter gatherers. Meat eating in hominines can be tracked at least back to 2.5 Mya. Few of the fossil discoveries are constant with scavenging practices, even though the relative relevance and timing of hunting vs aging in hominine evolution remain a source of debate. Meat supplies proteins, iron, zinc, vitamins, and other essential nutritional components in addition to high caloric benefits from animal-source foods (Keservani et al., 2017; Keservani et al., 2020). As raw meat is hard and slow to digest, preventing consumption in big numbers, the adaption of a large amount of meat eating may have required advanced processing methods, such as cooking (Luca et al., 2010).

### *13.2.3 COOKING*

In Israel, the earliest undeniable proof of control of fire by humans comes from 8,00,000 years ago. The rest of the possible sites date back as far as 1.5 Mya. Based on the reduction in the size of the tooth in *Homo erectus*, which indicates a switch to the processing of softer foods, Wrangham and colleagues argue that cooked food could have been part of human culture at least as far back as 1.9 Mya. Cooking, regardless of when it started, was most likely a major diet change because it improved the digestion of both meat and plant meals. Some fats and some ordinarily edible tubers are too hard to eat without being cooked. There is no solid proof for fire and cooking. If the early date they estimate is right, the resulting dietary shift could have assisted the growth in relative brain size shown at the same era in hominid evolution. The increase in brain size, which corresponded with a reduction in stomach size, suggests that food quality has improved. Whether such a change in diet quality was facilitated by cooking or simply by advancements in stone tool technology, and whether meat or starch-rich USOs were the most important and main food source in this scenario, are heavily debated topics. Because proof of control of fire by humans and sticks for digging up tubers does not always survive well in archaeological and fossil records, this dispute may not be resolved.

### *13.2.4 PLANTS AND ANIMALS DOMESTICATION*

The most recent major transition in human nutrition is the birth and expansion of agriculture and animal husbandry over the past 12,000 years, with centers of domestication in Asia, Europe, South America, and Africa. Food production and storage technologies connected with this dietary change result in population densities orders of magnitude larger than those achievable in hunter-gatherer subsistence civilizations. Overall, however, the growth of agriculture was connected to a startling fall in the diversity of nutritious intake. For example, starch accounts for 50–70% of the calories consumed in agricultural cuisine. Agricultural diets may have been connected with a nutritive availability that surpasses growth and energy requirements, as seen in the most developed modern agricultural economies, in addition to a decrease in nutritional diversity. Type-2 diabetes mellitus (T2DM) can be a current result of the ancient history of metabolic efficiency in terms of survival of the fittest in the scarce or unpredictably available food in a hunter-gatherer culture, as per Neel's "thrifty genotype" hypothesis. As per Brand–Miller and Colagiuri's "carnivore link" theory, insulin resistance might have been helpful in a previous diet that was rich in meat and poor in carbohydrates, but it is now hazardous due to a shift back to agriculture-associated high-carbohydrate diets. Estimate that plant food product plant food products made up about 65% of the diets of pre-agricultural paleolithic humans, based on information from anthropological studies of modern hunter-gatherer societies. Although it can be an exaggeration to recognize pre-agricultural hunter gatherers as "primarily carnivorous," a 35% or higher contribution of animal source foods would have resulted in significantly higher protein levels and lower dietary carbohydrate levels than their agricultural descendants.

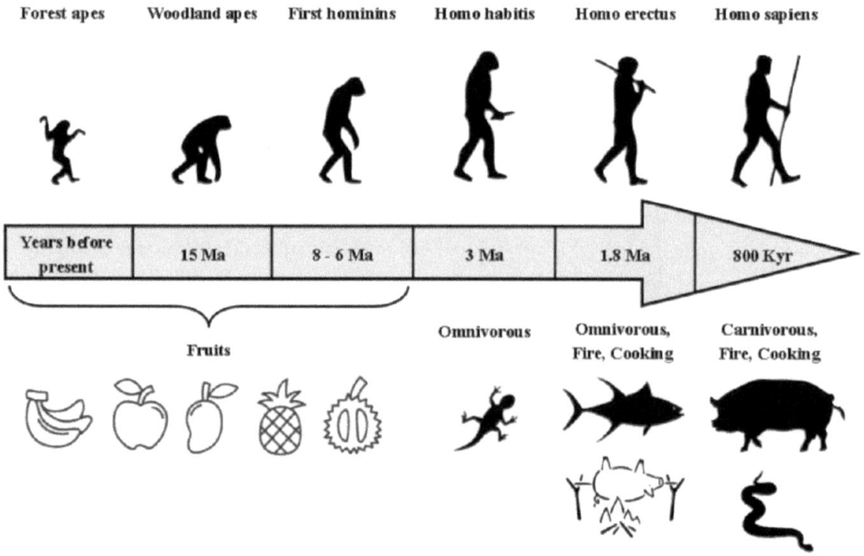


**FIGURE 13.1** Human diet evolution.

## 13.3 BIOLOGICAL PROCESSES INFLUENCED BY DIETARY CHANGES

During hominin evolution, changes in dietary abundance and diet composition presumably imposed severe selective pressures on several biological systems. Determining the genomic regions that are being targeted by nutrition-related selective factors could reveal information about our species' evolutionary past also the biological mechanisms that have a function in the impact of food risk factors for diseases such as diabetes mellitus, cancer, and hypertension. Many researchers have used evolutionary concepts to learn about the impact of selection on individual loci, and this interesting idea has aroused their interest. Numerous evolutionary research has mainly concentrated on genetic makeup and derivatives with well-established functional importance to resolve a few of the drawbacks stated above; by mixing multiple sources of evidence (functional and developmental), this plan aims to provide further interpretable results and improve the probability of real signals of alteration are identified. We concentrate on research that matches into categories listed below (Surana, et al., 2021; Ahire, et al., 2021):

a. Research of genetic loci about which strong prior assumptions refer to the biological activities have been developed;
b. Genetic variations that are significant candidates for evolutionary processes based on their influence on relevant phenotypes are studied; and
c. Research reporting robust selection signs. There is evidence of genetic modifications in response to nutrition modifications and the adoption of dietary specializations is examined in this section. Single-gene investigations are reviewed first for every biological mechanism and consist of a summary of the outcomes of genome-wide scans for shortlisting.

### 13.3.1 METABOLISM

Metabolism is the destruction and construction of molecular structures from nutrients for the generation of energy. Although several metabolic processes are important in all dietary programs, as populations specialize in specific dietary components, specific pathways, and responses may become serious. As a result, early studies to discover genetic adaptations to dietary specialization focused on enzymes with well-defined and highly precise functional roles in food metabolism. Patterns of variability in the coding of genes for the lactase (LCT) and amylase (AMY1) enzymes have provided a couple of the most interpretable instances of genetic adaptations to dietary specializations in human metabolism to date. After weaning, all nonhuman animals lose their capacity for lactose digestion, the major carbohydrate found in dairy products. This is due to lower quantities of the lactase-phlorizin hydrolase enzyme, which is generated in the gut. Lactose can be metabolized by certain humans in their adult years. The lactase persistence (LP) phenotype was discovered to have a genetic basis and to be transmitted as a dominant feature in the early 1970s. Only pastoral nomads and dairy products have a long history, such populations possess LP as a common trait. Several studies recommended the LCT gene as a chosen target in European cultures before any strong proof of choices in the lactase region was revealed, owing to the significant milk and dairy content in adult diets. The initial proof of positive selection at the LCT gene came from higher $F_{ST}$ values among European Americans, African Americans, and East Asians, as well as a large extended haplotype homozygosity (EHH) LCT for the haplotype containing an LP variation. However, while this mutation explains LP in European groups, it is only found in small numbers in African communities where the LP trait is widespread. In additional fact, other variations linked to LP in African pastoral nomad tribes were identified in a later study that found concrete proof that a minimum of one of these African variants had recently undergone positive selection, as evidenced by a substantial EHH score. Two more variations in the LCT gene's regulatory region have recently been discovered in Middle Eastern people from Saudi Arabia. In vitro investigation determined that these variations influence the transcriptional LCT gene control and are linked to a strong natural selection signature based on EHH. These results imply that the LP trait is categorized by genetic heterogeneity and is a remarkable example of human population convergent evolution. Messages of a recent selective sweep in European populations have been identified at the LCT locus in virtually all genome-wide scans for collection executed so far; in most studies, the LCT gene carries one of the strongest signals in the entire genome, with haplotype homogeneity extending up to 2 Mb, which is consistent with the studies cited above. This trend indicates that there had been a lot of evolutionary pressure related to adult milk consumption.

### *13.3.2 PERCEPTION OF THE SENSORY SYSTEMS*

Food is seen, handled, and smelled before it has been tested, making it a multisensory experience. The release of extra chemicals that stimulate the senses of smell and taste, and also the creation of noises that stimulate hearing, occur while food is processed in the mouth. Each of the five senses is enhanced to varying degrees in different species, reflecting their feeding patterns and demands, and also the overall complicated interactions with their surroundings. As a result, numerous investigations have been performed to

**TABLE 13.1** Ethnic and Ancestral Differences in Nutritional Processing and Metabolism.

| Nutrient | Ethnic differences | Dietary challenge or opportunity | Nutritional significance |
|---|---|---|---|
| Lactose | Asian and most African groups have lactose intolerance but many other social groups independently evolved the capacity to digest lactose | The capacity to domesticate cattle for milk production is found in many different parts of the world | Intolerance to high lactose intakes is usual after infancy but several societies relatively recently developed persistent lactase activity |
| Starch | Rainforest dwellers in Africa compared with pastoralists | Rainforest dwellers are particularly reliant on fruit so lower amylase digestive capacity is required but with more sparse forest and open country the availability of tubers and other crops increases starch availability markedly | Amylase 1 (AMY1) greater copy number improves digestion in groups on high starch diets vs pastoralists; a higher copy number may also help to overcome the impact of intestinal diseases |
| Long-chain fatty acids (LCFAs) | Multiple ethnic group differences across Africa, Europe, and Asia depend on traditional diet | Coastal populations have far more dietary LCFAs from fish than in inland communities; Animal sources provide some essential fats but otherwise, the main unsaturated fat source is from plants so desaturation and elongation to produce essential fatty acids become important. | Ethnic groups with long-standing diets low in animal foods have higher FADs enzymes so able to produce long-chain polyunsaturated fatty acids (LCPUFAS) more readily |
| Alcohol | Asia/Oceania societies intolerant | Alcohol tolerance was induced for about 10 Mya when the capacity to eat fermented fresh fruit became important in tropical forest conditions. Mitochondrial gene mutations relating to acetaldehyde dehydrogenase account for the loss of alcohol tolerance in Asians. | Low acetaldehyde dehydrogenase with an accumulation of acetaldehyde leading to hot flushing of face + intolerance of alcohol |
| Folate | Northern Eurasian groups have a higher prevalence of the methylenetetrahydrofolate reductase (MHTFR) C677T polymorphism associated with reduced activity and higher homocysteine levels | Marked differences in the ready availability of dietary folate throughout the year dependent on Climate, temperature, rainfall, and soil fertility. | Those with low activity of MHTFR C677T enzyme activity have higher needs for dietary folate with greater susceptibility to neural tube defects and possibly other conditions |

**TABLE 13.1** *(Continued)*

| Nutrient | Ethnic differences | Dietary challenge or opportunity | Nutritional significance |
|---|---|---|---|
| Vitamin D | Dark-skinned populations | Darker skin is more protective of skin from intense sun exposure in Africa; lighter skin develops as the sun becomes less intense away from the equator | Darker-skinned people need greater exposure to sunlight or more oral vitamin D2 to avoid bone changes, for example, rickets and to improve immune function |
| Fructose | Humans, great, and lesser Apes are more sensitive to the effects of fructose in increasing their liver fat and stimulating gluconeogenesis than other mammals due to a loss of uricase that occurred during the Miocene epoch | Tree-living primates dependent on fruit as an energy source with the capacity to induce energy reserves importantly | It may have had a role in protecting ancestral apes from starvation |

*Source*: Adapted from James et al. (2019).

better detect the changes in odor, taste, and vision in nonhumans and human primates, with crucial implications for understanding dietary adaptations. Chemical constituents from food trigger taste receptor cells, resulting in taste perception. Sweet, sour, bitter, salty, and umami are the five flavors that humans and most mammals can sense and distinguish. Different receptors are engaged in the perception of these diverse taste modalities. Because of its importance in preventing the consumption of hazardous food, the bitter taste has been a particular focus of evolutionary studies.

### *13.3.3 REGULATING APPETITE*

The frequency and quantity of eating episodes, the choice of high and low foods, the energy level, and the range of food consumed all contribute to appetite. Appetite regulation is a biological process that is closely related to food availability in the environment and lifestyle. As a result, genes involved in hunger regulation could be potential targets of selective pressure as humans encountered various nutritional obstacles. Hormonal, psychological, and neurological signals all influence feeding behavior. Genes encoding long-term energy storage and hunger-satiety hormones have been researched to better understand the genetic underpinnings of metabolic illnesses such as T2D. LEPR has increased haplotypic homozygosity, according to a search for selection signals across the entire genome. Furthermore, for this gene, there was significant link among climatic variables and allele frequencies discovered, similar to variants in candidate gene for metabolic syndrome. TCF7L2 is also one of those genes, which has been recognized as the gene with the highest signal in T2D whole-genome association studies. There was also evidence of spatially variable selection in two SNPs in this gene. The rate with which the allele is generated for SNP which defends against T2D is more in non-African people than in African people, and it is strongly linked to climate variables. Based on EHH, in Asia, Europe, and Africa, the protective allele at this identical SNP also has a positive selective signature. The era of the protective allele-tagged haplotype is estimated to be relatively recent, coinciding with the progress of the agricultural sector in each population. It is difficult to predict that whether the signals seen at TCF7L2 and LEPR are the consequence of adaptations to a cold climate or dietary and lifestyle modifications based on the proofs so far. Nonetheless, these genes are intriguing candidates for how short-term neuroendocrine hunger regulation and fat regulation over time and energy metabolism may have evolved during human evolution under positive natural selection.

### *13.3.4 THE DIGESTIVE SYSTEM'S MORPHOLOGICAL DEVELOPMENT*

Although the digestive tract structure is thought to have responded to dietary changes, there have been few conclusive signs of natural selection (positive) in this large number of genes. The thickness of the dental enamel is a particularly interesting phenotypic characteristic because it may represent an adaptation to dietary items with different fracture–resistance features. The production and elongation of enamel are thought to be assisted by enamelin peptides. Patients with amelogenesis imperfect, or a genetic deficiency in dental enamel production, have been found to have mutations in the coding of gene for enamelin (ENAM). Additionally, in human populations, the thickness of enamel is a polymorphic trait, with African and American populations having thicker enamel than European Americans. In genome-wide selection scanning, the gene ENAM was discovered as likely the focus of current positive selection in humans. The ENAM gene was resequenced in 10 populations of humans and 12 species of mammals by Kelley and Swanson with varying diets in follow-up research. In non-African population samples, patterns of gene variation revealed an inadequacy of variety and a tilt toward rare variants, showing a complete selection sweep in these groups. The developed allele is essentially stable in non-Africans, it happens in a conserved site across mammal species, and it results in the transition of polar to nonpolar residues.

## 13.4 GENES THAT ARE SUBJECTED TO DIET-RELATED GENETIC ADAPTATION

Although genetic modifications to dietary specializations are considered to have served as favorable evolutionary solutions in humans, the limit to which dietary considerations have operated as selective pressure on genes that play a role in food-related metabolic pathways is yet unknown. Recent research has discovered genetic markers of adaptations that are presumably generated by dietary stresses. Additionally, the candidate genes technique had previously yielded strong proof for genetic modifications to nutrient consumption disparities, such as at the lactase and amylase genes. Other metabolic-related genes, such as AGXT, code for alanine: glyoxylate aminotransferase (AGAT), this enzyme is liable for the glyoxylate transamination into glycine, has been hypothesized as dietary adaptation. MTRR, encoding methionine synthase reductase, an enzyme acting in the complex folate pathway; NAT2, encoding N-acetyltransferase 2, a phase-II enzyme that is liable for the detoxifying of a large variety of xenobiotics; and CYP3A5, encoding cytochrome P-450 3A5, a component of the CYP3A enzyme that is a component of oxygen metabolism and xenobiotics.

### *13.4.1 AGXT*

We examined the polymorphism c.32C > T in AGXT gene, where the developed allele T had previously been hypothesized to play an adaptability role in a population that ate a meat-rich diet (Ye and Gu, 2011). Studied the concept in detail, finding on frequency data that supported the model, an inference that accounted for the highest number of the developed gene in the Sweden Saami, who has a long backstory of ingesting huge amounts of animal items. Later, when reworked the question with a larger study of Central Asian populations, they would be unable to find increased genotype frequencies in populations with meat-rich diets compared to those with less meat-rich diets, casting aspersions on the adaptive model proposed for the variation. However, when evaluating the question with a larger sample of Central Asian populations, they were unable to find increased allele frequencies in populations with meat-rich diets especially in comparison to those with less meat-rich diets, raising doubts about the adaptive model proposed for the alteration. For example, it was eliminated from both Angolan farmers and Khoisan hunter-gatherers, even though the fact that the former comprises fewer meat users than the latter. The derived allele reached 19.15% in the sample from Portugal, which is regarded to be a farming population with a mixed diet that is reasonably balanced in terms of animal and plant food resources. This is a higher frequency than any of the African populations, regardless of their mode of subsistence or reliance on meat. Furthermore, while the diverse regional distribution of c.32C > T in Africa and Eurasia corresponded well to random neutral biomarkers used widely to identify big population groups, two populations with extreme results were recorded in Asia, where gene frequencies are dispersed widely, but with distinct traditional lifestyles: In the Tajiks, a group of sedentary agriculturalists from Western Tajikistan; and in the Tajiks, a group of sedentary agriculturalists from Western Tajiki (1.7%).

### 13.4.2 PLRP2

We focused on c.1074G > A, a variation that causes the pancreatic lipase-associated associated protein 2 to be prematurely turned specialized, resulting in the enzymes' more active version identified this variant as a convincing signal of adaptation to a dietary specialized knowledge, since the developed allele was considered, a genomic sequence search for selection in populations was considered to be very frequent in populations dependent on diets high in cereals (farmers) than in other societies. In comparison, the two hunter-gatherer groups, notably the Ju/hoansi, showed lower frequencies (2.42%). With 51.1%, the sample from Portugal had the highest number in this analysis. Overall, our findings strengthen the idea that the widespread of c.1074G > A is associated with the number of cereals consumed in diets, in the sense that, at least within Africa, farmers have greater frequencies of the developed allele than hunter gatherers who eat fewer cereals. These observations were then introduced in a broader perspective, with data on c.1074G > A for African and Eurasian groups gathered from a variety of sources, and the categorization was divided into populations that specialized in cereals and those who do not when first presented. The incomplete allele was found to be very common in communities with cereal-rich nutrition (average frequency 35.8% in Africa; 49.1% in Eurasia + Africa) than in communities that were less dependent on cereals (average frequency 22.7% in Africa; 22.9% in Africa + Eurasia), with statistically significant differences in Africa [$p = 0.0050$, one-way analysis of variance (ANOVA)] and Eurasia + Africa ($p = 0.0001$, one-way ANOVA). When comparing both, the herders and hunter gatherers, who were included in the cereal-poor population group; the mean frequency was 41.3% in Africa and 18.6% in African + Eurasian countries, respectively, with both differences statistically meaningful (one-way ANOVA: $p = 0.0060$ and $p = 0.0001$ for Africa and Africa + Eurasia, respectively). Overall, these results indicate that PLRP2 diversity was generated by selective forces that changed depending on the lifestyle of communities.

### 13.4.3 MTRR

We looked at the common variable c.1130A > G altering amounts of enzymatic activity within MTRR because it was another possible adaptive genetic variation found in the genome-wide analysis. MTRR had previously attracted a lot of attention in association studies, having been linked to things like the chance of spina bifida. However, just one study looked at its adaptive role in dietary specialization, finding that c.1130A > G was substantially associated with diets containing primarily folate-poor food roots and tubers. The allele derived was found to be relatively abundant in most of the groups in Africa, Angolan and Mozambican agronomists had the highest ratings of 0.529 and 0.550 respectively, in this survey. Both values are close to those reported in a recent survey on the Yoruba (0.548), the only African population with a diet based mostly on roots and tubers. As a result, a large number of these genes can be observed in populations without such dietary specialization, at least in Africa. Furthermore, there was no suggestion that the dispersion of c.1130A > G was associated with dietary folate availability, which is usually believed to be less in nonforager cultures (pastoral and agricultural) than in hunter-gatherer populations. Despite the disparities in the manner of food production, the developed allele was discovered at a similar frequency in the Baka hunter gatherers, Ugandan herders, and Equatorial Guinean farmers. As a result, the present patterns of diversity for the mutation c.1130A > G in MTRR do not indicate that it might indicate an adaptation to the human population's style of subsistence.

### *13.4.4 NTA2*

The availability of folates in the diet has previously been proposed as a genetic variant modulator at the NAT2 (N-acetyltransferase 2) gene. Individuals can be classed as slow, intermediate, or fast acetylating agents based on their haplotypic makeup, which is determined by NAT2 locus' genetic variants. Individuals are classified as slow, moderate, or quick acetylators based on their haplotypic structure, which is determined by gene changes at the NAT2 locus. The same researchers' subsequent findings, based on a more extensive examination of NAT2 global genetic variation, were also consistent with a scenario positing that slow acetylation phenotypes were preferentially chosen among populations reliant on low-folate diets, while the quick acetylators were even or neutral advantageous in the company of dietary regimens with enhanced folate supply. The number of acetylator phenotypes and NAT2 haplotypes was also assessed in this investigation, to the extent that prior studies' population coverage allowed. The haplotype distribution was highly variable across African populations, but the prevalence of the slow acetylators was markedly smaller in the two hunter-gatherer groups (Khoisan, 1.6%; Baka Pygmies, 13.5%) than in the three agriculturalist groups or Ugandan pastoralists, all of whom had values up to 37.4% ($p = 0.0139$, one-way ANOVA). The slow acetylator phenotypes accounted for a high proportion of 52.2% in the Portuguese, which is equal to other European groups. Following that, we examined our findings in comparison to those previously reported results for Eurasian and African populations, emphasizing c.590G > A, which describes allele NAT2, as it was the variant of the most information for groups representing the three ways of subsistence. In summary, our findings support prior evidence that NAT2 evolved under the effect of a selective factor affected by human food (Luca et al., 2010).

### *13.4.5 CYP3A5*

In the case of CYP3A5, we screened for the intronic variation c.219-237G > A, also known as the CYP3A5 polymorphism, where the derived allele A induces a premature stop codon, limiting protein expression. It has been found that the variation has a very peculiar global distribution, with the frequency of CYP3A5 being significantly associated with latitude. CYP3A5 is thought to affect retention of salt and water also the likelihood of salt-sensitive hypertension, with interactions with dietary salt consumption determining blood pressure. We wondered if CYP3A5 diversity could be linked to population diets because anthropological evidence suggests that hunting and gathering people's diets are often characterized by a low level of salt intake, trying to make them a surrogate for the preagricultural human diet, which has recently been appreciated as a model of well-balanced dietary patterns. As a result, analysis of variance in six African communities where the allele derived was only marginally represented, but suggestively, the smallest and the largest number were found in two groups of a farmer (11.7% in Mozambique and 24.0% in Angola), discrediting any link between lifestyle and differences in allele frequency across populations. The allele in the Portuguese achieved a high of 90.2%, which is typical of European populations where CYP3A5 fluctuates relatively narrowly, with most countries being near-fixation. It was confirmed that the number of the low expression allele increased significantly with the equatorial distance (SRCSC = 0.7540; $p = 0.0001$) while the data was combined with that acquired from the literature, and with expanded exposure of Eurasian and African populations. When the link was examined in all three continents separately, there was no statistically significant rank correlation in Africa or Europe (SRCSC = 0.1058; P = 0.2438) or (SRCSC = 0.4183; $p = 0.1310$), respectively, still the correlation coefficient was statistically significant in Asia (SRCSC = 0.5724; $p = 0.0002$). The remarkable relationship can be explained in Asia, where the mean frequency of allele was 0.793 because the smallest values are regularly present in groups from the continent's southern part, positioned extremely close to or already a component of the intertropical zone. These results concluded the conclusion that CYP3A5 was the object of a selective factor induced by human population geography.

## 13.5 METHODS FOR IDENTIFYING GENETIC ADAPTATION SIGNALS

Understanding the basic concepts of evolutionary analyses targeted at detecting signs of natural selection is important for a critical examination of the proof of genetic modifications to diet adaptations. In this part, we look at the main natural selection concepts, with a concentration on others more likely to be subject to dietary changes. We also give a summary of the most generally used selection procedures and statistical tests, as well as a consideration of statistical challenges surrounding the perception of selection signals.

### *13.5.1 NATURAL SELECTION*

The natural selection process is through which heritable good qualities become more frequent with time, whereas heritable undesirable traits become less common. This process can be governed by several models, each of which generates unique signs in variation of genetic patterns. When one among the two alleles at a polymorphic location is preferred over the other, the allele preferred will rapidly grow in frequency; if the selection is stable for a long enough time, the favorable allele will achieve fixation. Two or more genetic loci, in the population are maintained at a steady-state frequency for as long as the selective pressure is available in balancing selection; in one model, heterozygous individuals are more fit than both the homozygotes, also a new gene is expected to rise in frequency quickly until it reaches equilibrium. Several more models with regionally and temporally changing choices may be subject to human dietary adaptations. Humans believe in a diverse range of ecosystems, each with its own set of nutrients and resources, reducing the range of dietary choices available in any specific region. This indicates that diet-related selective pressures have differed throughout geographic space as populations moved across the globe and adapted to the nutritional components available in various habitats. Furthermore, hominid diets have altered considerably throughout time, especially recently, resulting in novel selective pressure at various points in evolutionary history. New alleles generated by mutation or transferred from a nearby population may have been utilized in adaptive responses to new selective pressures, or existing alleles that were not favorable before the environmental shift may have been utilized. The latter scenario allows for a faster adaptation response and, as a result, may be more common among species like humans who have experienced significant dietetic and environmental changes.

### 13.5.2 NATURAL SELECTION'S SIGNATURES

Most lots of the population's genetic approaches to identifying advantageous mutations rely on the trait of positive natural selection that leads the advantageous allele to the high frequency at a considerably faster rate than a neutrally evolving allele. A new beneficial gene must be related to a certain haplotype background if it is introduced into a population via mutation. Only some of the recombination events will occur because the allele is fast driven to moderate or high frequency, and as a result, the neutral alleles which are firmly related to the targeted site will tend to increase in frequency along with the favorable allele. EHH is the result of this mechanism, which results in extended identical haplotypes in a local pattern. An incomplete or partial selective sweep is a term that describes this process, and various statistical test methods have been designed to detect the structure that results. The beneficial allele will move to fixation if the selection is directed and constant over a long period. All variants near the selected site will be fixed in this instance, which is also referred to as a complete selective sweep, whereas new mutations originating during the sweep will occur at a low frequency. As a result, near a fixed favorable allele, a decrease in levels of polymorphism and a relatively large quantity of uncommon variations in the spectrum of allele frequencies are expected. Summary statistics of the frequency spectrum can be used to evaluate a tilt toward rare variants; the most generally used of these statistics is Tajima's D (TD), although others are also widely utilized. In contrast, with a large amount of intermediate variation numbers, as predicted by some balanced selection models, the TD value will be positive. Under these models, several statistical assessments have been designed to detect the influence of positive natural selection.

When selection works on a variant that is helpful in just a subset of populations, the frequency of that variant may differ among populations more than random drift would indicate. Several methods for detecting such adaptations to local surroundings have been devised. Historically, the most often utilized technique has been based on the $F_{ST}$ statistics, which highlights differences in allele frequencies between two groups or across numerous populations. Variants with extremely high $F_{ST}$ values are normally thought to be the targets of selective pressure related to certain habitats, as may be anticipated in populations that specialize in specific dietary components. Another method is to put up several independent contrasts among closely related groups with differing diets; this approach is exemplified in the specific gene for salivary amylase. Small-to-moderate variations in allele frequencies attributable to particular selective pressure acting in one of the populations but not in another one can be detected by comparing genetically similar populations. This method, which was recently developed, compares allele frequencies among populations classified based on an environmental variable, such as main dietary component or subsistence while regulating the link between human populations due to history of populations and sample size differences across populations.

### *13.5.3 GENOME SCANS FOR SELECTION SIGNALS: INTERPRETING THE RESULTS*

It is not easy to determine the statistical significance of the results of the above-mentioned tests. This is because existing genomic variation patterns in humans are the result of both selection and population history processes, and disentangling the effects of these processes is difficult. To get through the problem, many researchers have used an empirical approach that avoids making assumptions about population history and focuses on identifying the loci with the most interesting patterns in large-scale datasets of genetic variation. For example, variability patterns across the genomes can be described using one or more statistical tests are used to grade loci in the DNA. Loci that fall over an arbitrary cut-off are identified as outliers and are referred to as such. These outliers indicate candidate targets of high selective forces, provided that the majority of human DNA loci change neutrally.

If selection is weak, however, favorable alleles may show variability patterns that are unique but not identical severe in comparison to patterns found across the entire genome. Investigators assessed the degree of evidence for modifications in different gene groupings in different biological processes in a try to discover loci with a finer selection. A significant excessive number of signals in single group, even though the signals at specific genes may not achieve genome-wide relevance, they show that the biological mechanism itself was a focus of selection. Another factor to consider when examining evidence for dietary adaptations is if the selection was new mutations, it's more likely that standing was involved. This selection model may be useful for explaining how humans evolved to their ever-changing diets since standing variations that become favorable under new environmental conditions let populations adjust more promptly to new selective pressures. Theoretical studies have found that the variation pattern expected in this scenario may be unrecognizable from that discovered in neutrally developing genome regions, resulting in a huge loss of power.

**TABLE 13.2** Specifying Selection Signals in Humans.

| Signal of selection | Target variant | Normal biological function | Phenotype variantion | Other variation in the pathway | Evolutionary adaptation |
|---|---|---|---|---|---|
| Chr11, Africa | HBB, Coding | Oxyten carrying in RBCs | Sickle cell, Malaria resistance | CD36, HBA, SLCA1, DARC, HBB* | Protection from Malaria |
| Chr2, Europe, Africa | LCT, Regulatory | Enzymelactase to digest mil | Prolonged expression of LCT in adulthood | LCT* | Milk digestion as adults after cattle domestication |
| Chr15, Durope | SLC24A5, Coding | Pigment Formation | Decreased pigment formation | KITLG, SLC42A2, OCA2 | Lighter skin as humans moved to less sunlight |
| Chr2, Asia | EDAR, Coding | Development of hair, sweat, and teeth | Linked to thicker hair on head, largely unknown | EDA2R | |
| Chr16, Asia | ABCC11, Coding | Ear wax type Determinant | Dry ear wax | | |
| Chr18, Europe and Asia | DTNA, Unknown | | | | |

*Source*: Data adapted from Sabeti et al. (2006).

## 13.6 HUMAN ADAPTATION TO A COOKED DIET: GENETIC EVIDENCE

**FIGURE 13.2** Adaptation of Humans to a cooked diet.

Beginning around 2 Mya, ancestral humans experienced significant gains in volume of brain and body size, as well as decreases in size of gut and tooth. These biological properties are thought to represent a stronger reliance on animal foods and improved methods of food processing, including cooking, and imply the ingestion of an easier-to-digest diet with increased caloric density. Cooking improves nutritional digestibility and decreases diet-induced thermogenesis, allowing essential hominin foods like meat and tubers to provide significantly more energy. Evidence that modern humans cannot obtain enough energy from raw wild diets, whether or not they include meat, has led to the hypothesis that hunter gatherers are biologically committed to the benefits of cooking, such as the provision of enough fuel from energy. The idea that food that is cooked is essential for modern people expects genetic proof of human adaption to a diet that has been cooked. Direct testing of this hypothesis is urged by indirect information of such adaptation, such as pseudogenization of the masticatory (MYH16) a myosin gene, and (TAS2R62 and TAS2R64) two bitter taste receptor genes after chimpanzees separated from a common origin, but even before Neandertals and Denisovans split from a common origin. Several populations with a dairying history have developed the ability lactose digestion in adults due to lactase's ability to persist, a feature that has altered multiple times under strong positive selection over the previous 7000 years. Moreover, it has been suggested that populations with a background of consuming starch-rich diets had increased copy numbers of the gene encoding salivary amylase, the catalytic enzyme involved in the breakdown of starch in the mouth. Because amylase is ineffective at digesting starch until it has been gelatinized by heat, this adaptive covariation is regarded to be partly due to the adoption of cooking. Diet-induced genetic adaptations have been found in existing populations, implying that dietary changes with long evolutionary histories and large systemic consequences may result in more widespread gene mutation (Carmody, et al., 2016).

Even though the fact that Homo fossilized anatomic evidence implies that cooking is thought to have begun in the Lower Paleolithic, archaeological proof for fire control does not exist until the Middle Paleolithic. The fire was most likely regulated 2,50,000 years ago, though evidence of manmade fire goes back only to 4,00,000 years on rare occasions; the oldest widely accepted date of anthropogenic fire is 1 Mya from Wonderwerk Cave in South Africa. Although control of fire does not necessarily imply cooking, great apes' strong preferences for cooked foods, combined with their willingness to wait for raw food to be cooked, suggest that cooking would have occurred soon after. The suggested split between contemporary humans and the last common origin of Neandertals and Denisovans, dated to between 2,75,000 and 7,65,000 years ago, overlaps with later putative dates for the origins of cooking, leaving it unclear whether cooking was present in our clade's last common ancestor. By 50,000 years ago, granules of gelatinized starch coated in Neandertal tooth calculus indicated that they were eating cooked plant foods. However, rare evidence of the fire use in cold temperatures has led others to speculate that early Neandertals utilized fire as needed but did not have control over it. Determining whether cooked meal adaptation occurred before or after the divide between the modern human and Neandertal–Denisovan lineages could thus help determine when fire control was created. Because of the strict tissue samples and testing controls required, studies on the compatibility of a cooked diet genetically are hard to undertake in people. To find potential genes that could have been influenced by dietary change during human evolution, analyzed gene expression alterations in a model organism because of uncooked and cooked diets.

## 13.7 FUTURE DIRECTION

As previously said, there is now a large database of interesting observations that may reflect hominin adaptations to nutritional changes and allow the formation of new hypotheses to be evaluated in follow-up research. However, the functional and phenotypic outcome of the implicated genetic variants or modifications are frequently undiscovered when molecular genetic evidence for these adaptations is reported. Except for a few digestive enzymes (such as lactase and amylase), which have very specific roles, most genes involved in food processing and digestion have multiple functions and are expressed in different tissues. As a result, it's often tough to exclude the chance that selection signals are caused by adaptations unrelated to food. Furthermore, due to the complexity of molecular processes involved in gene regulation, severe functional work is required to clarify the functional and evolutionary consequences of gene regulatory alterations affecting the digestive system. Despite these challenges, this research has the ability to considerably improve our knowledge of hominin dietary evolution, specifically, if it is based on systems biology methods (Luca et al., 2010). The search for diet-adaptive genes is still in its beginning. In the not future, fast sequencing technologies advancements and falling sequencing expenses will result in an explosion of genetic data. Indigenous genome sequences will be particularly beneficial in this attempt because of their genetic homogeneity and distinct eating patterns. For example, indigenous groups who have traditionally eaten an animal-based diet, such as the Inuit and the Maasai, will be ideal candidates for investigating adaptation to a high-fat/low-carbohydrate diet (Luca et al., 2010).

The rising number of genetic variants, together with the massive number of dietary and phenotypic data, will present a once-in-a-lifetime opportunity to discover even minor frequency shifts in response to environmental influences, such as dietary components (Ye and Gu, 2011). The increasing incidence of population genomics data in nonhuman primates will also improve the understanding of early human adaption activities. The study of the gut microbiome is another area of particular interest. The study of the symbiotic interaction between the gut biome and the human host and will not only expose human nutritional history but also their dietetic adaptations to the human diet (Ye and Gu, 2011). Evolutionary responses to diet modifications in humans are highly important for human health, especially in the age of personal genomics. By combining this knowledge with genome-wide association studies and functional investigations, researchers will be able to gain a better understanding of the evolutionary and molecular roots of complex metabolic illnesses. Individualized nutritional practices will thereafter be available to tailor optimal health management strategies by considering personal genomic information into consideration (Ye and Gu, 2011).

## 13.8 CONCLUSION

Our knowledge of the diversity of human evolution is rapidly growing, but an investigation into how dietary choices may have influenced the gene function of specific metabolic traits and people's vulnerability to disease is still in its initial stages. In this study, it is identified as evidence that PLRP2, NAT2, and CYP3A5 are three genes that are considered to play a role in compound metabolism (including xenobiotics), which enter the body through dietary food, could be examples of human gene–culture coevolution. In the case of PLRP2, more studies are needed to see if the signal recorded is a hitching effect of its PLRP1 neighbor. To study the essence of the selective forces that contributed to generating existing genetic variation patterns at specific loci, it is also required to create which biological mechanisms, environmental factors, and their interactions were involved. Furthermore, detailed systems are required to confirm the possible biological impact of the changes considered, so excluding the possibility that the discovered indications of selection are due to other changes in linkage disequilibrium than some of those examined here.

## ACKNOWLEDGMENT

We thank MET's Institute of Pharmacy, Bhujbal Knowledge City, Adgaon, Nashik. We also thank Principal Dr. Sanjay J. Kshirsagar and our Guide Dr. Neelam L. Dashputre for their constant support and providing all facilities to complete this work. Also, many thanks from us to all the known and unknowns who helped us right from the writing till the publishing of this chapter.

## KEYWORDS

- **allele**
- **apes**
- **bipedalism**
- **dietary shift**
- **evolutionary genetics**
- **hominin**
- **evolution**

## REFERENCES


Ahire, E. D.; Sonawane, V. N.; Surana, K. R.; Talele, G. S. Drug Discovery, Drug-Likeness Screening, and Bioavailability: Development of Drug-Likeness Rule for Natural Products. In *Applied Pharmaceutical Practice and Nutraceuticals*; Apple Academic Press: 2021; pp 191–208.

Allison, A. C. Protection Afforded by Sickle-cell Trait Against Subtertian Malarial Infection. *Br. Med. J.* **1954,** *1* (4857), 290.

Andrews, P.; Johnson, R. J. Evolutionary Basis for the Human Diet: Consequences for Human Health. *J. Int. Med.* **2020,** *287* (3), 226–237.

Arjamaa, O.; Vuorisalo, T. Geme-Culture Coevolution and Human Diet: Rather than Acting in Isolation, Biology and Culture have Interacted to Develop the Diet We Have Today. *Am. Sci.* **2010,** *98* (2), 140–147.

Babbitt, C. C.; Warner, L. R.; Fedrigo, O.; Wall, C. E.; Wray, G. A. Genomic Signatures of Diet-related Shifts During Human Origins. *Proc. R Soc. B: Biol. Sci.* **2011,** *278* (1708), 961–969.

Barrett, J. C.; Fry, B.; Maller, J. D. M. J.; Daly, M. J. Haploview: Analysis and Visualization of LD and Haplotype Maps. *Bioinform.* **2005,** *21* (2), 263–265.

Brain, C. K.; Sillent, A. Evidence from the Swartkrans Cave for the Earliest Use of Fire. *Nature* **1988,** *336* (6198), 464–466.

Brunet, M.; Guy, F.; Pilbeam, D.; Mackaye, H. T.; Likius, A.; Ahounta, D.; et al. A New Hominid from the Upper Miocene of Chad, Central Africa. *Nature* **2002,** *418* (6894), 145–151.

Carmody, R. N.; Dannemann, M.; Briggs, A. W.; Nickel, B.; Groopman, E. E.; Wrangham, R. W.; et al. Genetic Evidence of Human Adaptation to a Cooked Diet. *Genome Biol. Evol.* **2016,** *8*(4), 1091–1103.



Carmody, R. N.; Dannemann, M.; Briggs, A. W.; Nickel, B.; Groopman, E. E.; Wrangham, R. W.; et al. Genetic Evidence of Human Adaptation to a Cooked Diet. *Genome Biol. Evol.* **2016,** *8* (4), 1091–1103.



Danpure, C. J. Variable Peroxisomal and Mitochondrial Targeting of Alanine: Glyoxylate Aminotransferase in Mammalian Evolution and Disease. *Bioessays* **1997,** *19* (4), 317–326.

Dominy, N. J.; Vogel, E. R.; Yeakel, J. D.; Constantino, P.; Lucas, P. W. Mechanical Properties of Plant Underground Storage Organs and Implications for Dietary Models of Early Hominins. *Evol. Biol.* **2008,** *35* (3), 159–175.

Exploring the Relationship Between Lifestyles, Diets and Genetic Adaptations in Humans. *BMC Genet. 16* (1), 1–15.

Fernández, M. H.; Vrba, E. S. Plio-Pleistocene Climatic Change in the Turkana Basin (East Africa): Evidence from Large Mammal Faunas. *J. Hum. Evol.* **2006,** *50* (6), 595–626.

Gomes, V.; Sánchez-Diz, P.; Alves, C.; Gomes, I.; Amorim, A.; Carracedo, Á.; et al. Population Data Defined by 15 Autosomal STR Loci in the Karamoja Population (Uganda) using AmpF/STR Identifiler Kit. *Forensic Sci. Int.: Genet.* **2009,** *3* (2), e55–e58.

Goren-Inbar, N.; Alperson, N.; Kislev, M. E.; Simchoni, O.; Melamed, Y.; Ben-Nun, A.; et al. Evidence of Hominin Control of Fire at Gesher Benot Yaaqov, Israel. *Science* **2004,** *304* (5671), 725–727.

Haile-Selassie, Y. Late Miocene Hominids from the Middle Awash, Ethiopia. *Nature* **2001,** *412* (6843), 178–181.

Hanchard, N. A.; Rockett, K. A.; Spencer, C.; Coop, G.; Pinder, M.; Jallow, M.; et al. Screening for Recently Selected Alleles by Analysis of Human Haplotype Similarity. *Am. J. Hum. Genet.* **2006,** *78* (1), 153–159.

Hancock, A. M.; Witonsky, D. B.; Ehler, E.; Alkorta-Aranburu, G.; Beall, C.; Gebremedhin, A.; et al. Human Adaptations to Diet, Subsistence, and Ecoregion are Due to Subtle Shifts in Allele Frequency. *Pro. Nat. Acad. Sci.* **2010,** *107* (Suppl 2), 8924–8930.

Hancock, A. M.; Witonsky, D. B.; Ehler, E.; Alkorta-Aranburu, G.; Beall, C.; Gebremedhin, A.; et al. Human Adaptations to Diet, Subsistence, and Ecoregion are Due to Subtle Shifts in Allele Frequency. *Pro. Nat. Acad. Sci.* **2010,** *107* (Suppl 2), 8924–8930.

Hehemann, J. H.; Correc, G.; Barbeyron, T.; Helbert, W.; Czjzek, M.; Michel, G. Transfer of Carbohydrate-active Enzymes from Marine Bacteria to Japanese Gut Microbiota. *Nature* **2010,** *464* (7290), 908–912.

Henn, B. M.; Gignoux, C. R.; Jobin, M.; Granka, J. M.; Macpherson, J. M.; Kidd, J. M.; et al. Hunter-gatherer Genomic Diversity Suggests a Southern African Origin for Modern Humans. *Pro. Nat. Acad. Sci.* **2011,** *108* (13), 5154–5162.

Hohmann, G.; Robbins, M. M.; Boesch, C., Eds.; *Feeding Ecology in Apes and Other Primates,* Vol. 48; Cambridge University Press: 2012.

Innan, H.; Zhang, K.; Marjoram, P.; Tavaré, S.; Rosenberg, N. A. Statistical Tests of the Coalescent Model Based on the Haplotype Frequency Distribution and the Number of Segregating Sites. *Genetics* **2005,** *169* (3), 1763–1777.

James, W.; Johnson, R.; Speakman, J.; Wallace, D.; Frühbeck, G.; Iversen, P.; et al. Nutrition and Its Role in Human Evolution. *J. Int. Med.* **2019,** *285* (5), 533–549.

Jones, B. L.; Raga, T. O.; Liebert, A.; Zmarz, P.; Bekele, E.; Danielsen, E. T.; et al. Diversity of Lactase Persistence Alleles in Ethiopia: Signature of a Soft Selective Sweep. *Am. J. Hum. Genet.* **2013,** *93* (3), 538–544.

Keservani, R. K.; Sharma, A. K.; Kesharwani, R. K. An Overview and Therapeutic Applications of Nutraceutical and Functional Foods. *Recent Adv. Drug Deliv. Technol.* **2017,** 160–201. ISBN:9781522507543.

Keservani, R. K.; Sharma, A. K.; Kesharwani, R. K. Eds.; *Nutraceuticals and Dietary Supplements: Applications in Health Improvement and Disease Management*; CRC Press: 2020. ISBN: 9781771888738.

Khan, N.; Pande, V.; Das, A. NAT2 Sequence Polymorphisms and Acetylation Profiles in Indians. *Pharmacogenomics* **2013,** *14* (3), 289–303.

Kosuge, K.; Chuang, A. I.; Uematsu, S.; Tan, K. P.; Ohashi, K.; Ko, B. C.; et al. Discovery of Osmosensitive Transcriptional Regulation of Human Cytochrome P450 3As by the Tonicity-responsive Enhancer Binding Protein (Nuclear Factor of Activated T Cells 5). *Mol. Pharmacol.* **2007,** *72* (4), 826–837.

Lachance, J.; Vernot, B.; Elbers, C. C.; Ferwerda, B.; Froment, A.; Bodo, J. M.; et al. Evolutionary History and Adaptation from High-coverage Whole-genome Sequences of Diverse African Hunter-gatherers. *Cell* **2012,** *150* (3), 457–469.

Lederberg, J. JBS Haldane on Infectious Disease and Evolution. *Genetics* **1999,** *153* (1), 1–3.

Leonard, W. R. Human Nutritional Evolution. *Hum. Biol.: An Evol. Biocultural Perspect.* **2012,** 251–324.

Ley, R. E.; Peterson, D. A.; Gordon, J. I. Ecological and Evolutionary Forces Shaping Microbial Diversity in the Human Intestine. *Cell* **2006,** *124* (4), 837–848.

Luca, F.; Perry, G. H.; Di Rienzo, A. Evolutionary Adaptations to Dietary Changes. *Annu. Rev. Nutr.* **2010,** *30*, 291-314.

Magalon, H.; Patin, E.; Austerlitz, F.; Hegay, T.; Aldashev, A.; Quintana-Murci, L.; et al. Population Genetic Diversity of the NAT2 Gene Supports a Role of Acetylation in Human Adaptation to Farming in Central Asia. *Eur. J. Hum. Genet.* **2008,** *16*(2), 243–251.

McBrearty, S.; Jablonski, N. G.. First Fossil Chimpanzee. *Nature* **2005,** *437* (7055), 105–108.

Middleton, J., Ed.; *Africa: An Encyclopedia for Students,* Vol. 1; Gale Cengage: 2002.

Novembre, J.; Di Rienzo, A. Spatial Patterns of Variation Due to Natural Selection in Humans. *Nat. Rev. Genet.* **2009,** *10* (11), 745–755.

Patillon, B.; Luisi, P.; Poloni, E. S.; Boukouvala, S.; Darlu, P.; Génin, E.; et al. A Homogenizing Process of Selection has Maintained an "Ultra-slow" Acetylation NAT2 Variant in Humans. *Hum. Biol.* **2014,** *86* (3), 185–214.

Ragir, S. Diet and Food Preparation: Rethinking Early Hominid Behavior. *Evol. Anthropol: Issues News Rev.* **2000,** *9* (4), 153–155.

Reed, K. E. Early Hominid Evolution and Ecological Change through the African Plio-Pleistocene. *J. Hum. Evol.* **1997,** *32* (2–3), 289–322.

Sabeti, P. C.; Reich, D. E.; Higgins, J. M.; Levine, H. Z.; Richter, D. J.; Schaffner, S. F.; et al. Detecting Recent Positive Selection in the Human Genome from Haplotype Structure. *Nature* **2002,** *419* (6909), 832–837.

Sayers, K.; Lovejoy, C. O. The Chimpanzee has No Clothes: A Critical Examination of Pan Troglodytes in Models of Human Evolution. *Curr. Anthropol.* **2008,** *49* (1), 87–114.

Schaffner, S. F. P. C. S.; Sabeti, P. Evolutionary Adaptation in the Human Lineage. *Nat. Educ.* **2008,** *1* (1), 14.

Schlebusch, C. M.; Skoglund, P.; Sjödin, P.; Gattepaille, L. M.; Hernandez, D.; Jay, F.; et al. Genomic Variation in Seven Khoe-San Groups Reveals Adaptation and Complex African History. *Science* **2012,** *338* (6105), 374–379.

Senut, B.; Pickford, M.; Gommery, D.; Mein, P.; Cheboi, K.; Coppens, Y. First Hominid from the Miocene (Lukeino formation, Kenya). *C R Acad Sci. III IIA-Earth Planet Sci.* **2001,** *332* (2), 137–144.

Strait, D. S.; Weber, G. W.; Neubauer, S.; Chalk, J.; Richmond, B. G.; Lucas, P. W.; et al. The Feeding Biomechanics and Dietary Ecology of Australopithecus Africanus. *Pro. Nat. Acad. Sci.* **2009,** *106* (7), 2124–2129.

Surana, K. R.; Ahire, E. D.; Sonawane, V. N.; Talele, S. G.; Talele, G. S. Informatics and Methods in Nutrition Design and Development. *Nat. Food Prod. Waste Recovery: Healthy Foods Nutr. Des. Extr. Valuable Compounds*, **2021,** 33.

Valente, C.; Alvarez, L.; Marks, S. J.; Lopez-Parra, A. M.; Parson, W.; Oosthuizen, O.; et al. 2015.

Voight, B. F.; Kudaravalli, S.; Wen, X.; Pritchard, J. K. A Map of Recent Positive Selection in the Human Genome. *PLoS Biol.* **2006,** *4* (3), p.e72.

Weiner, S.; Xu, Q.; Goldberg, P.; Liu, J.; Bar-Yosef, O. Evidence for the Use of Fire at Zhoukoudian, China. *Science* **1998,** *281* (5374), 251–253.

White, T. D.; Asfaw, B.; Beyene, Y.; Haile-Selassie, Y.; Lovejoy, C. O.; Suwa, G.; et al. *Ardipithecus ramidus* and the Paleobiology of Early Hominids. *Science* **2009,** *326* (5949), 64–86.

Wrangham, R.; Conklin-Brittain, N. Cooking as a Biological Trait. *Comp. Biochem. Physiol. Part A: Mol. Integr. Physiol.* **2003,** *136* (1), 35–46.

Wrangham, R. W.; Jones, J. H.; Laden, G.; Pilbeam, D.; Conklin-Brittain, N. The Raw and the Stolen: Cooking and the Ecology of Human Origins. *Curr. Anthropol.* **1999,** *40* (5), 567–594.

Yang, Z. Likelihood Ratio Tests for Detecting Positive Selection and Application to Primate Lysozyme Evolution. *Mol. Biol. Evol.* **1998,** *15* (5), 568–573.

Ye, K.; Gu, Z. Recent Advances in Understanding the Role of Nutrition in Human Genome Evolution. *Adv. Nut.* **2011,** *2* (6), 486–496.

Zeggini, E.; Weedon, M. N.; Lindgren, C. M.; Frayling, T. M.; Elliott, K. S.; Lango, H.; et al. Replication of Genome-wide Association Signals in UK Samples Reveals Risk Loci for Type 2 Diabetes. *Science.* **2007,** *316* (5829), 1336–1341.

## CHAPTER 14

# NUTRITIONAL EPIGENOMICS

RANI KANKATE,[1] TAMBE P. B.,[1] SHAIKH M. A.,[1] and SANJAY KSHIRSAGAR[2]

[1]*Department of Pharmaceutical Chemistry, MET's Institute of Pharmacy, Adgaon, Nashik, India*

[2]*Principal, Department of Pharmaceutics, MET's Institute of Pharmacy, Bhujbal Knowledge City, Adgaon, Nashik, India*

## ABSTRACT

Nutritional epigenomics provides a comprehensive overview of nutritional epigenomics as a mode of study as well as nutrition's role in the epigenomics regulation of disease, health and developmental processes. The epigenome, in addition to the genome, stores heritable information. The genome is supposed to be stable throughout one's life, whereas the epigenome is very dynamic and is influenced by environmental factors such as dietary molecules. It is expected that epigenetics will eventually explain the missing classical genetic heritability of susceptibility to complex diseases and traits. Persons born with low birth weight, for example, have a high risk of developing type-2 diabetes later in life, implying that epigenetic programing during embryogenesis and the intrauterine environment both contribute to type-2 diabetes risk. Furthermore, the human individual's lifestyle, particularly the daily choice of diet, appears to create a metabolic memory within the epigenome. This theory proposes that lifestyle changes such as personalized diet, increased physical activity, and subsequent weight loss can have a beneficial effect on the epigenome and thus reduce the risk of developing metabolic syndrome. In this chapter, we will define various epigenetic mechanisms that process information provided by dietary molecules, such as posttranslational

Nutrigenomics and Nutraceuticals. Raj K. Keservani, Eknath D. Ahire, Shubham J. Khairnar, Sanjay J. Kshirsagar, & Rajesh K. Kesharwani (Eds.)

histone modifications and DNA methylation. We will discover that many chromatin-modifying enzymes are sensitive to changes in the levels of intermediary metabolites that act as cosubstrates and cofactors as well as changes in nutrient intake and metabolism. We will gain insight into the concepts of nutritional epigenetic programing by understanding prenatal supplementation in mouse models. This will lead us to the thrifty hypothesis and we will talk about the various approaches to epigenetic epidemiology, including the concept of "epigenetic drift" during adulthood.

## 14.1 INTRODUCTION

The study of food nutrients and their effects on human health through epigenetic alterations is thought as biological process epigenomics. Obesity, upset, diabetes, high blood pressure, and cancer area unit all connected to biological process imbalances throughout gestation and lactation, in step with a growing body of information. If metabolic disruptions occur throughout vital organic process periods, the next epigenetic changes will cause lasting changes in tissue and organ type and performance furthermore as incline folks to unwellness (Desai et al., 2015).

Epigenetics is bothered with hereditary changes in sequence operations that occur severally with changes within the primary polymer sequence. Polymer methylation and simple protein modification area unit two major epigenetic mechanisms involved in nutriepigenomics. Polymer methylation in sequence promoter regions usually silences genes and influences organic phenomena (Hotamisligil et al., 2010).

Whereas this kind of sequence silencing is crucial for development and cellular differentiation, abnormal polymer methylation is harmful and has been coupled to a spread of unwellness processes as well as cancer. As a result of the alkyl teams employed in polymer methylation area unit derived from dietary sources (Sharma et al., 2020; Keservani et al., 2020), such as vitamin B and vitamin B complex, diet will have a big impact on methylation patterns and organic phenomenon (Huang et al., 2009).

Gene silencing may also be bolstered by recruiting simple protein deacetylases to scale back transcriptional activation. Simple protein acylation, on the opposite hand, causes transcriptional activation and therefore will increase the organic phenomenon. Dietary elements will influence these epigenetic events, sterilization of organic phenomena, and disrupting functions such as craving management, metabolic balance, and fuel utilization. Various genetic sequences are targeted for epigenetic modification. A transcriptome-wide analysis in mice discovered that a protein-restricted (PR) diet throughout gestation resulted in the differential organic phenomenon in more or less I Chronicles of the vertebrate genes analyzed. Multiplied expression was seen in genes concerned with the p53 pathway, apoptosis, negative regulators of cell metabolism, and epigenetic management (Keane et al., 2013; Ahire, et al., 2021)

## 14.2 NUTRI EPIGENOMICS AND DEVELOPMENT

The step of the organic process when the nutritional imbalance occurs is critical in influencing disease-related genes. Totally different organs undergo essential organic process stages and therefore the purpose at that they fail predisposes folks to specific diseases. Counting on the operation of the cistron, epigenetic changes that occur throughout development might not be expressed till later in life. Whereas the bulk of studies proposes the prenatal and perinatal periods as essential time windows, some analysis suggests that nutritionary intake throughout adulthood may influence the epigenome.

## 14.3 FACTORS AFFECTING NUTRITIONAL EPIGENOMICS

### *14.3.1 PRENATAL*

The process by that fetuses adapt to their surroundings is thought of as organic process of malleability. Environmental cues, as well as dietary parts, gift within the utero setting will cause vital changes within the expression of the ordering via epigenetic modifications. Fetal organic process plastic responses will cause changes in lean body mass, medical specialty, blood flow associate degreed vascular loading moreover as an inflated risk of varied diseases in adulthood (Weiss et al.1997).

#### *14.3.1.1 LOW BIRTH WEIGHT*

Birth weight has been coupled to fetal exposure to metallic element, folate, magnesium, high or low supermolecule and metal. Varied studies are conducted to analyze the connection between birth weight and illness risk and it has been discovered that low birth weight is considerably related to a coronary heart condition, stroke, and kind a pair of polygenic disorders. Most significantly, these associations persisted when dominant for way factors, indicating a genetic basis for illness onset. Low birth weight is related to impaired hypoglycemic agent secretion, which might cause hypoglycemic agent resistance as babies gain body fat. In mistreatment intrauterine growth backward (IUGR) rats, researchers discovered that growth inhibition may result in minimized expression of the Pdx1 transcription issue that is needed for exocrine gland β-cell differentiation and performance. Reduced simple protein acylation at Pdx1s proximal promoter is liable for minimized Pdx1 expression that ends up in a cascade of simple protein deacetylation and methylation events that may cause sort a pair of polygenic disorders.

#### 14.3.1.2 OBESITY

Obesity (Jarouliya and Keservani, 2019) throughout gestation and high-fat maternal diets has each been coupled to childhood fat. Because the range of overweight reproductive-age girls rises, therefore will the quantity of overweight youngsters and infants. Maternal fat is assumed to cause associate degree accumulation of fat in fetal fatty tissue (adiposity) and predisposes babies to fat in childhood and adulthood. Maternal deficiency disease has been shown in animal studies to own an impression on brain development and cause disruptions in neural structure programing. Offspring of mothers with WHO Greek deit—a high-fat or high-calorie diet—had higher levels of hypoglycemic agent, aldohexose, and leptin. These will increase area units thought to be caused by disruptions within the complicated somatic cell network that features the neuropeptide Y (NPY) and proopiomelanocortin (POMC) pathways. This altered somatic cell signal has the potential to influence food intake behavior and cause diet-induced fat in adults. Whereas epigenetic changes area unit presumably concerned in fat development, the particular target genes have nonetheless to be known. Adipogenesis-related genes with multiple CpG islands in their promoters, such as formative cell growth factor-2, enzyme and tensin homolog, cyclin-dependent enzyme matter 1A, and steroid hormone receptor-α, could act as epigenetic targets. What is more, prenatal exposure to a hypomethylating agent, such as bisphenol A (BPA), has been coupled to inflated weight, implying that altered deoxyribonucleic acid methylation could be a mechanism for inflated condition to fat (House et al.2018).

#### 14.3.1.3 FOLATE

Maternal folic acid intake throughout gestation has long been coupled to fetal development and growth and might scale back the danger of great birth defects. Folic acid could be a supply of S-adenosyl essential amino acid

(SAM) that is employed to produce alkyl teams to deoxyribonucleic acid methyltransferases. As a result, changes in folic acid offer have a big impact on deoxyribonucleic acid methylation patterns. Low folic acid levels area unit coupled to associate degree inflated risk of preterm labor, poor placental and female internal reproductive organ growth, and intrauterine growth retardation. In line with animal studies, low folic acid intake could increase the danger of exoderm defects by reducing the number of alkyl radical deoxyribonucleic acids throughout so exoderm closure. It had been recently discovered that folic acid protects against innate heart defects via epigenetics and Wnt signals. Throughout embryogenesis, multiple environmental factors target the Wnt signal pathway, which might cause pathway misregulation. Vitamin B metabolism produces a guided missile, which alters the methylation states of histones H3K9, H3K4, and H3K27 moreover as genetically neutering Wnt signal. A Recent double-blind placebo-controlled trial of high dose folinic acid (leucovorin calcium) incontestable effectivity in rising verbal communication in unfit youngsters (Finnin et al., 1990).

### 14.3.2 PERINATAL

The perinatal amount, that happens now before and when birth, is another essential organic process time window. It has been incontestable that maternal diet in late gestation moreover as associate degree infants diet within the early weeks, will all have a big impact on the organic phenomenon. As a result, perinatal nutrition encompasses each late stage in utero nutrition and lactation.

#### 14.3.2.1 BONE HEALTH

Perinatal nutrition has been coupled to bone mass and therefore the development of pathology. Once researching perinatal nutrition, it is essential to think about whether or not the baby was breastfed or formula-fed. Nursed babies have additional bone mass than non-breast-fed babies and this tiny increase in bone mass throughout an essential organic process amount might probably program the skeleton to continue on a healthy growth flight. Maternal cholecalciferol deficiency throughout late gestation has additionally been coupled to the minimized bone size and mineral mass in late childhood. Peak bone mass could be a smart predictor of fracture and pathology risk, with even a little increase leading to a far lower risk of bone fracture. In line with analysis, genetic markers make a case for solely a little portion of the variation in bone mass and fracture risk. As a result, varied epigenetic mechanisms, such as learning of the growth-promoting genes IGF-2 or changes to the hypothalamic–pituitary–adrenal axis, area unit presumably influencing healthy bone programing (HPA).

#### *14.3.2.2 NEURODEVELOPMENT*

Maternal nutrition imbalances may have a big impact on fetal neurodevelopment (Keservani et al., 2016; Keservani et al., 2017). Brain development is most speedy throughout the fetal development and infancy, and analysis has shown that bound environmental conditions will have long-run effects on knowledge. N-3 fatty acids, iodine, iron, and vitamin B, particularly, are shown to influence brain development, psychological feature ability, and behavior. The foremost compelling proof for a link between nutrition associate degree neurodevelopment comes from studies that link low birth weight to an occasional intelligence quotient and an inflated risk of psychosis. Breast-feeding seems to push long-run neurodevelopment by providing the nutrients needed for correct brain development, in line with many studies. A mouse study found that choline-deficient diets throughout late gestation impaired fetal brain development, as well as minimized cell proliferation and visual–spatial and sensory system memory (Waterland et al.2003). These psychological feature changes were found to be caused by altered simple protein and deoxyribonucleic acid methylation patterns within the fetal hippocampus, establishing a link between maternal nutrition, epigenetics and early brain development.

#### *14.3.2.3 TYPE-1 DIABETES*

Breastfeeding might also defend against sort one polygenic disorder, in line with analysis, with formula-fed infants at a better risk of developing isle autoantibodies. People with sort one polygenic disorder undergo a preclinical polygenic disorder section marked by pathology to exocrine gland islets. When put next to babies WHO area unit exposed to solid foods later in life, the introduction of bound foods within the initial few months of life, such as berries and cereal, is considerably related to associate degree inflated risk of isle antibody development. Whereas the pathologic process of autoantibodies remains mostly unknown, an associate degree epigenetic link probably exists between perinatal diet and therefore the risk of sort one polygenic disorder.

### 14.3.3 ADULTHOOD

The majority of nutrigenomics analysis has targeted nutritionary imbalances throughout gestation and lactation. Foods consumed throughout adulthood, on the opposite hand, will influence organic phenomenon and illness pathologic process. Cancer is that the most typically associated illness with adult nutrition and epigenetic changes. Cancer progression is power-assisted by deoxyribonucleic acid hypomethylation that permits for inflated cistron transcription, whereas hypermethylation will silence growth suppressor genes and promote uncontrolled biological process and growth formation. Food compounds, such as genistein and tea polyphenols, will regulate deoxyribonucleic acid methyltransferases and simple protein acylation in cultivated cancer cells and should defend against bound kinds of cancer. In clinical trials, different dietary compounds, such as diallyl disulfide found in garlic and sulforaphane found in dilleniid dicot family vegetables, are coupled to cancer hindrance. This might be due to their ability to inhibit simple protein deacetylase (HDAC) enzymes, preventing the silencing of vital restrictive genes.

## 14.4 MECHANISMS OF EPIGENETICS

Chromatin, a three-dimensional complicated of genomic deoxyribonucleic acid and nuclear proteins, is split into two types: Less densely packed euchromatin, that is well accessible to transcription factors and compact heterochromatin that represents a functionally pent-up state. Epigenetics is that the study of functionally relevant chromatin granule modifications that do not involve an amendment within the ordination sequence itself. Epigenetic changes is stable throughout cell divisions and last for multiple generations. The most effective example is that the method of cellular differentiation, within which at one time ability stem cells endure epigenetic changes to become the embryos numerous pluripotent cell lines that successively are the precursors of terminally differentiated cells. The two most vital epigenetic mechanisms are posttranslational modifications of nucleosome-forming simple protein proteins and pyrimidine methylation inside genomic deoxyribonucleic acid (Ameres et al., 2013). Mechanisms of nutritionary epigenetics mentioned in the following sections.

### 14.4.1 HISTONE PROTEINS AND THEIR MODIFICATIONS

The nucleosome is that the basic, each 200 bp continuation unit of chromatin granule, consisting of 147 bp genomic deoxyribonucleic acid wrapped nearly doubly around Associate in nursing octamer of four pairs of the simple protein proteins H2A, H2B, H3, and H4. Simple protein proteins contain tons of essential amino acid residues, particularly at their amino-terminal tails that protrude from the nucleosome. Posttranslational, covalent, reversible chemical modifications to lysines in addition as essential amino acid and amino acid residues, such as acetylations, methylations, and phosphorylations are distributed by a range of chromatin-modifying enzymes. These simple protein modifications, referred to as epigenetic marks, are a sort of chromatin granule assortment. The bulk of marks are allotted to practical chromatin granule regions such as promoters, enhancers, or heterochromatin, the principles that are summarized within the simple protein code. This code could be a larger determinant of the accessibility of transcription issue binding sites and their associated cofactors within the ordination, ultimately leading to increased or shriveled organic phenomenon. Several chromatin granule modifiers and alternative nuclear proteins share a group of common domains that acknowledge specific chromatin granule modifications, permitting these proteins to "read" the simple protein code (Bernstein et al., 2007).

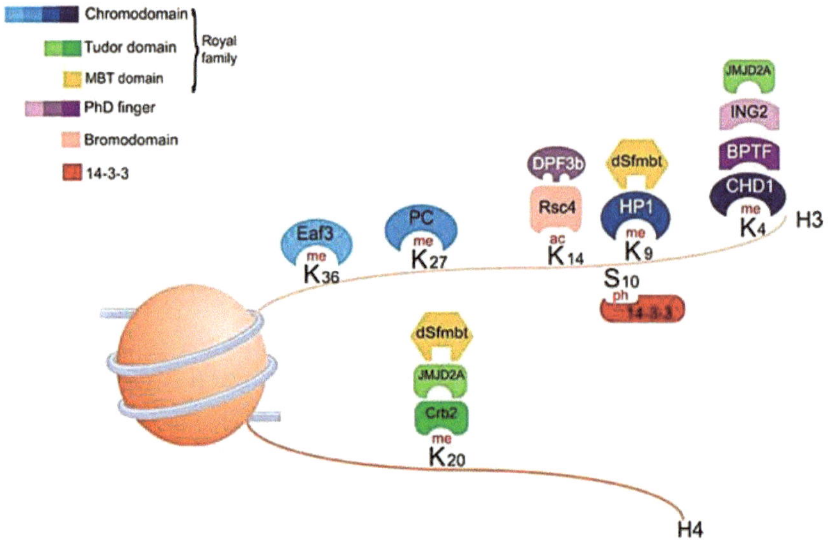


**FIGURE 14.1** Regulation of chromatin by histone modifications.

*Source*: Reprinted with permission from Bannister and Kouzarides, 2011. © 2011 Shanghai Institutes for Biological Sciences, Chinese Academy of Sciences.

### *14.4.2 DNA METHYLATION*

Only the cytosine-containing ester of the four DNA-forming nucleotides is alkyl, and solely at the dinucleotide CpG, that is, wherever C and G are found on a similar deoxyribonucleic acid strand and are connected by a phosphodiester bond. CpG islands are genomic regions that are a minimum of 200 bp long and have a CG content of a minimum of 55th (compared to the 42% average within the human genome). Solely 3–6% of all cytosines within the order of traditional human cells are alkyl, implying that CpG islands are for the most part unmethylated which the genes in their neck of the woods will still be activated by transcription factors. DNMTs use S-adenosylmethionine (SAM) as an alkyl radical donor to methylate the carbon within the cytosines. The enzymes DNMT3A and DNMT3B, on the opposite hand, primarily operation throughout embryogenesis and cellular differentiation and have an American state novo methylation rate of 3–5% per cell division. Ten-eleven translocation (TET) proteins management the de-methylation of genomic deoxyribonucleic acid by changing 5-methylcytosine to 5-hydroxymethylcytosine, 5-formylcytosine and 5-carboxylcytosine. These modifications are then removed throughout deoxyribonucleic acid replication via repetitive oxidization and base excision repair (Kanherkar et al., 2014).

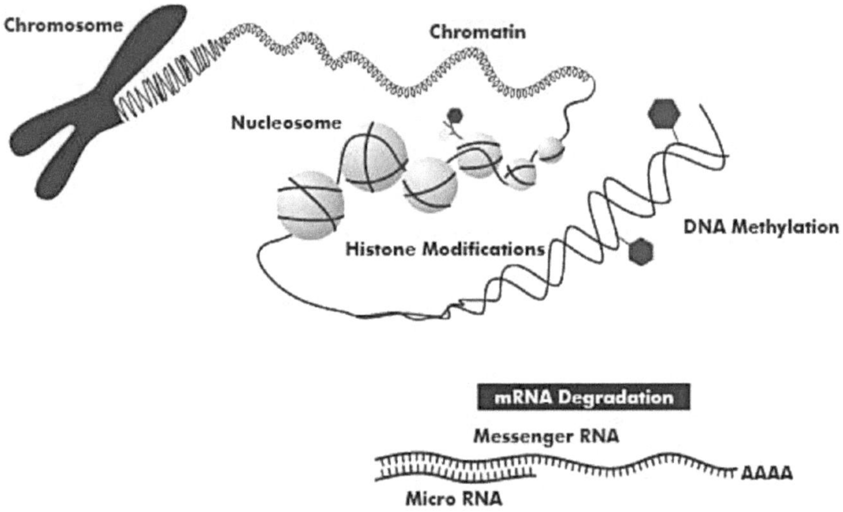


**FIGURE 14.2** DNA methylation in epigenetics.

*Source*: Reprinted courtesy of h ttps://www.labclinics.com/2018/11/08/role-dna-methylation-disease/?lang=en

The encrypt project and different giant biology collaborations offer genome-wide maps of epigenetic marks in over 100 human cell lines, primary cells, and tissues. These findings recommend that in open, active chromatin granule, simple protein proteins area unit acetylated (e.g., at essential amino acid fourteen of simple protein three, H3K14ac) whereas genomic deoxyribonucleic acid remains unmethylated (Lillycrop et al., 2014). In distinction, histones area unit alkyl (e.g., at essential amino acid 27 of simple protein three, H3K27me) and deoxyribonucleic acid is alkyl in inhibited, closed chromatin granule. The modification in deoxyribonucleic acid methylation throughout development begins with demethylation throughout animate being cell divisions, followed by first-state novo methylation once implantation. Attributable to this epigenetic reprograming throughout development, the intrauterine amount is thought to be crucial for semi-permanent health and sickness risk. However, deoxyribonucleic acid methylation not solely regulates the expression of specific genes throughout tissue development and differentiation; still it is additionally needed for the silencing of imprinted genes, the second feminine sex chromosome and retrotransposons. Significantly, simple protein modifications occur before or once deoxyribonucleic acid demethylation, implying that each epigenetic processes act. HDACs may be recruited to alkyl regions of genomic deoxyribonucleic acid by methyl-CpG binding proteins, for instance (Millar et al., 2002).

The accessibility of genomic binding sites for transcription factors is controlled by chromatin granule density, which plays a crucial role in regulation organic phenomenon. An outsized set of enzymes that either covalently modifies simple protein proteins, referred to as chromatin granule modifiers or move, reconfigure or eject nucleosomes, referred to as chromatin granule remodelers, influence the dynamic competition between nucleosomes and transcription factors for vital binding sites. This method determines the density, composition and positioning of nucleosomes relative to the transcription issue binding sites in every genomic region (Hendrich et al., 1999).

Secondary metabolites derived from fruits, vegetables, teas, spices, and ancient healthful herbs (Rane et al., 2020a,b,c), such as genistein, resveratrol, curcumin, and polyphenols derived from tea leaf, low, and cocoa, will modulate the activity of transcription factors and chromatin granule modifiers. Next-generation sequencing technologies change genome- and transcriptome-wide assessments of the specificity and effectualness of those compounds, for instance, within the hindrance and treatment of cancer (Bellacosa et al., 1999; Petronzelli et al., 2000; Wong et al., 2002).

### 14.4.3 GENOMIC IMPRINTING

Because of the development of genomic acquisition in diploid cells, some genes are expressed in a very parental-origin-specific manner, that is, either the father's or the mother's factor copy is employed. The acquisition method is needed for traditional class growth and development, as proven by the inactivation of females' second X chromosome throughout early embryogenesis. The pent-up copy of imprinted genes' regulative region is hyperalkyl and placed inside heterochromatin (Rotondo et al., 2013). This repression is passed down through corporal cells via mitotic division. The bulk (but not all) of the genomic acquisition is erased within the next generation's primeval germ cells and reset as they differentiate into parental gametes.

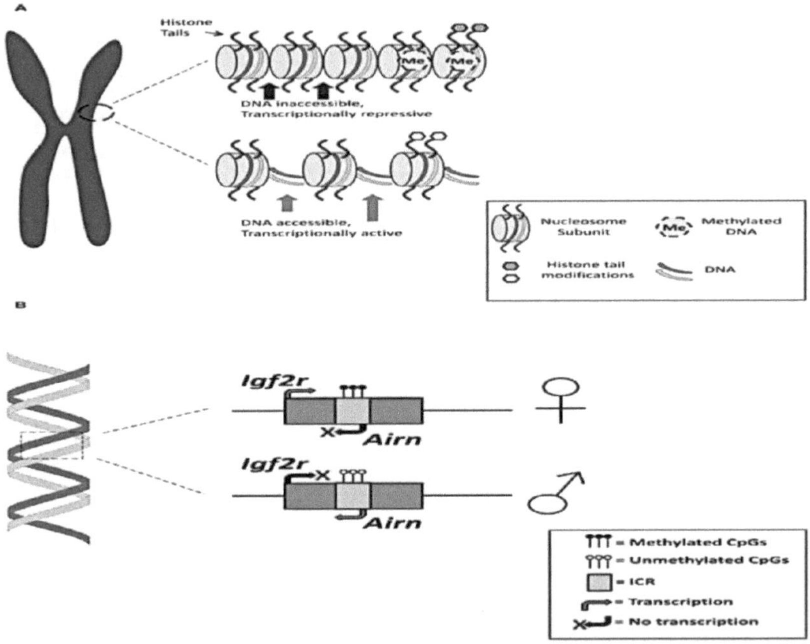


**FIGURE 14.3** Epigenetic mechanisms associated with genomic imprinting.

*Source*: Reprinted from O'Doherty et al., 2015. Copyright © 2015 authors. https://creativecommons.org/licenses/by/4.0/

### *14.4.4 LEVELS OF ORGANIC PHENOMENON*

Cells will adapt to changes in environmental signals due to the precise coordination of transcriptional programs. As a result, sequence ribonucleic acid expression is tightly controlled on three completely different levels. The primary level, the deoxyribonucleic acid code, is that the human ordering sequence, that contains ≥200,000 protein-coding exons that has to be transcribed and combined by splice into mature ribonucleic acid molecules.

What is more, the human ordering encodes ≥20,000 non-coding RNAs," the bulk of that square measure used for finetuning and feedback management of organic phenomenon starting from body substance accessibility to macromolecule translation. The second level, the epigenetic code, is ruled by a mix of the simple protein code and deoxyribonucleic acid methylation rules. The transcription issue program, the third level, represents the relative quantity and activity state of the transcription factors expressed in an exceedingly given cell Relton et al. (2012). Transcription issue expression, in turn, is ruled by each the deoxyribonucleic acid code and therefore the epigenetic code. What is more, most transcription factors' activity is regulated by a spread of processes, together with posttranslational modification, substance binding, dimerization and translocation from the cytoplasm to the nucleus. What is more, small RNAs (miRNAs), the organic process oldest regulators of organic phenomenon, primarily counteract the activities of transcription factors, finetuning the soundness of mRNAs.

In general, organic phenomenon is regulated by (1) the sequence of genomic deoxyribonucleic acid (the deoxyribonucleic acid code), (2) the accessibility of genomic deoxyribonucleic acid (the epigenetic code) and (3) the abundance and activity of transcription issues (the transcription factor program). The bulk of trait-associated genomic deoxyribonucleic acid variations, such as the most important SNP dominant the persistence of LCT organic phenomenon when substitution, square measure found outside of protein-coding regions. These regulative SNPs enable researchers to raise perceive the mechanistic perform of SNPs that do not alter the organic compound composition of a macromolecule. However, as a result of regulative SNPs square measure a part of the deoxyribonucleic acid code, they are doing not fill the void in classical genetic heritability. As a result, the latter is explained by the epigenetic code or the transcription issue program.

## 14.5 INTERMEDIARY METABOLISM AND EPIGENETIC SIGN

Within a cell, signal transduction pathways activate organic phenomenon programs that integrate environmental inputs, such as the provision of energy substrates, to mediate homeostasis-targeting responses. There square measure varied links famous between go-between metabolism merchandise and chromatin-modifying enzymes. Many body substance modifiers that interpret ("read"), add ("write") or take away ("erase") posttranslational simple protein modifications square measure expressed tissue-specifically within the human ordering. Bromodomains square measure found all told proteins that acknowledge acetylated residues, together with HATs, HMTs and body substance transforming enzymes, coactivators and general transcription factors. In distinction, "chromo domains" square measure way more specific for a given body substance modification; for instance, chromodomain-containing nuclear proteins acknowledge their genomic targets with so much larger accuracy than bromodomain proteins. the majority of those chromatin-modifying enzymes square measure critically passionate about intracellular levels of essential metabolites such as acetyl-CoA, uridine diphosphate (UDP)-glucose, -ketoglutarate, NAD+, ketone purine dinucleotide (FAD), ATP or SAM. As a result of the cellular concentrations of many of those metabolites represent the metabolic standing of the cell, the body substance modifies activities replicate the go-between metabolism. The enzymes perform as sensors, changing metabolic data into dynamic post-translational simple protein modifications that coordinate adaptive transcriptional responses. As a result, such metabolic signal sensing modulates the activity of sequence regulative networks, that successively management cell fate selections. Crucial disruptions in energy metabolism may additionally lead to stable epigenetic changes that square measure passed down through the germ line and should have a control on the health of future generations. Acetyl-CoA is created once nutrients square measure eaten and enter the catabolic pathways of go-between metabolism. This implies that simple protein acylation via HATs is primarily passionate about cytosolic acetyl-CoA pools. Simple protein acylation throughout feeding, that is, beneath high-energy conditions, ends up in transcriptionally open body substance, permitting the expression of genes that regulate cellular proliferation, lipogenesis, and adipocyte differentiation. Simple protein deacetylation via HDACs, in turn, is joined to intracellular energy levels. For instance, the useful effects of calorie restriction on metabolic parameters square measure passionate about sirtuin-type HDACs sensing the energy carrier coenzyme. Apparently, coenzyme + levels fluctuate daily, implying that the peripheral clock is joined to transcriptional regulation of metabolism via epigenetic mechanisms involving sirtuins. Significantly, sirtuins deacetylate non-histone proteins such as the transcription issue p53 and therefore the coactivator peroxisome proliferator-activated receptor gamma, coactivator one alpha (PPARGC1A).

Butyrate, a short-chain carboxylic acid that's created within the lumen of the colon from dietary fibers, may be a potent matter of many HDACs. In humans, a fiber-rich diet will stop rubor and carcinoma, which may well be explained partially by butyrate-mediated inhibition of HDAC-dependent transcriptional programs of colonocyte proliferation. The addition and removal of N-acetylglucosamine (O-GlcNAc) to amino acid and essential amino acid residues of histones by the enzymes O-GlcNAc enzyme (OGT) and O-GlcNAcase (OGC) provides another link of the epigenetic code to go-between metabolism (OGA). UDP-glucose, the substrate of this glycosylation reaction, comes from the hexosamine pathway and reflects changes in aldohexose levels (Bellacosa et al., 2007).

## 14.6 ONE-CARBON METABOLISM AND DEOXYRIBONUCLEIC ACID METHYLATION

During physiological state, alkyl group donor's square measure essential and dietary excess still as deficiency, could have a sway on epigenetic programing in mice and humans. SAM, an alkyl group donor substrate, links deoxyribonucleic acid methylation to negotiant metabolism. Within the one-carbon metabolism pathway, SAM is synthesized from the aminoalkanoic acid essential amino acid and adenosine triphosphate. S-adenosylhomocysteine (SAH) is recycled back to SAM once an alkyl of SAM is transferred to deoxyribonucleic acid or a simple protein. SAH, apparently, acts as a feedback regulator for HMTs, implying that the SAM/SAH quantitative relation, conjointly called the "methylation index," is vital for simple protein and deoxyribonucleic acid methylation. Tetrahydrofolate, a by-product of the water-soluble vitamin M, acts as an alkyl donor for the cyclic one-carbon pathway. Another example of an immediate link between nutrition and epigenetics is that the pathway's reliance on vitamin M and different micronutrients. The role of vitamin M in traditional ectoblastic closure throughout early gestation (21–28 days once human conception) is well understood and maternal vitamin M supplementation is usually recommended to forestall ectoblastic defects. Elevated homocysteine could be a well-studied indicator of one-carbon metabolism disruption and is related to low levels of vitamin M, vitamins B12 and B6, B-complex vitamin and alkaloid. What is more, a mutation within the human MTHFR factor that codes for the catalyst methylenetetrahydrofolate enzyme, that catalyzes the conversion of five, 10-methylenetetrahydrofolate to 5-methyltetrahydrofolate, reduces the enzyme's activity by regarding 500th. State for the missense SNP C677T (rs1801133) leads to the substitution of essential amino acid for amino acid at position 222 of the supermolecule. This genetic variation is found in 10–15% of Europeans. People with the TT genotype square measure additional tormented by low vitamin M intake than those with the CCor CT gene. As a result, hyperbolic plasma homocysteine concentrations increase the danger of premature birth, low birth weight, and ectoblast defects (Kucharski et al., 2008).

## 14.7 NUTRITION-TRIGGERED TRANSGENERATIONAL EPIGENETICS IN MICE

In the mouse, the factor rodent signal supermolecule (Asip) encodes a paracrine signal molecule that causes follicle melanocytes to provide pheomelanin, a yellow pigment, instead of eumelanin, a black or brown pigment. The insertion of the retro {transposon, jumping factor, deoxyribonucleic acid, desoxyribonucleic acid and DNA} Intracisternal A particle (IAP) into the Asipgene regulative region leads to the expression of a dominant gene of the gene (called A vy) that's passionate about the methylation standing of the IAP part. Some retro transposons, such as IAP, exhibit random epigenetic variability, which can contribute considerably to composition variability in mammals. This leads to coat colors starting from yellow to mottle to the wild-type dark coat color A vy/mice, indicating that A vy could be a stability epiallele. Yellow coat mice, apparently, become rotund and hyperinsulinemic. Once the IAP part is alkyl group, the synthesis of the yellow pigment is suppressed, leading to pseudo rodent dark coat color. Unmethylated IAP, on the opposite hand, permits omnipresent Asip expression, leading to each yellow coat color and fatness. Another intriguing feature of the A vy mouse model is its ability to transmit epigenetic data, which suggests that once A vy/ animals inherit the A vy gene from their mothers, Asip expression and coat color correlate with the maternal constitution. What is more, the patterned constitution suggests that IAP methylation is mosaic, implying that the Asip factor isn't expressed altogether cells. This implies that the IAP element's methylation pattern is established early in development and therefore the coat color provides a straightforward composition readout of the element's methylation and expression standing throughout life. All of those characteristics create the A vy Asip mouse a perfect in vivo model for learning the mechanistic relationship between environmental stimuli, such as nutrition, and epigenetic states of the order (Bishop et al., 2015).

The following experiment substantiated this: Feminine wild-type a/a mice were either supplemented or not supplemented with alkyl group donors such as vitamin M, B-complex vitamin and alkaloid fortnight before pairing with male A vy/mice. Throughout physiological state and lactation, the supplementation was continuing. Whereas the F1 generation of non-supplemented mothers had the expected variety of yellow coat color phenotypes, the offspring of supplemented mothers shifted toward a brown coat color constitution. This means that maternal alkyl group donor supplementation will increase A vy methylation within the offspring. The transmission of epigenetic programing to future generation suggests that a minimum of stability epialleles, such as IAP, will face up to world demethylation of the order before preimplantation. Once, A vy mice were fed a soy polyphenol diet, which caused changes in their deoxyribonucleic acid methylation patterns, their offspring were protected against polygenic disease, fatness and cancer for multiple generations. It was incontestable during a mouse model of maternal underneath nutrition that leads to low birth weight and aldohexose intolerance in male and feminine F1 offspring that exposure to suboptimal nutrition throughout craniate development causes changes within the germ-cell deoxyribonucleic acid methylome of child even once these males were nourished commonly once ablactation. These composition variations square measure passed right down to the F2 offspring via the paternal line. During a genome-wide analysis of deoxyribonucleic acid methylation, quite 100 genomic regions in F1 sperm cell from motherly malnourished child were hypo alkyl group compared to the equivalent regions on top of things offspring. The presence of novel hypo alkyl group regions indicated that early germ cells from nutritionally restricted fetuses didn't utterly remethylate their deoxyribonucleic acid. Taken along, the assorted gnawing animal models show that epigenetic memory are often passed down from generation to generation by heritable identical categorization of chromatin granule marks (Campión et al., 2009).

However, whereas short "day-to-day" epigenome responses square measure primarily mediate by non-inherited changes in simple protein acylation levels, deoxyribonucleic acid methylation seems to be specifically designed for long cell memory (Campión et al., 2009).

## 14.8 EPIGENETIC PROGRAMING IN HUMANS

The findings from gnawing animal models raise the question of whether or not the construct of metabolic memory is additionally applicable to humans. However, there are not any comparable natural human mutants and embryonal feeding experiments do not seem to be underneath moral standards. However, the comparatively new field of epigenetic medical specialty has discovered many various approaches to responsive this question. The Dutch Hunger Families Study piqued the public's interest the foremost. The study looked into people at large World Health Organization had been exposed in utero to severe deficiency disease within the European country throughout the winter of 1944–1945. The themes among the cohort were weedy at birth, but were overweight later in life associated had AN-accumulated incidence of hormone resistance. what's additional, even six decades once birth, lower deoxyribonucleic acid methylation levels were found among the restrictive region of the imprinted gene insulin-like supermolecule 2 (IGF2), that has been connected to associate accumulated risk of avoirdupois, dyslipidemia, and hormone resistance. This points to a link between prenatal nutrition and epigenetic changes in humans, such as what has been shown in rodents (Bernstein et al., 2007).

## 14.9 METHODS OF EPIGENETIC MEDICAL SPECIALTY

Natural experiments square measure studies during which the topic is exposed to a condition while not being subjected to experimental management. For instance, the Dutch Hunger Winter Families Study (www.dutchfamine.nl) follows people World Health Organization were exposed as embryos to their mothers' famine throughout the winter of 1944–1945. Longitudinal birth cohort studies examine epigenetic changes over time and link them to environmental exposures and malady onset. Within the UK, for instance, the Avon Longitudinal Study of oldsters and youngsters (www.bristol.ac.uk/alspac) monitors the health of 14,500 families with youngsters born in 1991–92. Longitudinal twin studies as a result of monozygotic twins share identical oldsters, birth date, and gender, they're ideal cross-sectional references for learning individual epigenetic variations. For instance, the Peri or postpartum Epigenetic Twins Study (www.mcri.edu.au/research/researchprojects/pets) in Australia investigates epigenetic variation between twins at birth. The prenatal biological process amount is thought to be the foremost crucial amount for epigenetic programing. Maternal and intrauterine influences, genotype and postpartum atmosphere will all be coupled to health and malady once people square measure tracked from conception forward. Within the UK, for instance, the Southampton Women's Survey (www.mrc.soton.ac.uk/sws) follows twelve, 500 ladies and their youngsters throughout their pregnancies.

The Danish National IVF Cohort Study investigated the prevalence of learning diseases in youngsters born once in vitro fertilization (IVF). When human mothers square measure exposed to AN adverse atmosphere throughout embryonic development, such as underneath nutrition or placental pathology leading to impaired blood flow, nutrient transport or drive, their offspring square measure additional doubtless to develop metabolic syndrome symptoms such as fatness, impaired aldohexose tolerance, and eventually sort a pair of polygenic disease throughout adulthood. These findings fashioned the premise of the "thrifty hypothesis," that planned that poor nutrition in childhood causes permanent changes in glucose-insulin metabolism. Craniate deficiency disease impairs craniate growth and low birth weight favors a thrifty constitution that's epigenetically programmed to use biological process energy with efficiency, that is, to be ready for AN adult atmosphere with restricted resources. Once these folks square measure exposed to AN obesogenic atmosphere, they're additional doubtless to develop fatness and its consequences for a range of diseases, indicating that the metabolic memory fashioned throughout craniate programing in utero has a sway on adult life (Altucci et al., 2016).

The thrifty hypothesis has been extended to a broader construct of "developmental origins of health and malady," implying that early-life development is very sensitive to poor nutrition. Different environmental factors (Buettner et al., 2008). during this context as a result of permanent changes in metabolism, which might increase status to advanced diseases such as polygenic disease, obesity, upset, cardiovascular disease, respiratory disorder, and cancer. As a result of their transmitted over much more consecutive mitotic divisions, epigenetic changes throughout embryonic development have a way larger impact on the general epigenetic standing of the organism than adult stem cells or corporal cells. These adolescence exposures square measure recorded and hold on as cellular memory via persistent variations in cellular functions with long consequences. This suggests that a human's constitution is that the results of advanced order atmosphere interactions that end in life-long epigenome transforming. The idea behind epigenetic programing via biological process compounds is that dietary interventions in human adults, such as calorie restriction, the Mediterranean or Nordic diet, will all have an effect on the chromatin granule standing of the themes and result in the expression of genes that square measure useful to metabolic health. This suggests that epigenetic states that square measure mounted throughout embryogenesis will amendment in response to intrinsic and environmental factors such as biological process compounds. This epigenetic drift of stability epialleles has been discovered in monozygotic twins World Health Organization begin with AN indistinguishable epigenome however reach completely different stages throughout their lives. Changes in diet or metabolism related to fatness could end in AN epigenetic drift that's passed right down to future generations. If this theory is correct, the world's growing fatness and metabolic malady epidemic could end in a major epigenetic predisposition to the metabolic syndrome in consequent generations, leading to a positive feedback. The construct of familial genetic drift proposes that epigenetic changes nonheritable throughout written account aging scale back the capability for physiological condition responses to biological process stress, such as over nutrition, leading to a decrease in overall lifespan. The goal of nutrigenomics is to realize a genome-wide understanding of the results of nutrients and their metabolites on negotiant metabolism and their impact on the epigenome.

Ideally, this could embrace not solely simply accessible tissues such as blood, however, conjointly metabolic organs just like the liver, pancreas, fatty tissue and muscle. For instance, the deoxyribonucleic acid methylome of duct gland islets from folks with sort a pair of polygenic disease from healthy people differed at over 800 factor loci. Notably, a number of these methylation changes were associated with ribonucleic acid levels, implying variations in cell perform. During this means, nutrigenomic approaches can maintain well-being, promote health, and open the door to new therapeutic methods, such as the doable reprograming of the epigenome of metabolic organs via a customized diet, which has natural compounds that modulate the activity of chromatin granule modifiers and transcription factors (Lee et al., 2013).

**TABLE 14.1** A Summary of Some Dietary Components.

| Sr. No | Nutrients | Food origin | Epigenetic role |
|---|---|---|---|
| 1. | Methionine | Sesame seeds, Brazil nuts, fish, peppers, and spinach | SAM synthesis |
| 2. | Folic acid | Leafy vegetables, sunflower seeds, baker's yeast, and liver | Methionine synthesis |
| 3. | Vitamin B12 | Meat, liver, shellfish, and milk | Methionine synthesis |
| 4. | Vitamin B6 | Meats, whole grain products, vegetables, nuts | Methionine synthesis |
| 5. | SAM-e(SAM) | Popular dietary supplement pill; unstable in food | Enzymes transfer methyl group |
| 6. | Choline | Egg yolks, liver, soy, cooked beef, chicken, veal, and turkey | Methyl donor to SAM |
| 7. | Betaine | Wheat, spinach, shellfish, and sugar beets | Break down the toxic byproducts of SAM synthesis |
| 8. | Resveratrol | Red wine | Removes acetyl group from histones, improving health |
| 9. | Genistein | Soy and soy products | Increased methylation, cancer prevention, unknown mechanism |
| 10. | Sulforaphane | Broccoli | Increased histone acetylation turning on anti-cancer genes |
| 11. | Butyrate | A compound produced in the intestine when dietary fiber is fermented | Increased histone acetylation turning on protective genes, increased lifespan |
| 12. | Diallyl sulfides (DADS) | Garlic | Increased histone acetylation turning on anti-cancer genes |

*Source*: Reprinted from Tiffon, 2018. 

**Advantages**

- Increased stress on a healthy diet and manner.
- Increased awareness of the chance of sure conditions.
- Improved health quality of life.
- Put a stress on unwellness interference.
- Lower morbidity and mortality.
- Lowering health-care prices.
- A higher understanding of the mechanisms underlying unwellness susceptibleness.

**Disadvantages**

- Other modifiable risk factors are being disregarded.
- Only specific nutrients/foods ought to be prioritized.
- Misleading statements.
- Personalized diets and designer foods return at the next worth.

## 14.10 FUTURE VIEW

The selection of tissue kind is important for future epigenetic studies. Epigenome mapping is done a lot of simply with cells from blood, buccal tissue layer or intestines (harvested via stool). Comparisons of epigenetic marks inside a similar individual's tissues can generate a list of epigenetic marks which will be figure to inaccessible tissues such as the brain, heart, respiratory organ or exocrine gland. In these comparisons, it is necessary to differentiate between intraindividual variations and inter-tissue correlations related to true epigenetic variants. Moreover, cross-tissue comparison studies ought to be disbursed in the least stages of life. Within the future, we tend to might learn why some regulative variants are a lot of vital than others in terms of attribute variability or unwellness risk. Understanding the epigenome, together with non-coding polymer pathways, has huge potential within the interference or treatment of diseases such as cancer (Chatenoud et al., 2014).

## 14.11 CONCLUSIONS

The combination of varied environmental influences with one's heredity determines one's health. Epigenetic pathways are expected to change human health and offspring as a results of manner factors. Epigenetics that is influenced by our life experiences influences organic phenomenon variations and habits, together with as nutrition, behavior, and toxic exposures within the surroundings. The results of the surroundings are chronic. Exposures are thought to play a task within the rising rates of all malignancies, not simply those in youngsters. Not just for the mother, however, additionally for the fetus some stages of development are a lot of vulnerable than others. Moreover, the toxins' negative consequences—the toxin itself, dose, and important exposure all work along. It is necessary to think about the window, however, this could be troublesome to forecast. Epigenomics changes occur once poison exposure. Several organs are also littered with changes within the offspring. The surroundings influence epigenetic marks, and these changes is passed down between generations, leading to Transgenerational epigenetic inheritance. a number of our characteristics, actions, illnesses, positive and negative experiences leave epigenetic tags which will be transmissible, however, they will even be eliminated by dynamical our manner. Manner decisions and environmental exposures manufacture epigenetic markings on our polymer that influence factor expression; a number of these marks are useful, whereas others are harmful. Diets high in fruits and vegetables might have effects on polymer that are such as epigenetic medicines. A more robust understanding of epigenetic changes and also the signal pathways engaged by bioactive dietary parts would aid in determinant the role and potential price of nutrients in rising human health and lowering cancer risk. A similar is aforesaid for substances found within the surroundings. Chemicals with epigenetic effects became a heavy international public health concern that has got to be self-addressed because of their negative impacts on people, youngsters, and presumably, throughout generations. Chemicals that promote cancer formation are probably to focus on checkpoints or important molecules within the cell at the interface between the cell cycle and differentiation processes. Understanding the epigenetic impacts of endocrine-disrupting chemicals (EDCs) would facilitate assess their activity and mapping such marks may well be used for risk management exploitation pathway-based toxicity testing.

## ACKNOWLEDGMENTS

The authors are thankful to Dr. Sanjay J. Kshirsagar, Principal, MET's Institute of Pharmacy and the trusts of MET's Institute of Pharmacy for their constant support and providing all facilities to complete this work. The authors are also grateful to CRC Press and Apple Academic Press—a Taylor and Francis group for providing them this opportunity.

## KEYWORDS


- **acetyl–CoA**
- **agouti mice**
- **chromatin**
- **chromatin-modifying enzymes**
- **DNA methylation**
- **epigenome**
- **epigenetic programing**


## REFERENCES


Ahire, E. D.; Sonawane, V. N.; Surana, K. R.; Talele, G. S. Drug Discovery, Drug-Likeness Screening, and Bioavailability: Development of Drug-Likeness Rule for Natural Products. In *Applied Pharmaceutical Practice and Nutraceuticals*; Apple Academic Press: 2021; pp 191–208.

Altucci, L.; Rots, M. G. Epigenetic Drugs: From Chemistry via Biology to Medicine and Back. *Clin. Epigenetics* **2016,** *8*, 56.

Ameres, S. L.; Zamore, P. D. Diversifying microRNA Sequence and Function. *Nat. Rev. Mol. Cell Biol.* **2013,** *14*, 475–488.

Bannister AJ, Kouzarides T. Regulation of chromatin by histone modifications. Cell Res. 2011 Mar;21(3):381-95. doi: 10.1038/cr.2011.22. Epub 2011 Feb 15. PMID: 21321607; PMCID: PMC3193420.

Bellacosa, A.; Cicchillitti, L.; Schepis, F.; Riccio, A.; Yeung, A. T.; Matsumoto, Y.; et al. MED1, A Novel Human Methyl-CpG-binding Endonuclease, Interacts with DNA Mismatch Repair Protein MLH1. *Proc. Natl. Acad. Sci. USA* **1999,** *96*, 3969–3974.

Bernstein, B. E.; Meissner, A.; Lander E. S. The Mammalian Epigenome. *Cell* **2007,** *128* (4), 669–681. Crossref, Medline, CAS, Google Scholar.

Bishop, K. S.; Ferguson, L. R. The Interaction Between Epigenetics, Nutrition and the Development of Cancer. *Nutrients* **2015,** *7*, 922–947.

Campión, J.; Milagro, F. I.; Goyenechea, E.; Martínez, J. A. TNF-α Promoter Methylation as a Predictive Biomarker for Weight-loss Response. *Obesity* **2009,** *17* (6), 1293–1297.

Chatenoud, L.; Bertuccio, P.; Bosetti, C.; Malvezzi, M.; Levi, F.; Negri, E.; et al. Trends in Mortality from Major Cancers in the Americas: 1980–2010. *Ann. Oncol.* **2014,** *25*, 1843–1853.

Desai, M.; Jellyman, J. K.; Ross, M. G. Epigenomics, Gestational Programming and Risk of Metabolic Syndrome. *Int. J. Obes. (Lond.)*. **2015,** *39* (4).

Diamanti-Kandarakis, E.; Bourguignon, J. P.; Giudice, L. C.; Hauser, R.; Prins, G. S.; Soto, A. M.; et al. Endocrine-disrupting Chemicals: An Endocrine Society Scientific Statement. *Endocr. Rev.* **2009,** *30*, 293–342.

Druesne, N.; Pagniez, A.; Mayeur, C.; Thomas, M.; Cherbuy, C.; Duee, P. H.; et al. Diallyl Disulfide (DADS) Increases Histone Acetylation and p21(waf1/cip1) Expression in Human Colon Tumor Cell Lines. *Carcinogenesis* **2004,** *25*, 1227–1236.

Finnin M. S.; Donigian J. R.; Cohen A.; Richon V. M.; Rifkind R. A.; Marks, P. A.; et al. Structures of a Histone Deacetylase Homologue Bound to the TSA and SAHA Inhibitors. *Nature* **1999,** *401*, 188–193.

Gluckman P. D.; Hanson M. A. Developmental Origins of Disease Paradigm: A Mechanistic and Evolutionary Perspective. *Pediatr. Res.* **2004,** *56* (3), 311–317.

Hotamisligil, G. S. Endoplasmic Reticulum Stress and the Inflammatory Basis of Metabolic Disease. *Cell* **2010,** *140*, 900–917.

House, J. S.; Mendez, M.; Maguire, R. L.; Gonzalez-Nahm, S.; Huang, Z.; Daniels, J.; et al. Periconceptional Maternal Mediterranean Diet Is Associated with Favorable Offspring Behaviors and Altered CpG Methylation of Imprinted Genes. *Front Cell Dev. Biol.* **2018,** *6*, 107.

Huang P. L. A comprehensive Definition for Metabolic Syndrome. *Dis. Model Mech.* **2009,** *2*, 231–237.

Jarouliya, U.; Keservani, R. K. *"Pathways Leading to Child Obesity", Global Perspectives on Childhood Obesity: Current Status, Consequences and Prevention*, Second Edition. Debasis, B., Ed. Academic Press, Elsevier: 2019; Chapter 12, pp 137–146. ISBN: 9780128128404.

Jenke A. C.; Postberg J.; Raine T.; Nayak K. M.; Molitor M.; Wirth S.; et al. DNA Methylation Analysis in the Intestinal Epithelium – Effect of Cell Separation on Gene Expression and Methylation Profile. *PLoS ONE* **2013,** *8* (2), e55636.

Kanherkar R. R.; Bhatia-dey N.; Csoka A. B. Epigenetics Across the Human Lifespan. *Front Cell Dev. Biol.* **2014,** 2, 1–19.

Keane D.; Kelly S.; Healy N. P.; McArdle M. A.; Holohan K.; Roche H. M. Diet and Metabolic Syndrome: An Overview. *Curr. Vasc. Pharmacol.* **2013,** *11*, 842–857.

Keservani, R. K.; Sharma, A. K.; Kesharwani, R. K. Medicinal Effect of Nutraceutical Fruits for the Cognition and Brain Health. *Scientifica* **2016,** Article ID 3109254, 10. doi:10.1155/2016/3109254.

Keservani, R. K.; Sharma, A. K.; Kesharwani, R. K. An Overview and Therapeutic Applications of Nutraceutical and Functional Foods. *Recent Adv. Drug Deliv. Technol.* **2017,** 160–201. ISBN:9781522507543.

Keservani, R. K.; Sharma, A. K.; Kesharwani, R. K., Eds.; *Nutraceuticals and Dietary Supplements: Applications in Health Improvement and Disease Management*; CRC Press: 2020. ISBN: 9781771888738.

Kucharski, R.; Maleszka, J.; Foret, S.; Maleszka, R. Nutritional Control of Reproductive Status in Honeybees via DNA Methylation. *Science* **2008,** *319* (5871), 1827–1830.

Lee, H.-S.; Barraza-Villarreal, A, Hernandez-Vargas, H.; Sly P. D.; Biessy C.; Ramakrishnan U.; et al. Modulation of DNA Methylation States and Infant Immune System By Dietary Supplementation with ω-3 PUFA During Pregnancy in An Intervention Study. *Am. J. Clin. Nutr*. **2013,** *98* (2), 480–487.

Lillycrop, K. A.; Hoile, S. P.; Grenfell, L.; Burdge, G. C. DNA Methylation, Ageing and the Influence of Early Life Nutrition. *Proc. Nutr. Soc.* **2014,** *73*, 413–421.
Lusis A. J.; Attie A. D.; Reue, K. Metabolic Syndrome: From Epidemiology to Systems Biology. *Nat. Rev. Genet.* **2008,** *9*, 819–830.
Paustenbach, D. J. The US EPA Science Advisory Board Evaluation (2001) of the EPA Dioxin Reassessment. *Regul. Toxicol. Pharmacol.* **2002,** *36*, 211–219.
Rane, B. R.; Bharath, M. S.; Patil, R. R.; Keservani, R. K.; Jain, A. S. Novel Approaches in Nutraceuticals, In *Enhancing the Therapeutic Efficacy of Herbal Formulations through Novel Drug Delivery Systems*; Kesharwani R. K.; Keservani R. K.; Sharma A. K., Eds.; IGI Global International Publisher: Pennsylvania, USA, Chapter 11, 2020b; 241–266. ISBN13: 9781799844532.
Rane, B. R.; Patil, A. S.; Keservani, R. K.; Jain, A. S. Novel Approaches in Herbal Formulation. In *Enhancing the Therapeutic Efficacy of Herbal Formulations through Novel Drug Delivery Systems*; Kesharwani R. K.; Keservani R. K.; Sharma A. K., Eds., IGI Global International Publisher: Pennsylvania, USA, Chapter 2, 2020a; 43–68. ISBN13: 9781799844532.
Rane, B. R.; Tadavi, S. A.; Keservani, R. K. Naturopathy. In *Herbal Product Development*; Sharma A. K.; Keservani R. K., Eds.; Surya Prakash Gautam, Apple Academic Press, CRC Press, Taylor & Francis Group: Chapter 12, 2020c; 321–347. ISBN: 9781771888776.
Reamon-Buettner, S. M.; Mutschler, V.; Borlak, J. The Next Innovation Cycle in Toxicogenomics: Environmental Epigenetics. *Mutat. Res.* **2008,** *659*, 158–165.
Relton C. L.; Davey Smith G. Two-step Epigenetic Mendelian Randomization: A Strategy for Establishing the Causal Role of Epigenetic Processes in Pathways to Disease. *Int. J. Epidemiol.* **2012,** *41* (1), 161–176.
Rotondo, J. C.; Selvatici, R.; Di Domenico, M.; Marci, R.; Vesce, F.; Tognon, M.; Martini, F. Epigenetics. **September 2013,** *9* (18), 990–7.
Sharma, A. K.; Keservani, R. K.; Gautam, S. P. *Herbal Product Development*; Apple Academic Press, CRC Press, Taylor & Francis Group: 2020. ISBN: 9781771888776.
The Academy of Medical Sciences. Stratified, Personalised or P4 Medicine: A New Direction for Placing the Patient at the Centre of Healthcare and Health Education. https://acmedsci.ac.uk/file-download/38266–56e6d483e1d21
Tiffon C. The impact of nutrition and environmental epigenetics on human health and disease. International journal of molecular sciences. 2018 Nov 1;19(11):3425.
Waddington, C. H. The Epigenotype. *Int. J. Epidemiol.* **2012,** *41* (1), 10–13.
Waterland, R. A.; Jirtle, R. L. Transposable Elements: Targets for Early Nutritional Effects on Epigenetic Gene Regulation. *Mol. Cell Biol.* **2003,** *23* (15), 5293–5300. Crossref, Medline, CAS, Google Scholar
Weiss, H. A.; Potischman, N. A.; Brinton, L. A.; Brogan, D.; Coates, R. J.; Gammon, M. D.; et al. Prenatal and Perinatal Risk Factors for Breast Cancer in Young Women. *Epidemiology* **1997,** *8*, 181–187.

# INDEX

## D



## E



## F

## G



## H

## O



## P

## R



## S



## T



## U



## V



## X